HAWKING RADIATION

From Astrophysical Black Holes to Analogous Systems in Lab

Other Related Titles from World Scientific

The Encyclopedia of Cosmology
(In 4 Volumes)
Volume 1: Galaxy Formation and Evolution
(by Rennan Barkana)
Volume 2: Numerical Simulations in Cosmology
(edited by Kentaro Nagamine)
Volume 3: Dark Energy
(by Shinji Tsujikawa)
Volume 4: Dark Matter
(by Jihn E Kim)
editor-in-chief Giovanni G Fazio
ISBN: 978-981-4656-19-1 (Set)
ISBN: 978-981-4656-22-1 (Vol. 1)
ISBN: 978-981-4656-23-8 (Vol. 2)
ISBN: 978-981-4656-24-5 (Vol. 3)
ISBN: 978-981-4656-25-2 (Vol. 4)

Centennial of General Relativity: A Celebration
edited by César Augusto Zen Vasconcellos
ISBN: 978-981-4699-65-5

One Hundred Years of General Relativity: From Genesis and Empirical Foundations to Gravitational Waves, Cosmology and Quantum Gravity
(In 2 Volumes)
edited by Wei-Tou Ni
ISBN: 978-981-4635-12-7 (Set)
ISBN: 978-981-4678-48-3 (Vol. 1)
ISBN: 978-981-4678-49-0 (Vol. 2)

Introduction to the Theory of the Early Universe: Hot Big Bang Theory
2nd Edition
by Valery A Rubakov and Dmitry S Gorbunov
ISBN: 978-981-3209-87-9
ISBN: 978-981-3209-88-6 (pbk)

Hawking on the Big Bang and Black Holes
by Stephen Hawking
ISBN: 978-981-02-1078-6
ISBN: 978-981-02-1079-3 (pbk)

HAWKING RADIATION

From Astrophysical Black Holes to Analogous Systems in Lab

Francesco D Belgiorno
Politecnico di Milano, Italy

Sergio L Cacciatori
Università degli Studi dell'Insubria, Italy

Daniele Faccio
Heriot-Watt University, UK

World Scientific

NEW JERSEY · LONDON · SINGAPORE · BEIJING · SHANGHAI · HONG KONG · TAIPEI · CHENNAI · TOKYO

Published by

World Scientific Publishing Co. Pte. Ltd.
5 Toh Tuck Link, Singapore 596224
USA office: 27 Warren Street, Suite 401-402, Hackensack, NJ 07601
UK office: 57 Shelton Street, Covent Garden, London WC2H 9HE

Library of Congress Cataloging-in-Publication Data
Names: Belgiorno, Francesco, author. | Cacciatori, Sergio, author. | Faccio, Daniele, author.
Title: Hawking radiation : from astrophysical black holes to analogous systems in lab /
Francesco D. Belgiorno (Politecnico di Milano, Italy), Sergio L. Cacciatori
(Università degli Studi dell'Insubria, Italy), Daniele Faccio (Heriot-Watt University, UK).
Description: Singapore ; Hackensack, NJ : World Scientific Publishing Co. Pte. Ltd., [2018] |
Includes bibliographical references and index.
Identifiers: LCCN 2018017435| ISBN 9789814508537 (hardcover ; alk. paper) |
ISBN 9814508535 (hardcover ; alk. paper)
Subjects: LCSH: Black holes (Astronomy) | Radiation. | Space and time.
Classification: LCC QB843.B55 B46 2018 | DDC 523.8/875--dc23
LC record available at https://lccn.loc.gov/2018017435

British Library Cataloguing-in-Publication Data
A catalogue record for this book is available from the British Library.

For any available supplementary material, please visit
https://www.worldscientific.com/worldscibooks/10.1142/8812#t=suppl

Desk Editor: Ng Kah Fee

Printed in Singapore

To our parents

Contents

Introduction xiii

First Part 1

1. A short scrapbook on classical black holes 3
   - 1.1 Mathematical black holes 3
   - 1.2 Schwarzschild and Kerr black holes 6
     - 1.2.1 The Schwarzschild black hole 6
     - 1.2.2 The Kerr black hole 10
   - 1.3 Bifurcate Killing horizons 15
   - 1.4 Quantum fields and particles 19
     - 1.4.1 The Boulware state 20
     - 1.4.2 The Hartle-Hawking state 21
     - 1.4.3 The Unruh state 22
   - 1.5 BMS group for asymptotically flat spacetimes 23
   - 1.6 Further readings 26

2. The seminal paper 27
   - 2.1 Particle creation by black holes: The computation 31
   - 2.2 Particle creation by black holes: Dependence on the details 45
     - 2.2.1 Nonsymmetrical collapse 45
     - 2.2.2 The spin of the fields 47
     - 2.2.3 Massive fields 48
     - 2.2.4 Angular momentum 49
     - 2.2.5 Electric charge 50
     - 2.2.6 Back reaction 51

2.3 A very short list of further readings 55

3. Thermality of Hawking radiation: From Hartle-Hawking to Israel and Unruh 57

3.1 Hartle-Hawking approach to black hole radiance 58
3.2 Gibbons-Perry analysis for thermality 64
3.3 Thermofield dynamics and Hartle-Hawking-Israel state . . 66
3.3.1 Thermofield Dynamics 66
3.3.2 Israel contribution 68
3.4 Unruh's cornerstone 71

4. The tunneling approach 73

4.1 Damour-Ruffini 74
4.2 Hamilton-Jacobi tunneling method 76
4.2.1 Canonical invariance 78
4.2.2 The null geodesic method 78
4.2.3 The analyticity argument 81
4.2.4 A trick à la Nikishov 83
4.2.5 The 4D case: Role of the transverse coordinates . . 86
4.3 Parikh-Wilczek approach 87

5. The anomaly route to Hawking radiation 89

5.1 Christensen-Fulling way (1977) 89
5.2 Robinson-Wilczek (2005) and Iso-Umetsu-Wilczek (2006) . 93

6. The Euclidean section and Hawking temperature 101

6.1 Local diffeomorphism and extendability 102
6.2 Conical singularity and black hole horizon 104
6.2.1 The 2D case 104
6.2.2 The 4D case 105
6.3 The Hawking temperature from the extendability of the metric at $r = r_+$ 106
6.4 Example: Schwarzschild black hole 108
6.5 The failure of the argument for extremal black holes 108

7. Rigorous aspects of Hawking radiation 111

7.1 Local definiteness principle and local stability 112
7.1.1 The local definiteness 113

7.2 Hawking temperature from local definiteness and stability . . . 115
7.3 Existence of Hartle-Hawking state . . . 120
7.3.1 Almost free double quantum dynamical systems . . 120
7.3.2 The Minkowski vacuum . . . 124
7.3.3 The Hartle-Hawking state . . . 127
7.3.4 The Boulware one particle structure . . . 129
7.4 Hawking temperature for a spacetime with bifurcated Killing horizon . . . 131
7.4.1 Modular inclusion and Hawking temperature . . . 131
7.5 Further readings for black holes of the Kerr-Newman family 133

Second Part **135**

8. The roots of analogue gravity 137
8.1 Experimental black hole evaporation in water . . . 138
8.2 Analogue systems and dispersion: The Gospel according to Unruh . . . 141
8.3 A sample model for dispersive fluids: Essentials . . . 144
8.4 Analogue gravity in BEC . . . 147
8.5 Further readings concerning analogue gravity in presence of dispersion . . . 150

9. Hawking radiation in a non-dispersive nonlinear Kerr dielectric 153
9.1 Classical analysis of the effective spacetime geometry . . . 154
9.1.1 Wave rays geometry in the RIP frame . . . 157
9.1.2 Horizons in effective geometries . . . 159
9.1.3 Null geodesics in dispersionless and in dispersive dielectrics . . . 163
9.1.4 Trapping in dispersive regime . . . 168
9.2 Hawking radiation in a static dielectric black hole . . . 170
9.2.1 The “standard” Hawking prediction . . . 175
9.2.2 The electrodynamics point of view . . . 177
9.3 Effects of optical dispersion: Preliminary heuristics . . . 185
9.3.1 Horizons in dispersive media . . . 188

10. Hawking radiation in a dispersive Kerr dielectric 195

10.1 The relativistic Hopfield model 196
10.1.1 The covariant generalization of the Hopfield model 197
10.1.2 Quantum covariant Hopfield model 199
10.2 Uniformly moving RIP 214
10.3 Hawking radiation in the perturbative formulation 216
10.3.1 The $\varphi\psi$–model 217
10.3.2 Separation of variables 220
10.3.3 Quantization of the fields 221
10.4 An interlude: Semi-analytical and numerical calculations from Maxwell equations in the lab 228
10.4.1 Born approximation 229
10.4.2 Thermality for a Gaussian perturbation 231
10.4.3 A sample of numerical results 233
10.5 Calculation of thermality in the $\phi\psi$ model 234
10.5.1 Determination of the microscopic parameters in terms of the physical ones 236
10.5.2 Near horizon approximation: Solutions of equation (10.232) 237
10.5.3 Steepest descent approximation 238
10.5.4 Convergence regions 241
10.5.5 Decreasing mode inside the black hole $x < 0$ 241
10.5.6 Possible diagrams in the external region $x > 0$ . . . 241
10.5.7 Special configurations and thermality of pair-creation 242
10.5.8 Branch cuts along steepest descent paths 245
10.5.9 Vertical branch cuts 246
10.5.10 Near horizon: A different saddle point approximation 249
10.5.11 WKB solutions 252
10.5.12 A dimensionless parameter for the saddle point approximation 255
10.5.13 A further rescaling and insights for thermality . . . 257
10.5.14 Coalescence of branch points as $\omega \to 0$ 259
10.5.15 Horizons and dispersion 261
10.6 Recapitulation 263
10.7 Further readings 264
10.8 Hawking radiation in a dispersive nonlinear dielectric . . . 265

11. Hawking radiation in the lab 267

11.1 The Como experiment 267
11.2 The Vancouver experiment 276
11.3 The Technion experiment 278

Appendix A Algebraic methods in QFT 281

A.1 Araki-Haag-Kastler algebraic formulation of QFT 281
A.2 Haag-Hugenholz-Winnik formulation of quantum statistical systems 286
A.2.1 Structure of the statistical system 286
A.2.2 The Gibbs states 289
A.2.3 KMS condition and the infinite volume limit . . . 290
A.2.4 The thermal representations 292
A.3 The Tomita-Takesaki theorem 294
A.3.1 Polar decomposition 294
A.3.2 Some simple facts about Von Neumann algebras . 295
A.3.3 The Tomita-Takesaki theorem and the KMS condition 297

Bibliography 299

Index 321

Introduction

Hawking radiation represents one of the most fascinating predictions of quantum field theory. Its deduction by Steven Hawking in 1974 had very important consequences, and not only for black hole physics. Indeed, from one hand the prediction that black holes, which are classically stable and live till the end of the Universe (if any), are affected by a instability associated intrinsically with quantum mechanics, forcing them to evaporate with a temperature related to geometrical parameters of the black hole itself, gave rise to an unexpected link between the so-called laws of black hole mechanics and the laws of thermodynamics: as the black hole can be conceived as a thermodynamic object, due to its possibility to emit Hawking radiation, it became possible to interpret laws of black hole mechanics (1973), perfectly simulating laws of thermodynamics from a purely formal point of view, as actual laws of black hole thermodynamics. In particular, a thermodynamic notion of black hole entropy became available. On the other hand, Hawking calculation stimulated the framework of quantum field theory on curved spacetime, giving rise to a series of studies in the field.

Still, the deduction of Hawking radiation, albeit corroborated in several ways from a theoretical point of view, is far from being experimentally confirmed, due to the fact that black hole temperature is too low for astrophysical black holes to be actually detected: a solar mass black hole would be associated with a temperature order of 50 nanokelvin, so that it is quite impossible to detect it, and the temperature decreases as the inverse of the mass of the black hole, so for more massive black holes this goal is even harder. Primordial black holes have been suggested as possible more efficient sources of Hawking radiation, but no real possibility to detect Hawking radiation emerged so far.

In view of the aforementioned difficulties, at the beginning of eighties (1981) William Unruh proposed a different framework. The basic idea is that kinematics at the root of Hawking radiation does not require to dispose of an event horizon, but what really is important is that an horizon exists for some kind of particles. The original proposal was to consider an horizon for phonons (the particle-like side of sound waves). Then, according to Unruh, one can deduce an effective metric and an effective temperature which corresponds to the Hawking one. The so-called analogue gravity arose, and the possibility to detect the analogous Hawking radiation emerged.

The aim of the present book is to provide a survey on the methods for deducing both Hawking radiation from black holes and its analogue in dielectric media. The first part of the book is devoted to a survey of methods for deducing Hawking radiation in 'standard' black hole backgrounds. After recalling some basics of black hole spacetime, we start from the seminal paper by Hawking. Then, we consider the Hartle-Hawking masterpiece, together with some further papers concerning thermality of quantum propagators in black hole backgrounds. A further chapter is devoted to the so-called tunneling method, and its variants. The unexpected relation between Hawking radiation and quantum anomalies and rigorous quantum field theory results for Hawking radiation are considered in the two subsequent chapters.

The second part of the book starts from a discussion of the analogue framework of the seminal paper by Unruh. We remark that, in absence of dispersive (nonlinear) effects, all the methods of the first part of the book are applicable to the analogue gravity framework. Problems arise as far as dispersive effects come into play. Indeed, in that case the relation between Hawking radiation and geometry is automatically weakened, as from one hand there is a weaker link with the geometrical framework, and on the other hand one has to consider possible contributions of scattering effects weakening the Hawking radiation contribution. The main difficulty of the dispersive analogue case is that scattering methods have to be set up carefully, and analytical results are a quite hard task. We choose to limit ourselves to dielectric media, as they are the direct subject of our studies in the field. This is a limit of the book, but a detailed account would have required a less self-contained structure of the book itself, depleting the aim of our efforts to provide the reader an almost self-contained presentation of the various topics. Our hope is that book can stimulate curiosity and further efforts in order to made as clear as possible the original picture as well as the analogue gravity one, and can shift more attention by experts in the 'standard' Hawking effect towards the very interesting field of dispersive analogue gravity, for which we provide also a survey on recent experimental results.

We preliminarily apologize with many authors because we have surely missed some fundamental references (we hope to restore them in a future edition). We also point out what follows concerning publications to which we contributed as coauthors. In the fall of 2008 - spring of 2009 two of us, i.e. SC and FB, were invited to join Daniele Faccio and his group in studying Hawking radiation in dielectric media. This engagement gave rise to a very beautiful collaboration which was the origin of several studies and publications. It is correct to point out that we adopted two different attitudes in the ordering of the authors in our works. All works listed in the bibliography where DF is not present adopt an alphabetical order of authors, whereas the other works involving DF adopt mostly an ordering in which the first author is the main contributor, and the group leader DF appears as last author. There is an important exception to the aforementioned rule which FB wishes to underline: the Physical Review Letters entitled 'Hawking radiation from ultrashort laser pulse filaments' (2010) should have DF as first author as also the main contributor, but he preferred to appear as last author as the group leader. So we were able to cause confusion about that authorship.

Last, but not least, the present book displays the author list in rigorous alphabetical order.

Francesco Belgiorno
Sergio L. Cacciatori
Daniele Faccio

PART 1
First Part

Chapter 1

A short scrapbook on classical black holes

There is not a general definition of a black hole. The intuitive idea is that a black hole appears when there is a trapping region from which nothing can escape, bounded by an horizon, the event horizon. In this chapter we will recall the definition of black holes in a particular class of spacetimes, which are asymptotically flat and with a far future predictable from Cauchy surfaces, as introduced by Hawking and Ellis [Hawking and Ellis (2011)]. Next, we will illustrate the classical "physical" examples of Schwarzschild and Kerr black holes. One could expect that, given the present knowledge of cosmological data, the de Sitter versions of such solutions would be more physical. These can also be considered, but, then, one should include black holes in a eventually closed expanding universe. To our knowledge, no rigorous definition of such a black hole, generalizing the one of Hawking and Ellis, exists in the last class of spacetimes. Thus, in more general situations the intuitive definition of a black hole must be sufficient. In all the explicit known examples of black hole, the event horizon is also a Killing horizon. Since some properties of Hawking radiation (and more general phenomena) are related to this fact, we will end this chapter by defining Killing horizons and illustrating some of their properties.

1.1 Mathematical black holes

Following [Hawking and Ellis (2011)], we now will introduce a rigorous mathematical definition of a black hole. The main purpose is to collect some of the mathematical tools which will allow to get some exact results as illustrated in chapter 7. This general definition may be easily extended to the cases in presence of electromagnetic fields end/or a cosmological constant, but not to all possible physically interesting spacetimes, so the reader not interested in mathematically exact results can skip this section.

A D-dimensional *spacetime* is a D-dimensional smooth manifold M, endowed with a Lorentzian metric $\boldsymbol{g}$. We will indicate it with a pair $(M, \boldsymbol{g})$. It is said to be *time orientable* if it admits a continuous vector field everywhere non-spacelike (i.e. never spacelike) and future directed. In an orientable spacetime, a non-spacelike curve is *future (past) directed* if its tangent vector is everywhere future (past) directed. Given two subsets $A, B \subset M$, one defines

- the *chronological future of A relative to B* as the set $I^+(A, B)$ of all points of B that can be reached from A along future directed timelike curves;
- the *chronological past of A relative to B* as the set $I^-(A, B)$ of all points of B that can be reached from A along past directed timelike curves;
- the *causal future of A relative to B* as the set $J^+(A, B)$ obtained from the union of $A \cap B$ to all points of B that can be reached from A along future directed non-spacelike curves;
- the *causal past of A relative to B* as the set $J^-(A, B)$ obtained from the union of $A \cap B$ to all points of B that can be reached from A along past directed non-spacelike curves.

In particular, $I^{\pm}(A) := I^{\pm}(A, M)$ and $J^{\pm}(A) := J^{\pm}(A, M)$.

A set A is said to be a *future set* if it properly contains its chronological future, $I^+(A) \subset A$. A is said to be *achronal* if $A \cap I^+(A) = \emptyset$. It happens that the boundary of a future set is an achronal $(D-1)$-dimensional C^1 manifold. A spacetime is said to satisfy the *chronological condition* if it does not contain closed time-like curves. It satisfies the *causality condition* if it does not contain closed non-spacelike curves. We will say that such spacetimes are chronological or causal respectively. A causal or a chronological spacetime cannot be compact, so in general physically interesting universes are non compact. Chronological or causal conditions do not avoid for an observer to pass more and more times arbitrarily near to a given event. To this end, one introduces the concept of *strongly causal spacetime* in which any point admits an open neighborhood that is not crossed more than one time by any non-spacelike curve. This is not enough, since small perturbations of the metric could allow the existence of non-causal curves. If this does not happens, then the spacetime is said to be *causally stable.* $(M, \boldsymbol{g})$ is causally stable if and only if there is at least a function f over M whose gradient is everywhere timelike.

The initial value problem for the Einstein equations requires the selection of a Cauchy surface, a spacelike $(D-1)$-dimensional submanifold of M, where constrained initial condition for the metric and its first derivatives uniquely determine the metric on the whole spacetime. Not all spacetimes are completely specified by a Cauchy surface. The ones for which the Cauchy problem is well defined are said to be *globally hyperbolic*. It can be shown that a subset $A \subset M$ is globally hyperbolic if and only if $\forall p, q \in A$, it happens that $J^+(p) \cap J^-(q)$ is a compact subset of A.

$(M, \boldsymbol{g})$ is *asymptotically simple* if it is both time and space orientable, and it exists a nonphysical strongly causal spacetime $(N, \boldsymbol{h})$ and an embedding

$$\iota : M \hookrightarrow N, \tag{1.1}$$

satisfying the following conditions:

(1) $\iota(\partial M)$ is smooth in N;
(2) it exists a smooth positive function Ω over N, which is positive on $\iota(M)$ and satisfies $\boldsymbol{h}_{\iota(M)} = \Omega^2 \iota_*(\boldsymbol{g})$;
(3) $\Omega|_{\iota(\partial M)} = 0$, $d\Omega|_{\iota(\partial M)} \neq 0$;
(4) any null geodesic curve has two endpoints on ∂M.

An asymptotically simple spacetime is asymptotically empty if the Ricci tensor $R_{\mu\nu}$ is zero in an open neighborhood of $\iota(\partial M)$ in N. Such a spacetime is always globally hyperbolic, but such conditions are too strong in order to include black holes. They have all the same topology as $\mathbb{R}^D$.

In order to include black holes, one defines the *weakly asymptotically simple and empty* spacetime $(M, \boldsymbol{g})$ if it exists an asymptotically simple and empty space $(M', \boldsymbol{g}')$ and an open neighborhood U' of $\partial M'$ isometrically isomorphic to an open subset U of M which conformally embeds in $(N, \boldsymbol{h})$ including ∂M as a smooth submanifold.

It follows that for a weakly asymptotically simple and empty spacetime the boundary ∂M consists on the union $\partial M = \mathfrak{I}^+ \bigcap \mathfrak{I}^-$ of two null hypersurfaces called the future null infinity $\mathfrak{I}^+$ and the past null infinity $\mathfrak{I}^-$. These can be thought as the point reached in the far future or back to far past by non trapped light rays.

A spacelike hypersurface S with the property that it is crossed only one time by any non-spacelike curve is called a *partial Cauchy surface*. The *maximal Cauchy development* $D^+(S)$ of a surface S is the maximal subset of the spacetime M that is completely determined by Cauchy data on S. A weakly asymptotically simple and empty spacetime is *future asymptotically predictable* (fap) from a partial Cauchy surface S if $\mathfrak{I}^+ \subset \overline{D^+(S)}$. In order to avoid the appearance of naked singularities because of perturbations, one finally introduce the concept of a *strongly future asymptotically predictable* spacetime from a partial Cauchy surface S, a fap that satisfies $J^+(S) \cap \bar{J}^-(\mathfrak{I}^+, M \cup \partial M) \subset D^+(S)$.

We are now ready to define black holes: given a strongly fap spacetime $(M, \boldsymbol{g})$, its *black hole region* is the topological complement B of $J^-(\mathfrak{I}^+)$ in M. Now, assume that the closure of $M \cap J^-(\mathfrak{I}^+)$ in the unphysical fap is contained in a globally hyperbolic region, and let S be a Cauchy surface for this region. Then, each connected component of $B(S) := B \cap S$ is called a *black hole* in S or at the time S. Since $J^-(\mathfrak{I}^+)$ is open, B is closed and its boundary is called the *event horizon* $H = \partial B$ of B, whereas $H(S) = H \cap S$ are the event horizons at time S. Note that the event horizon is an achronal boundary with null generators.

This concludes the mathematical presentation of black holes. In the next section we present more concretely black hole spacetimes and illustrate their principal properties by very known explicit examples.

1.2 Schwarzschild and Kerr black holes

We now describe the most known classical solutions of the vacuum Einstein equations, that are the Schwarzschild solution and the Kerr solution. We will work in a traditional four dimensional spacetime.

1.2.1 *The Schwarzschild black hole*

The Schwarzschild black hole represents the general solution of the Einstein equations in vacuum, with spherical symmetry. In Schwarzschild coordinates (t, r, θ, ϕ) the metric tensor has expression

$$ds^2 = -\left(1 - \frac{2mG}{c^2 r}\right) c^2 dt^2 + \frac{1}{\left(1 - \frac{2mG}{c^2 r}\right)} dr^2 + r^2 d\Omega^2, \tag{1.2}$$

where

$$d\Omega^2 = d\theta^2 + \sin^2\theta \, d\phi^2 \tag{1.3}$$

is the standard metric on the two dimensional round sphere. The range of the coordinates t and r is $(t, r) \in (-\infty, \infty) \times (r_H, \infty)$ so that this local chart does not cover the horizon hypersurface $r = r_H := 2mG/c^2$. Staticity plus spherical symmetry allow for the existence of four Killing vectors

$$K_0 = \frac{\partial}{\partial t}, \tag{1.4}$$

$$K_1 = \sin\phi \frac{\partial}{\partial \theta} + \cot\theta \cos\phi \frac{\partial}{\partial \phi}, \tag{1.5}$$

$$K_2 = \cos\phi \frac{\partial}{\partial \theta} - \cot\theta \sin\phi \frac{\partial}{\partial \phi}, \tag{1.6}$$

$$K_3 = \frac{\partial}{\partial \phi}. \tag{1.7}$$

In particular, note that K_0 is timelike, but its norm tends to zero in reaching the boundary $r = r_H$ of the local chart. Indeed, this boundary is not a singularity as can be seen by changing coordinates.

Eddington-Finkelstein coordinates

In the Schwarzschild coordinates it is easy to determine the radial null geodesics, which are indeed described by the curves

$$ct + r + r_H \log \left| \frac{r}{r_H} - 1 \right| = \text{constant}, \tag{1.8}$$

$$ct - r - r_H \log \left| \frac{r}{r_H} - 1 \right| = \text{constant}, \tag{1.9}$$

which represent light rays moving from infinity toward the horizon, and from the horizon toward infinity respectively. The Eddington-Finkelstein coordinates consist in choosing the *advanced null coordinate*

$$v = ct + r + r_H \log \left| \frac{r}{r_H} - 1 \right|, \tag{1.10}$$

in place of the time t. In these coordinates the metric assumes the form

$$ds^2 = -\left(1 - \frac{2mG}{c^2 r}\right) dv^2 + 2dvdr + r^2 d\Omega^2. \tag{1.11}$$

We immediately see that the metric coefficients are now regular everywhere in $r > 0$, and, in particular, are well defined on the horizon. In these coordinates the timelike Killing vector K_0 is

$$K_0 = \frac{\partial}{\partial v}, \tag{1.12}$$

which becomes lightlike and tangent to the horizon on the horizon itself. The Eddington-Finkelstein coordinates thus define a prolongation of the

Schwarzschild solution down to $r = 0$, which is a true singularity, since, for example, the curvature invariant

$$R_{\mu\nu\rho\sigma}R^{\mu\nu\rho\sigma} = 48\frac{m^2G^2}{c^4r^6} \tag{1.13}$$

diverges there. A second prolongation can be obtained by employing the *retarded null coordinate*

$$u = ct - r - r_H \log\left|\frac{r}{r_H} - 1\right|, \tag{1.14}$$

in place of the advanced one. One can obtain the maximal extension of the solution by using both coordinates.

Kruskal-Szekeres coordinates

Using both the retarded and anticipated null coordinates, the metric becomes

$$ds^2 = -\left(1 - \frac{r_H}{r}\right) dudv + r^2(u,v)d\Omega^2, \tag{1.15}$$

where $r(u,v)$ is defined by the equation

$$v - u = 2r + 2r_H \log\left|\frac{r}{r_H} - 1\right|, \tag{1.16}$$

or, equivalently,

$$r = r_H + r_H\Lambda\left(e^{\frac{v-u}{2r_H}-1}\right), \tag{1.17}$$

Λ being the Lambert function (Usually for Λ it is used the symbol W). Since for $r > r_h$

$$\left(1 - \frac{r}{r_H}\right) = -e^{\frac{v-u-2r}{2r_H}}, \tag{1.18}$$

it is convenient to introduce the coordinates

$$U = -e^{-\frac{u}{2r_H}}, \qquad V = e^{\frac{v}{2r_H}} \tag{1.19}$$

and then

$$T = r_H\frac{V+U}{2c}, \qquad R = r_H\frac{V-U}{2}, \tag{1.20}$$

so that the metric assumes the form

$$ds^2 = \frac{8mG}{c^2r(T,R)}e^{-\frac{r(T,R)c^2}{2mG}}(-c^2dT^2 + dR^2) + r^2(T,R)d\Omega^2, \tag{1.21}$$

with

$$r(T,R) = r_H(1 + W(|c^2T^2 - R^2|/r_H^2) \tag{1.22}$$

In these coordinates, the Schwarzschild edge correspond to $U < 0$, $V > 0$, but it is clear that the metric remains regular by extending V to negative values and U to positive values. Such extensions define the maximal extension of the Schwarzschild solution. It is easy to check that this spacetime is globally hyperbolic and that contain two horizons

$$H^+ = \{(U,V,\theta,\phi)|U = 0\}, \tag{1.23}$$
$$H^- = \{(U,V,\theta,\phi)|V = 0\}, \tag{1.24}$$

called the *future horizon* and the *past horizon*, which correspond to $r = r_H$, $t = +\infty$ and $r = r_H$, $t = -\infty$ respectively. The horizons intersects in the spatial two dimensional surface Σ defined by $U = V = 0$. On this surface the Killing vector field

$$K_0 = \frac{1}{2r_H}\left(V\frac{\partial}{\partial V} - U\frac{\partial}{\partial U}\right) \tag{1.25}$$

vanishes, so that the Σ is left invariant by the Killing vector field. Moreover, the horizons are generated by the null geodesics $V = 0$ and $U = 0$ starting normally from the surface Σ. Indeed, the null vectors

$$\boldsymbol{\nu} = \frac{\partial}{\partial V}, \qquad \boldsymbol{\mu} = \frac{\partial}{\partial U} \tag{1.26}$$

are orthogonal to Σ, and along the horizons are parallel to the Killing vector field:

$$K_0|_{H_-} = -\frac{U}{2r_H}\frac{\partial}{\partial U}, \qquad K_0|_{H_+} = \frac{V}{2r_H}\frac{\partial}{\partial V}. \tag{1.27}$$

One says that the two event horizons form a bifurcate Killing horizon. We will say something of more general on bifurcate Killing horizons in the next section.

It is worth to note that, in particular, the coordinate V parametrizes the null geodesics on the future horizon $U = 0$, whereas U parametrizes null geodesics on the past horizon $V = 0$. Indeed, these are affine parameters for such geodesics. To see this, note that the metric has the form

$$\begin{aligned} ds^2 &= -4\frac{r_H^3}{r(U,V)}e^{-\frac{r(U,V)}{r_H}}dUdV + r^2(U,V)d\Omega^2 \\ &= f(U,V)dUdV + r^2(U,V)d\Omega^2, \end{aligned} \tag{1.28}$$

since

$$r(U,V) = r_H(1 + \Lambda(|UV|/e)). \tag{1.29}$$

From this, one sees that the equations for geodesic transversal to the surfaces U = const, V = const expressed in terms of an affine parameter λ are

$$\frac{d^2V}{d\lambda^2} + U\frac{f'}{f}\left(\frac{dV}{d\lambda}\right)^2 = 0, \tag{1.30}$$

$$\frac{d^2U}{d\lambda^2} + V\frac{f'}{f}\left(\frac{dU}{d\lambda}\right)^2 = 0. \tag{1.31}$$

Thus, along H^- we have $V = 0$ and $d^2U/d\lambda^2 = 0$, whereas on H^+ is $U = 0$ and $d^2V/d\lambda^2 = 0$, which proves our statement.

Carter-Penrose diagram

A way for better understand all properties just stated and, in general, the causal structure of the spacetime is via the Carter-Penrose diagram. This consists in conformally compactifying the spacetime, by introducing the coordinates

$$\tau = \frac{1}{2}(\arctan V + \arctan U), \qquad \rho = \frac{1}{2}(\arctan V - \arctan U). \tag{1.32}$$

The resulting picture in the (ρ, τ) plane is depicted in Fig. 1.1.

In particular, the singularity is doubled in a past singularity at $\tau = -\pi/4$ and a future singularity, at $\tau = \pi/4$.

1.2.2 *The Kerr black hole*

The Kerr solution represents a rotating black hole.[1] In the Boyer-Lindquist coordinates, which generalize the Schwarzschild ones, the metric is

$$\begin{aligned} ds^2 = & -\frac{\Delta(r)}{\rho^2(r,\theta)}[dt - a\sin^2\theta d\phi]^2 + \frac{\sin^2\theta}{\rho^2(r,\theta)}[(r^2+a^2)d\phi + adt]^2 \\ & + \frac{\rho^2(r,\theta)}{\Delta(r)}dr^2 + \rho^2(r,\theta)d\theta^2, \end{aligned} \tag{1.33}$$

where, as usual

$$\rho^2(r,\theta) = r^2 + a^2\cos^2\theta, \qquad \Delta(r) = r^2 - 2mr + a^2. \tag{1.34}$$

Here, in order to simplify the expressions, we have set $c = G = 1$. The asymptotically flat region is for $r > r_+$, where

$$r_\pm = m \pm \sqrt{m^2 - a^2} \tag{1.35}$$

[1] But in general, differently from the Schwarzschild solution, it is not the exterior of a rotating star except for the approximate cases when the multipole momenta are very weak.

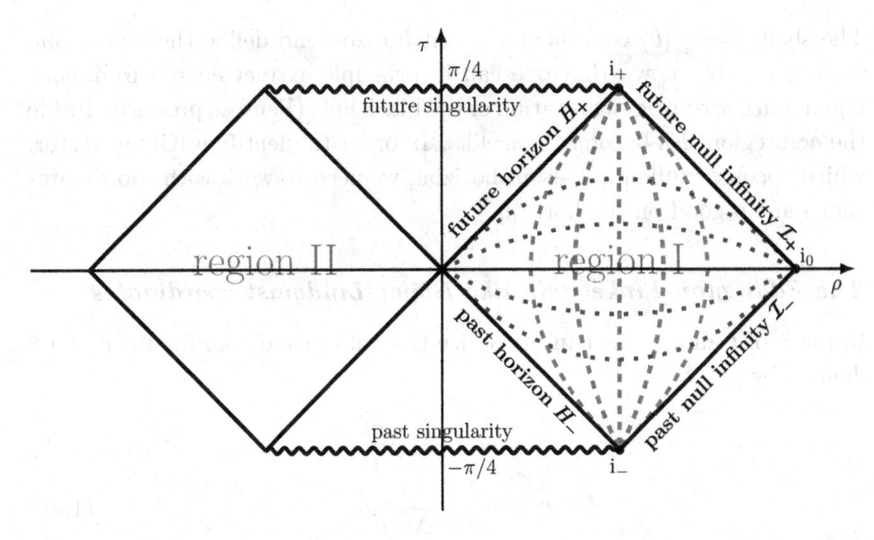

Fig. 1.1 Dotted lines correspond to t = *const.*, whereas dashed lines correspond to r = *const.* in the Schwarzschild patch. The "points" i_+, i_-, i_0 represent the future time infinity, the past time infinity and the space infinity. Notice that each point in the diagram is a sphere S^2.

are solutions of the equation $\Delta(r) = 0$, and are real only for $a \leq m$. a is related to the angular momentum of the black hole, so we say that such black holes are slowly rotating (or extremal if $m = a$). $r = r_-$ is a Cauchy horizon, whereas $r = r^+$ is an event horizon. Note that for $a \neq 0$ are both positive. There is also a true singularity at $r = 0$, $\theta = \pi/2$ (see below), which is called a ring singularity. When $a > m$, for fast rotating spacetimes, there are not real solutions to $\Delta(r) = 0$, and this is no more a black hole, but a rotating space time that exhibits a naked singularity.

The Kerr black holes have two commuting Killing fields

$$K_0 = \frac{\partial}{\partial t}, \tag{1.36}$$

$$K_1 = \frac{\partial}{\partial \phi}. \tag{1.37}$$

Differently to the Schwarzschild case, the Killing vector K_0 doesn't have vanishing norm on the event horizon. Indeed

$$K_0 \cdot K_0 = -1 + \frac{2mr}{r^2 + a^2 \cos^2\theta}, \tag{1.38}$$

which becomes null for

$$r_e^{\pm} = m \pm \sqrt{m^2 - a^2 \cos^2\theta}. \tag{1.39}$$

The shell $r = r_e^+(\theta)$ contains the event horizon end define the *ergoregion*, since it can be shown that one can in principle extract energy from such region until arresting the rotation of the black hole (Penrose process). Inside the ergoregion, K_0 becomes spacelike. In order to identify a Killing vector, which becomes null on the event horizon, we need to work with coordinates that remain good on the horizon.

The Eddington-Finkelstein like Boyer-Lindquist coordinates

In place of t and ϕ, one can introduce the anticipated coordinates v and ξ defined by

$$dv = dt + \frac{r^2 + a^2}{\Delta(r)} dr, \tag{1.40}$$

$$d\xi = d\phi + \frac{a}{\Delta(r)} dr, \tag{1.41}$$

so that, for example,

$$v = t + r + \frac{1}{\kappa_+} \log \left| \frac{r}{r_+} - 1 \right| + \frac{1}{\kappa_-} \log \left| \frac{r}{r_-} - 1 \right|, \tag{1.42}$$

$$\xi = \phi + \frac{a}{2\sqrt{m^2 - a^2}} \log \left| \frac{r - r_-}{r - r_+} \right|, \tag{1.43}$$

where

$$\kappa_\pm = \pm \frac{r_+ - r_-}{4mr_\pm} \tag{1.44}$$

are the surface gravities on the horizons. More physically, one can compute κ_+ as the acceleration an astronaut on the event horizon should keep in order to avoid falling in. Note that κ_+ is positive, whereas κ_- is negative, and

$$\frac{\kappa_+}{\kappa_-} = -\frac{r_-}{r_+}. \tag{1.45}$$

In these coordinates, the metric takes the form

$$\begin{aligned} ds^2 = &- \left(1 - \frac{2mr}{\rho^2(r,\theta)}\right) dv^2 + 2dvdr - 2a\sin^2\theta d\xi dr - \frac{4amr\sin^2\theta}{\rho^2(r,\theta)} dvd\xi \\ &+ \frac{(r^2+a^2)^2 - \Delta a^2 \sin^2\theta}{\rho^2(r,\theta)} \sin^2\theta d\xi^2 + \rho^2(r,\theta) d\theta^2. \end{aligned} \tag{1.46}$$

From this expression we see that $\partial/\partial r$ is a null direction, and the curves $\xi = \xi_0,\ \theta = \theta_0,\ v = v_0$ individuate radial ingoing geodesics.

In these coordinates the metric tensor is well defined on the horizons, and is singular only where $\rho(r,\theta)$ vanishes, which means $r=0$, $\theta=\pi/2$. This singularity cannot be eliminated by a change of coordinate, since the quadratic invariant (the *Kretschmann scalar*)

$$R_{\mu\nu\alpha\beta}R^{\mu\nu\alpha\beta} = \frac{48m^2(r^2-a^2\cos^2\theta)}{(r^2+a^2\cos^2\theta)^6} \times[(r^2-a^2\cos^2\theta)^2-16a^2r^2\cos^2\theta] \tag{1.47}$$

diverges in $r=0$, $\theta=\pi/2$. This is called the *ring singularity*, as it corresponds to a ring for $\phi\in[0,2\pi)$.

On the horizon $r=r_\pm$, the metric takes the form

$$ds^2_\pm = \rho^2(r_\pm,\theta)d\theta^2+\frac{\sin^2\theta}{\rho^2(r_\pm,\theta)}[adv-(r^2_\pm+a^2)d\xi]^2, \tag{1.48}$$

from which it is immediate to see that the Killing vector

$$K=a\frac{\partial}{\partial\xi}+(r^2_\pm+a^2)\frac{\partial}{\partial v}=aK_0+(r^2_\pm+a^2)K_1, \tag{1.49}$$

is a null vector on the horizons.

Exactly like in the Schwarzschild case, one can obtain a second prolongation by introducing the retarded coordinates

$$du=dt-\frac{r^2+a^2}{\Delta(r)}dr, \tag{1.50}$$

$$d\psi=d\phi-\frac{a}{\Delta(r)}dr, \tag{1.51}$$

so that, for example,

$$u=t-r-\frac{1}{\kappa_+}\log\left|\frac{r}{r_+}-1\right|-\frac{1}{\kappa_-}\log\left|\frac{r}{r_-}-1\right|, \tag{1.52}$$

$$\psi=\phi-\frac{a}{2\sqrt{m^2-a^2}}\log\left|\frac{r-r_-}{r-r_+}\right|, \tag{1.53}$$

in place of the advanced ones. By using a mixing of these coordinates one can define the maximal extension of the Kerr solution.

The Kruskal-Szekeres like Boyer-Lindquist coordinates

In order to construct the maximal extension one considers the coordinates

$$u=t-r-\frac{1}{\kappa_+}\log\left|\frac{r}{r_+}-1\right|-\frac{1}{\kappa_-}\log\left|\frac{r}{r_-}-1\right|, \tag{1.54}$$

$$v=t+r+\frac{1}{\kappa_+}\log\left|\frac{r}{r_+}-1\right|+\frac{1}{\kappa_-}\log\left|\frac{r}{r_-}-1\right|, \tag{1.55}$$

$$\phi^\pm=\phi-\frac{a}{r^2_\pm+a^2}t, \tag{1.56}$$

$$\theta=\theta, \tag{1.57}$$

where the choice of plus or minus sign is convenient when one is working around the event or the Cauchy horizon respectively. Next, like in the Schwarzschild case, one introduces the coordinates

$$U_\pm = -e^{-\kappa_\pm u}, \qquad V_\pm = e^{\kappa_\pm v}. \tag{1.58}$$

From these one easily see that r is a function of $U_\pm V_\pm$ only, implicitly defined by

$$U_\pm V_\pm = (1 - r/r_\pm) G_\pm(r)^{-1}, \tag{1.59}$$

$$G_\pm(r) = e^{\kappa_\pm r} \left| \frac{r}{r_\mp} - 1 \right|^{\frac{r_\mp}{r_\pm}}. \tag{1.60}$$

In these coordinates the metric takes the form

$$\begin{aligned} ds^2 = & \frac{r_\pm^2}{\rho^2(r,\theta)} \left(\frac{\rho^2(r,\theta)}{r^2+a^2} + \frac{\rho^2(r_\pm,\theta)}{r_\pm^2+a^2} \right) \frac{(r_\mp - r)(r_\pm + r) a^2 \sin^2\theta}{(r^2+a^2)(r_\pm^2+a^2)} \\ & \times \frac{G_\pm(r)^2}{4\kappa_\pm^2} (V_\pm^2 dU_\pm^2 + U_\pm^2 dV_\pm^2) \\ & - \frac{1}{\rho^2(r,\theta)} \left(\frac{\rho^4(r,\theta)}{(r^2+a^2)^2} + \frac{\rho^4(r_\pm,\theta)}{(r_\pm^2+a^2)^2} \right) G_\pm(r) \frac{(r_\mp - r) r_\pm}{2\kappa_\pm^2} dU_\pm dV_\pm \\ & + \rho^2(r,\theta) d\theta^2 + \frac{\sin^2\theta}{\rho^2(\theta,r)} \Bigg[a\Delta(r) \sin^2\theta d\phi^\pm \\ & - \frac{\rho^2(r_\pm,\theta)}{r_\pm^2+a^2} (r_\mp - r) r_\pm G_\pm(r) \frac{U_\pm dV_\pm - V_\pm dU_\pm}{\kappa_\pm} \Bigg] d\phi^\pm \\ & + \frac{\sin^2\theta}{\rho^2(r,\theta)} \Bigg[- (r^2+a^2) d\phi^\pm \\ & + a G_\pm(r) \frac{(r_\pm + r) r_\pm}{r_\pm^2 + a^2} \frac{U_\pm dV_\pm - V_\pm dU_\pm}{2\kappa_\pm} \Bigg]^2 . \end{aligned} \tag{1.61}$$

From this expression, we see easily that we can continue the solutions around the event horizon and the Cauchy horizon simply by extending the values of the coordinates $U_\pm$, $V_\pm$, which are defined initially for positive V and negative U, to all positive and negative values. Indeed, it follows that these patches can be repeated with infinitely many copies glued together in a maximal solution, as can be seen by drawing the Carter-Penrose diagram (see below).

Because of the presence of the Cauchy horizon the maximal solution is not globally hyperbolic, but the first region generated by the V_+, U_+ coordinates is. We will refer to this globally hyperbolic region as the Kerr black hole. It contains the future event horizon H^+ defined by $U_+ = 0$

and the past event horizon H^- defined by $V_+ = 0$. On these horizons the Killing vector

$$K = (r_\pm^2 + a^2)\left(V_+ \frac{\partial}{\partial V_+} + U_+ \frac{\partial}{\partial U_+}\right) \tag{1.62}$$

becomes null. Indeed, it vanishes on the spatial surface Σ defined by $V_+ = U_+ = 0$. The horizons $H^\pm$ are generated by null geodesics with tangent vectors $\partial/\partial V_+$ and $\partial/\partial U_+$, starting from the surface Σ. The Killing vector restricted to the horizons is parallel to the null generators:

$$K|_{H^+} = (r_\pm^2 + a^2)V_+ \frac{\partial}{\partial V_+}, \qquad (r_\pm^2 + a^2)U_+ \frac{\partial}{\partial U_+}. \tag{1.63}$$

Like in the case of Schwarzschild, the parameters V and U are affine parameters for the null geodesics generating the future and the past horizon respectively. To see this, let us consider for example the geodesic $(V^+(\lambda), 0, \phi_0^+, \theta_0)$, where λ is an affine parameter. Then

$$\frac{d^2 V_+}{d\lambda^2} + \Gamma^{V_+}_{V_+ V_+} \frac{dV_+}{d\lambda}\frac{dV_+}{d\lambda} = 0. \tag{1.64}$$

Now, since in this case the lightlike condition implies $g_{V_+ V_+} = 0$ and $\partial g_{V_+ V_+}/\partial V_+ = 0$, we have

$$\Gamma^{V_+}_{V_+ V_+} = \frac{1}{2} \sum_{i:x^i \neq V_+} g^{V_+ i}\left(2\frac{\partial g_{V_+ i}}{\partial V_+} - \frac{\partial g_{V_+ V_+}}{\partial x^i}\right). \tag{1.65}$$

Using that r is a function of $V_+ U_+$ and the way U^+ and V^+ appear in the coefficients of the metric, it is immediate to check that on $U_+ = 0$ it holds $\Gamma^{V_+}_{V_+ V_+} = 0$. Then, V^+ is affine.

Carter-Penrose diagram

Compactifying in a way similar to the Schwarzschild case we get a periodic diagram.

1.3 Bifurcate Killing horizons

The crossing horizons H^+ and H^-, present in both the Schwarzschild and the Kerr black holes, have several common characteristics: they are generated by null geodesics starting from a spatial two dimensional surface on which a Killing vector field vanishes. Moreover, the Killing vector field restricted to the horizons is tangent to the generating geodesics. These

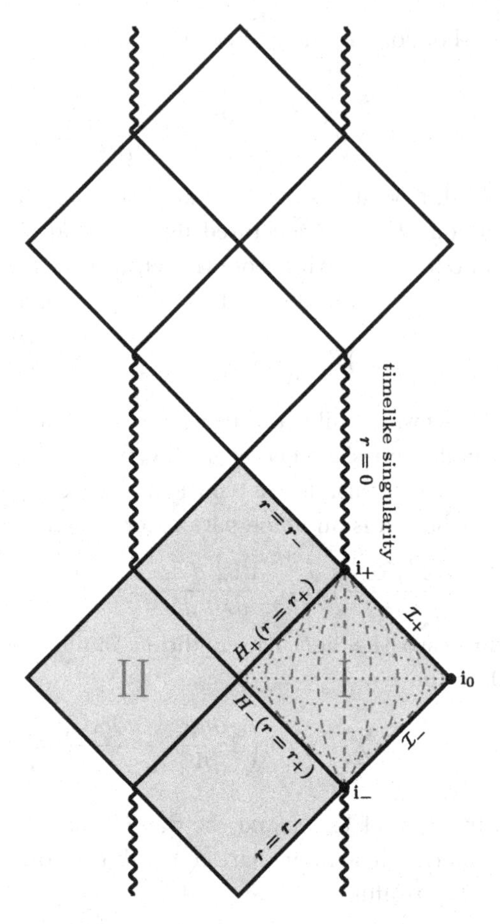

Fig. 1.2 A piece of the periodic continuation. The shaded region is a globally hyperbolic portion with past and future Cauchy horizons in $r = r_-$.

properties define in general what is called a *bifurcate Killing horizon*: first of all, it consists of a spatial two dimensional surface Σ of which any point is left invariant by a one parameter group of isometries, generated by a Killing vector field K. In particular, this means that K vanishes on Σ. In any point of Σ one can take two future directed lightlike vectors. Consider the lightlike geodesics generated by these vectors (as Cauchy conditions). It easily follows that the isometry must leave the geodesics invariant. Moreover, the covariant gradient of the Killing vector cannot vanish at any point of the geodesic different from the initial points. This means that necessarily the Killing vector field is tangent to this geodesic along them.

These kinds of horizons not necessarily coincide with an event or a Cauchy horizon, as the examples in the previous section. For example, in the Minkowski spacetime one can consider the isometries generated by the boosts, say, in the z direction, which leave the $x - y$ plane invariant. The motivation of studying Killing horizons is that most of the relevant properties of the event horizons are common to all Killing horizons.

In a four dimensional spacetime $(\mathcal{M}, g)$ assume that there exists a two dimensional spacelike surface Σ such that each of its point is left invariant by a one parameter isometry group generated by a Killing vector field ξ. This means that the Killing vector field vanishes along the surface. By using the identity

$$\nabla_\mu \nabla_\nu \xi_\rho = \xi_\sigma R^\sigma{}_{\mu\nu\rho}, \tag{1.66}$$

true for Killing vector fields, it is easy to prove that if both ξ and its gradient vanish in a point, then ξ vanishes everywhere. Since we are assuming that the isometry group is nontrivial, then the gradient of ξ cannot vanish at any point of Σ and thus ξ is different from zero at least in all points out of Σ at least in a small open neighborhood of Σ. Obviously the covariant derivatives of ξ along the directions tangent to Σ must vanish. For a fixed $p \in \Sigma$, apart from a normalization, there are exactly two independent null vectors χ and η in p, such that $\chi^\mu \eta_\mu = 1$, orthogonal to Σ. Let γ_χ and γ_η the null curves passing through p with direction χ and η respectively.

Since the isometry leaves fixed p, it will also act on each of these vectors by a possible multiplicative factor (indeed, it must send null vectors into null vectors with continuity). It follows that the curves γ_χ and γ_η are transformed each one into itself by the isometry, so that the Killing vector along each of these curves if does not vanishes is tangent to the curve. Now suppose that in $p \in \Sigma$ it holds $\chi^\mu \nabla_\mu \xi^\nu = 0$. Since the gradient of ξ cannot vanish, then, necessarily $\eta^\mu \nabla_\mu \xi^\nu \neq 0$ must be parallel to η^ν. Since it does not vanishes, it follows that $\eta^\mu \nabla_\mu \xi^\nu \chi_\nu \neq 0$ and then $0 \neq \chi^\nu \nabla_\mu \xi_\nu = -\chi^\nu \nabla_\nu \xi_\mu$, contradicting the initial hypothesis. Thus, we see that the gradient of ξ along the null directions is different from zero in p and then in at least in a small neighborhood of p along γ_χ and γ_η. Assuming both $\mathcal{M}$ and Σ are orientable, we can choose χ and η as continuous (future directed) vector fields over Σ, and prolong them along the null geodesics they generate. We now recognize the set so generated having the same structure as the future and past horizons we met in the previous sections. Such a set is called a *bifurcate Killing horizon.* It is the union of the horizon H_χ generated by χ and the horizon H_η generated by η.

Let us fix, for example, the horizon H_χ. For what we have seen, on H_χ there must exist a (continuous and differentiable) real valued function α such that

$$\xi|_{H_\chi} = \alpha\chi. \tag{1.67}$$

The quantity

$$\kappa = \xi^\mu \nabla_\mu \ln|\alpha| \tag{1.68}$$

is called the *surface gravity* on the horizon. It coincides with the surface gravity defined in the previous sections for the black holes. We will show that κ is constant on the whole bifurcate Killing horizon. Following [Kay and Wald (1991)], after taking the Lie derivative of (1.67) w.r.t. ξ, after using the geodesic equation and the Killing equation we get

$$\kappa\xi_\mu = -\frac{1}{2}\nabla_\mu(\xi^\nu\xi_\nu). \tag{1.69}$$

By taking the Lie derivative of this expression we get

$$\mathcal{L}_\xi\kappa = 0, \tag{1.70}$$

which means that the surface gravity is constant along each geodetic generator of the horizon. This is a weak version of our statement. We see as a consequence that ξ can never vanish along the generating geodesic in points out of Σ.

In a similar way, on H_η one finds that it must exist a function β satisfying relations analogous to the ones we found for α, from which one finds a surface gravity k' such that

$$k'\xi_\mu = \frac{1}{2}\nabla_\mu(\xi^\nu\xi_\nu). \tag{1.71}$$

Thus

$$\kappa^2 = -\frac{1}{2}\nabla^\mu\xi^\nu\nabla_\mu\xi_\nu, \qquad \kappa'^2 = -\frac{1}{2}\nabla^\mu\xi^\nu\nabla_\mu\xi_\nu, \tag{1.72}$$

from which it follows that $\kappa = \kappa'$ for geodesics meeting at the same point p. Taking the derivative of the last expression along any tangent direction to Σ and using again (1.66), we find that indeed κ is constant everywhere on the horizons.

Finally, in order to complete the comparison with the particular cases analyzed in the previous sections, we can find the general relation among affine null parameters and Killing parameters on the horizons. Let H_χ^F (H_χ^P) the portion of H_χ in the causal future (past) of Σ. With V we indicate the null affine coordinates of the lightlike geodesics, so that $\chi^\mu\nabla_\mu V = 1$

($\chi^\mu \nabla_\mu V = -1$), with $V = 0$ on Σ. The Killing parameter v is defined by $\xi^\mu \nabla_\mu v = 1$ ($\xi^\mu \nabla_\mu v = -1$) and is normalized so that $v = 0$ on the surface $V = 1$ ($V = -1$). From the definition of κ we see that

$$\kappa = \frac{\partial \log \alpha}{\partial v}, \tag{1.73}$$

so that, being κ constant,

$$\alpha(v) = A e^{\kappa u} \tag{1.74}$$

for some constant κ. On Σ α vanishes and then Σ correspond to $v \to -\infty$. On the other hand, since $\xi = \alpha\chi$, it must hold

$$\frac{dU}{du} = \alpha, \tag{1.75}$$

and using the fact that $V = 1$ ($V = -1$) on $v = 0$, we get

$$V = e^{\kappa v} \qquad (V = -e^{\kappa v}). \tag{1.76}$$

In a similar way, one defines the affine parameter U and Killing parameter u on H^F_η (H^P_η), which are related by $U = e^{-\kappa u}$ ($U = -e^{-\kappa u}$).

1.4 Quantum fields and particles

In order to speak about particles creation in a black hole background, one has to specify what one means about what a particle is. Whereas there is not any ambiguity at classical level, in a quantum theory of fields the concept of particle is not independent from the notion of the field: particles appear as excited configurations of the field in a given perturbative representation. Their properties are specified by the particle spectrum, which can be completely characterized in the free theory, and for the interacting theories in a perturbative regime, on a flat Minkowski space time. The energy is defined after fixing any global inertial frame. A Fourier expansion of the free fields in the inertial time allows to select the positive energy modes and, correspondingly, the creation and annihilation operators, and then a vacuum state, which generates the Fock space of the particle representation. Such space supports a unitary representation of the Poincaré group, so that the choice of the inertial time is irrelevant in order to define the concept of particle. Moreover, the Minkowski spacetime supports the Poincaré structure, which allows to complete the particle interpretation by completely characterizing the spectrum and the relevant representation of the Poincaré group, specified by the mass m and the spin s of the field.

On a curved background things are different. First of all, one has not any general global symmetry, which can help in identifying a preferred class of time parameters. Indeed, in general the Poincaré group is no more a symmetry of the spacetime, and the Lorentz subgroup characterize the tangent space only. Moreover, independently from the existence of symmetries, there not exist in general any global time coordinate, so that there is no way of globally defining a vacuum state or to characterize particle states. A possible way out, followed, for example, by [Araki *et al.* (1977)], could be to define a local algebra of observables, on any small open set in space time, in such the way that in the point limit they match with a well defined quantum field theory on the tangent space at the given point. We will discuss this strategy in the next sections, here we want to consider some simplest situations, which include the examples of asymptotically flat black holes we are considering here. In general it is possible to define positive modes on a globally hyperbolic spacetime, or in a spacetime containing a timelike or a null Killing vector field, or, further, a spacetime admitting an asymptotically flat region. In our case of asymptotically flat spacetimes, we have several choices for a defining a vacuum state, but there are three relevant possibilities. Interestingly, different concepts of vacuum states give rise to inequivalent concepts of particles. Let us look at the example of the Schwarzschild black hole.

1.4.1 *The Boulware state*

For both the Schwarzschild black hole and the Kerr black hole, as well as for the electrically charged solutions, the asymptotically flat time coordinate t correspond, at least in the external region, to a Killing vector field $\frac{\partial}{\partial t}$. Then, it looks quite natural to consider it as a good time parameter in order to define the evolution parameter with respect to which define the positive energy modes. This choice of time coincides with the time measured by the asymptotically inertial observers, the observers very far from the horizon. To be more precise, one in general prefers to adopt null coordinates $w = ct - r$, $v = ct + r$ such that a field ψ can be expressed in terms of positive energy modes as

$$\psi(t, r, \theta, \phi) = \sum_{l,m} \sum_{s} \int_0^\infty \frac{d\omega}{\sqrt{4\pi\omega}} \frac{Y_{lm}^{(s)}(\theta, \phi)}{\sqrt{2\pi}\, r} (\hat{a}_{lm,\omega}^{(s)}(r) e^{-i\omega u} + \hat{a}_{lm,\omega}^{(s)\dagger}(r) e^{i\omega u}$$
$$+ \hat{b}_{lm,\omega}^{(s)}(r) e^{-i\omega v} + \hat{b}_{lm,\omega}^{(s)\dagger}(r) e^{i\omega v}), \tag{1.77}$$

where $Y_{lm}^{(s)}$ are the spinorial spherical harmonics (depending on the spin of the field). The $a^\dagger$ operators generate the modes originating from the past infinity, whereas the $b^\dagger$ operators generate the modes originating from the past horizon. This expansion defines the Boulware vacuum state [Boulware (1975)] $\Omega_B = |0_B\rangle$, defined by

$$\hat{a}_{lm,\omega}^{(s)}(r)|0_B\rangle = \hat{b}_{lm,\omega}^{(s)}(r)|0_B\rangle = 0. \tag{1.78}$$

Such a state corresponds to a *vacuum of particles*, since, evidently, asymptotically inertial observers do not see particles in this state. However, despite appearing to be a natural choice for the vacuum state the t coordinate is ill-defined on the horizon and, as a consequence, it does not allow to construct a state that remains well defined on the horizon (and, consequently, in the Kruskal-Szekeres extension of the black hole).

1.4.2 *The Hartle-Hawking state*

In order to get a state that is well defined on the horizon, it is convenient to expand the fields with respect to the Kruskal-Szekeres like coordinates U and V, which are also null generators of the future and past horizons (and the null infinities), so that on the horizons the field is expressed in terms of the oscillating modes $e^{\mp i\omega U}$ and $e^{\mp i\omega V}$. Such expansion needs to be analytically extended everywhere in the extended solution. These oscillating modes are not the exact forms to be adopted as expanding functions, since U and V are far from selecting Killing parameters. On the other end, the right functions approximating the oscillating modes on the horizons must be well defined in an open neighborhood of the intersection between the past and the future horizon. Consequently, they can be expressed as combinations not only of modes well defined on the first region external to the horizon, but also of analogue modes defined on the second external region of the extended solution (the region II in the Carter-Penrose diagrams).

Thus, one has to consider the expansion

$$\begin{aligned}
\psi(t,r,\theta,\phi) = \sum_{l,m}\sum_{s}\int_0^\infty \frac{d\omega}{\sqrt{4\pi\omega}} \frac{Y_{lm}^{(s)}(\theta,\phi)}{\sqrt{2\pi}\, r} \\
\times \Big[({}_L\hat{a}_{lm,\omega}^{(s)}(r)\, {}_L\phi_{lm,\omega}(U,V) + {}_L\hat{a}_{lm,\omega}^{(s)\dagger}(r)\, {}_L\phi_{lm,\omega}(U,V)^* \\
+ {}_L\hat{b}_{lm,\omega}^{(s)}(r)\, {}_L\varphi_{lm,\omega}(U,V) + {}_L\hat{b}_{lm,\omega}^{(s)\dagger}(r)\, {}_L\varphi_{lm,\omega}(U,V)^*) \\
+ ({}_R\hat{a}_{lm,\omega}^{(s)}(r)\, {}_R\phi_{lm,\omega}(U,V) + {}_R\hat{a}_{lm,\omega}^{(s)\dagger}(r)\, {}_R\phi_{lm,\omega}(U,V)^* \\
+ {}_R\hat{b}_{lm,\omega}^{(s)}(r)\, {}_R\varphi_{lm,\omega}(U,V) + {}_R\hat{b}_{lm,\omega}^{(s)\dagger}(r)\, {}_R\varphi_{lm,\omega}(U,V)^*) \Big],
\end{aligned} \tag{1.79}$$

where R and L refer to the regions I and II in the Penrose diagrams, and

$$ {}_L\phi_{lm,\omega}(U,0) = e^{-i\omega U}, \qquad {}_L\varphi_{lm,\omega}(0,V) = e^{-i\omega V}, \tag{1.80} $$

and similar for the R modes, and are all of the form $\chi_{lm,\omega}(r^*(U,V))$, where χ satisfies the wave equation

$$ \left[-\frac{r(U,V)}{2r_H^3} e^{\frac{r(U,V)}{r_H}} \frac{\partial}{\partial V}\frac{\partial}{\partial U} + \frac{1}{2r_H r}\left(V\frac{\partial}{\partial V} + U\frac{\partial}{\partial U} \right) + \frac{l(l+1)}{r^2(U,V)} \right] \chi(U,V) = 0, \tag{1.81} $$

plus suitable boundary conditions for the different modes. They naturally prolong analytically to the whole Kruskal spacetime.

The Hartle-Hawking vacuum state $\Omega_{HH} = |0\rangle_{HH}$ [Hartle and Hawking (1976)] is thus defined by

$$ {}_L\hat{a}^{(s)}_{lm,\omega}|0\rangle_{HH} = {}_R\hat{a}^{(s)}_{lm,\omega}|0\rangle_{HH} = {}_L\hat{b}^{(s)}_{lm,\omega}|0\rangle_{HH} = {}_R\hat{b}^{(s)}_{lm,\omega}|0\rangle_{HH} = 0. \tag{1.82} $$

1.4.3 *The Unruh state*

The Hartle-Hawking state is a thermal state describing the *eternal black hole*, the global solution described by the Kruskal extension of a stationary asymptotically flat black hole. This suffices for explaining the thermalization of the radiation. However, a special role in the production of the radiation, as we will see, is played by the collapse of the star generating the black hole. For this reason, the collapsing vacuum should be considered as a state approximating the flat space time, including the auto gravitating matter, for very early times, whereas for very late times the state must be well defined on the created horizon. So, a good approximation should be obtained by a mixing of the boundary conditions of the previous examples. This gives rise to the *Unruh state*, or *collapsing state* [Unruh (1976)]: the field is expanded in terms of modes that approximate the asymptotically flat Fourier modes at very early times, whereas they approximate the horizon modes at very late times. With obvious notations:

$$ \begin{aligned} \psi(t,r,\theta,\phi) = \sum_{l,m}\sum_s \int_0^\infty \frac{d\omega}{\sqrt{4\pi\omega}} \frac{Y^{(s)}_{lm}(\theta,\phi)}{\sqrt{2\pi}\, r} & \\ \times \Big[\big({}_R\hat{b}^{(s)}_{lm,\omega}(r)\, {}_R\varphi_{lm,\omega}(U,V) + {}_R\hat{b}^{(s)\dagger}_{lm,\omega}(r)\, {}_R\varphi_{lm,\omega}(U,V)^* & \\ + {}_L\hat{b}^{(s)}_{lm,\omega}(r)\, {}_L\varphi_{lm,\omega}(U,V) + {}_L\hat{b}^{(s)\dagger}_{lm,\omega}(r)\, {}_L\varphi_{lm,\omega}(U,V)^* \big) & \\ + \big(\hat{a}^{(s)}_{lm,\omega}(r) e^{-i\omega w} + \hat{a}^{(s)\dagger}_{lm,\omega}(r) e^{i\omega w} \big) \Big]. & \end{aligned} \tag{1.83} $$

The Unruh state $\Omega_U = |0\rangle_U$ is defined by

$$\hat{a}^{(s)}_{lm,\omega}|0\rangle_U =_L\hat{b}^{(s)}_{lm,\omega}|0\rangle_U =_R\hat{b}^{(s)}_{lm,\omega}|0\rangle_U = 0. \tag{1.84}$$

The last two states can be interpreted physically by expressing them in terms of the modes with positive energy with respect to the Killing vector field η, which allows to understand what asymptotically flat observers see. One then finds that these states are thermal states with a temperature

$$T = \frac{\hbar\kappa}{2\pi k_B c}, \tag{1.85}$$

where k_B is the Boltzmann constant and κ the surface gravity. Proving this in several ways is the task of the next chapters.

Here we simply note that the Hartle-Hawking state is the most symmetric state over the Kruskal-Szekeres spacetime, and, thus, it is the most suitable for representing the vacuum state. Nevertheless, a more realistic situation includes the collapse of the star, and this case is more likely described by the Unruh vacuum.

1.5 BMS group for asymptotically flat spacetimes

The role of the whole Poincaré group for particle physics is evident in establishing the microscopical degrees of freedom. In General Relativity the translation invariance is lost but for asymptotically flat spacetime one can reasonably expect to be able to recover the whole Poincaré symmetry at least compatibly with the approximated flatness. In a series of paper devoted to the study of gravitational waves in General Relativity, in particular [Bondi *et al.* (1962)], the analysis of asymptotically flat spacetimes has been included exactly with the aim of recovering the Poincaré group at least as an asymptotic symmetry. The main surprise, however, is that the asymptotic group of transformations respecting the approximated flatness form an infinite dimensional group of transformations, containing an infinite dimensional abelian normal subgroup such that the quotient group is isomorphic to the proper orthocronous Lorentz group. In other words, it essentially looks like the Poincaré group where the translations are substituted with an infinite dimensional abelian group, called *super translations*. The whole group is called the Bondi Metzner Sachs group or *BMS group*.

In [Bondi *et al.* (1962)], Bondi, van der Burg and Metzner given a suitable definition of locally flat condition in terms of coordinates for axially symmetric Lorentzian manifolds. Then, they considered the set of diffeomorphisms that conserve the asymptotic expression of the metric (including the boundary conditions for the metric coefficients). A little more general situation has been considered by Sachs in [Bondi *et al.* (1962)] where no symmetries are assumed a priori. There, by introducing a suitable choice of coordinates (u, r, θ, ϕ), where u is a null retarded coordinate, r a "luminosity distance", and θ and ϕ "spherical coordinates", it is shown that the local form for an asymptotically flat metric should be

$$ds^2 = \frac{V}{r} e^{2\beta} du^2 - 2e^{2\beta} du dr + r^2 \sum_{\alpha,\beta=2}^{3} h_{\alpha\beta}(dx^\alpha - U^\alpha du)(dx^\beta - U^\beta du), \tag{1.86}$$

where

$$2 \sum_{\alpha,\beta=2}^{3} h_{\alpha\beta} dx^\alpha dx^\beta = (e^{2\gamma} + e^{2\delta}) d\theta^2 + 4 \sin\theta \sinh(\gamma - \delta) d\theta d\phi + \sin^2\theta (e^{-2\gamma} + e^{-2\delta}) d\phi^2, \tag{1.87}$$

$x^2 = \theta$, $x^3 = \phi$ and V, β, γ, δ, U^2, U^3 are six arbitrary functions of all the coordinates, which are constrained to satisfy the following global conditions:

(1) for at least a value of u it is possible to go to the limit $r \to \infty$ for all values of θ and ϕ;
(2) for some fixed choice of θ and ϕ and for a fixed value of u chosen as above, one has that

$$\lim_{r\to\infty} V/r = 1, \quad \lim_{r\to\infty} rU^\alpha = \lim_{r\to\infty} \beta = \lim_{r\to\infty} \gamma = \lim_{r\to\infty} \delta = 0. \tag{1.88}$$

Notice that in the case of General Relativity, in vacuum, the first condition is a consequence of the remaining ones;
(3) In the range of coordinates $u_0 \leq u \leq u_1$, $r_0 \leq r \leq \infty$, $0 \leq \theta \leq \pi$, $0 \leq \phi \leq 2\pi$, all components of the metric can be expanded in power of r^{-1} with at most a finite pole at $r = \infty$, Such power series can be added, multiplied, derived etc., freely.

The last requirement is technical and is expected to be relaxable. The other two make clear the meaning of asymptotic flatness. Under these assumptions Sachs showed that the coordinate transformations preserving all these conditions constitute a group acting on the coordinates as

$$u = k^{-1}(\tilde{\theta}, \tilde{\phi}, \lambda_1, \ldots, \lambda_6)[u' + F(\tilde{\theta}, \tilde{\phi})], \tag{1.89}$$

$$x^\alpha = G^\alpha(\tilde{\theta}, \tilde{\phi}, \lambda_1, \ldots, \lambda_6), \tag{1.90}$$

where F is an arbitrary function and for $\alpha = 0$ the transformation is a Lorentz homogeneous transformation parameterized by the six parameters $\lambda_1, \ldots, \lambda_6$. The infinite dimensional subgroup generated by $F(\theta, \phi)$ is the group of super translations. After developing F in terms of spherical harmonics,

$$F(\theta, \phi) = \sum_{l=0}^{\infty} \sum_{m=-l}^{l} c_{lm} Y_{lm}(\theta, \phi), \tag{1.91}$$

the super translation group is parameterized by the infinite coefficients c_{lm}, of which the first four $c_{0,0}$, $c_{1,-1}$, $c_{1,0}$, $c_{1,1}$ individuate standard translations. In [Sachs (1962)], a further simpler deduction of the transformation has been obtained by looking for the infinitesimal ones, that is the Lie algebra of diffeomorphisms preserving the above conditions. In particular, it is shown in the same paper that *the only four dimensional normal subgroup of the BMS group is the translation group.* This statement seems to solve the problem of recognizing the Poincaré group as an asymptotic symmetry. However, there is not a canonical way to individuate the translations inside the super translation group and, indeed, the Poincaré group can be embedded in the BMS group in infinite inequivalent ways. Moreover, the given analytic derivation does not led to a geometrical interpretation of the BMS group.

A geometrical viewpoint has been proposed by Penrose in [Penrose (1963)]. There, a general notion of "points at infinity" is given in terms of the conformal properties of the spacetime. If $g_{\mu\nu}$ is the physical metric on the spacetime $\mathcal{M}$, then we assume the existence of a conformal factor Ω^2 so that the unphysical metric $\tilde{g}_{\mu\nu} = \Omega^2 g_{\mu\nu}$ is everywhere defined on $\mathcal{M}$. The boundary $\mathcal{I}$ of $\mathcal{M}$ is defined by the points where $\Omega = 0$ and $\nabla_\mu \Omega \neq 0$. The asymptotic flatness is equivalent to say that the Riemann tensor goes to zero on $\mathcal{I}$. This implies $\nabla_\mu \Omega \nabla^\mu \Omega = 0$, so that $\mathcal{I}$ is a null surface. Actually we know that $\mathcal{I}$ has five components: $\mathcal{I}^\pm$, $i_\pm$ and i_0. We say that $(\mathcal{M}, g)$ is asymptotically flat if $(\mathcal{M}, \tilde{g})$ as above exists and $\tilde{g}$ is regular everywhere with, possibly, the exception of $i_\pm$ and i_0. This way, Penrose identifies the BMS asymptotic group as a group of conformal self transformations of the three dimensional manifold $\mathcal{I}^+$ (or $\mathcal{I}^-$). This group preserves both finite angles and null angles (angles between pairs of tangent vectors of which $n^\mu = \nabla^\mu \Omega$ is linear combination). The inhomogeneous Lorentz group can be singled out only if $\tilde{g}$ can be chosen regular also in $i_\pm$ an i_0. But generically this is not the case [Bergmann (1961)].

1.6 Further readings

There exist several texts on black hole physics, and a few texts on quantum field theory in curved spacetime and in external field. We indicate some of them in the following list:

- In the book [Frolov and Novikov (1998)] one can find both classical aspects of black hole physics and an extensive discussion about quantum effects in black hole manifolds, with particular reference to the Hawking effect. Further discussions can be found in [Frolov and Zelnikov (2011)].
- Classical aspects of black hole physics can be found in standard textbooks on General Relativity like e.g.[Misner *et al.* (1973); Wald (1984)]. See also [Raine and Thomas (2014); Poisson (2004); O'Neill (2014); Thorne *et al.* (1986)]. In [Wald (1984)] one can find also a chapter dedicated to quantum field theory in curved spacetime.
- Quantum field theory in curved spacetime, with reference also to the black hole case, is discussed in [Birrell and Davies (1984); Wald (1995)]. Further references are [Fulling (1989); Grib *et al.* (1994); Fabbri and Navarro-Salas (2005); Mukhanov and Winitzki (2007); Parker and Toms (2009); DeWitt (2003); Dimock (2011)]. See also the renowned report by DeWitt [DeWitt (1975)].
- Miscellanea of contributions, discussing also topics we don't deal with in the present book, are found in [Fre *et al.* (1999); Hehl *et al.* (1998); Calmet (2014)]. See also for further topics [Padmanabhan (2005)].

Chapter 2

The seminal paper

Here we summarize and discuss the famous paper of S. Hawking [Hawking (1975)]. See of course also [Hawking (1974)]. We will use his notations.

The first part of Hawking's analysis is devoted to heuristics for black hole pair-creation. Indeed, he starts by giving an idea of why one should expect particle production as a quantum effect in presence of a nontrivial curved background or, more precisely, when the background passes from a flat configuration to another one through a curved background transient. As a prototype of matter field let us consider a massless real Klein-Gordon field. At the quantum level it can be described as the superposition

$$\phi = \sum_i (f_i a_i + \bar{f}_i a_i^\dagger), \tag{2.1}$$

where f_i are an orthonormal set of complex valued solutions of the wave equation

$$f^{;a}_{;a} = 0, \tag{2.2}$$

chosen to have positive frequencies (energies) $\omega > 0$. The choice of positive frequencies appears to be natural, but it is related to the existence of a (unique) vacuum state, which is the only state, up to normalization, that is invariant under the action of translations. Since the latter are not symmetries of a curved background, flatness of spacetime is important for implementing the spectral condition. However, as Hawking argues, asymptotic flatness of certain regions should be sufficient in order to select a vacuum state and then positive and negative frequencies. Thus, one can think about the history of a black hole, which is born from the collapse of star, as divided essentially in three epochs:

(1) initially the matter is diluted with a very low density, which is just a weak perturbation of the flat spacetime. In this situation the symme-

tries of spacetime are the ones of the Minkowski one, and the vacuum state $|0\rangle_1$ is well defined;

(2) after long time, under the continuous action of gravity with an increasing amount of mass in a more and more reduced spatial region, it collapses in a black hole forming an event horizon. This epoch is strongly dynamical, in general there are not well defined symmetries and there is not any natural way of defining a vacuum state;

(3) after a longer time, when all matter is fallen inside the horizon and all asymmetries of the horizon have been radiated away, spacetime ends in a stationary configuration described by a Kerr rotating black hole solution or a static Schwarzschild spacetime. In both cases, there exists an asymptotic flat region, the external space far from the horizon. Observers in that region are expected to be able to define a vacuum state $|0\rangle_3$.

Since the vacuum states $|0\rangle_1$ and $|0\rangle_3$ do not coincide, the initial state $|0\rangle_1$ will appear to the asymptotic flat observers of the third epoch as decomposed into a $|0\rangle_3$ component plus a non vanishing complement. This means that particles have been created presumably during the collapse.

The particle creation should not be surprising in the presence of gravity, t.i. in presence of a non vanishing curvature. Fix a point p, where the curvature tensor $|R_{abcd}(p)|$ is non vanishing and bounded by B. In that point a (super local) inertial observer will be described by a four-velocity time-like vector v^a and will define an inertial system in a region U of radius $B^{-\frac{1}{2}}$, since she will experience a local metric tensor

$$g_{ab}(x) = \eta_{ab} - \frac{1}{3}R_{acbd}x^c x^d + \dots, \tag{2.3}$$

in coordinates x^a in U centred in p, $x^a(p) = 0$, or, equivalently, for the Vermeil theorem [Vermeil (1917)]. For very large frequencies, $\omega \gg B^{\frac{1}{2}}$, one expects a discrepancy among positive and negative frequencies of the order $B^{\frac{1}{2}}$. In such a situation the phase of the states is essentially undetermined, and the number operator is perfectly well defined. But for small frequencies, $\omega < B^{\frac{1}{2}}$, the phase is finely determined, so that the number operator $a_i^\dagger a_i$ has an ambiguity of order $\pm 1/2$. So the number of particles fluctuates and creation of particles is expected to be detected. If the field is massless, or the particle mass is much smaller than $\sqrt{B}$, then the number of modes in the unit volume having energy in the interval $\omega - \omega + d\omega$ can be estimated to be of order $\sim \omega^2 d\omega$. The estimate of the uncertainty on the energy because of the inability of exactly defining the excited modes with wavelength larger then $B^{-\frac{1}{2}}$ will be of the order

$$\int_0^{B^{\frac{1}{2}}} \omega\, \omega^2 d\omega \sim B^2, \tag{2.4}$$

and particle creation in the given region is thus expected to be of order B^2. Higher is the curvature, higher is the number of particles expected to be created. Obviously the production should be to the detriment of the curvature and back reaction on the background spacetime should be included. A complete treatment of the problem would then probably require a control of the quantum aspects of gravity. Nevertheless, if the curvature is not too large, the above estimate suggests that the fluctuations will not influence the Einstein equation too much, and back reaction can be neglected at a zeroth order approximation. If $B^{-\frac{1}{2}}$ is much larger than the Planck length, then gravity may be tractable classically. For astrophysical black holes, for example, the construction where the spacetime black hole is classical and the back reaction intervenes adiabatically is expected to work well everywhere far away from the physical singularity. In particular, it is expected to be a good approximation in the region external to the horizon. Surely there are other length scales where the detailed description of the evolution is expected to change, like, for example, the Compton length of massive particle. However, the most significative change expected on physical grounds, when the curvature is of such order of magnitude, is a sudden enhancement of the production of these particles, which will be exponentially suppressed at larger length scales. For astrophysical black holes a relevant particle production is thus expected only for massless fields, like photons and neutrinos (if at least one family is massless, and, in any case, the possible mass is very low).

Let us consider the case of a Schwarzschild black hole. If the particle creation occurs, then a flux of negative energy enters the horizon. The possible particle production by black hole could be also understood as follows. Crossing the horizon, from outside to inside, the temporal Killing vectors change signature, then becoming space-like. In this way, states having negative energy, coupled with states of positive energy during quantum fluctuations, may become real inside the horizon thus giving rise to real particles of positive energy outside the horizon. The separation of such modes depends on the surface gravity κ, which is the gradient of the amplitude of the Killing vector, and measures the velocity at which the last becomes space-like (see Chapter 1). Thus, the black hole is expected to lose mass and the area of its horizon will decrease. Such emission of particles makes the second law of thermodynamics holding true also in the presence of black holes, according to the proposals of Bekenstein.

The negative energy flux would give account of the violation of the classical theorems predicting the increasing of the area of the horizon as the only possibility for physical processes involving black holes. On the other hand, one could interpret the ingoing negative flux as an outgoing positive flux of particle tunneling out through the horizon, with the tunneling probability governed by the surface gravity. However, this is a too much naive picture, which should not considered too seriously, since the concept of particle is not even well defined far from flat backgrounds.

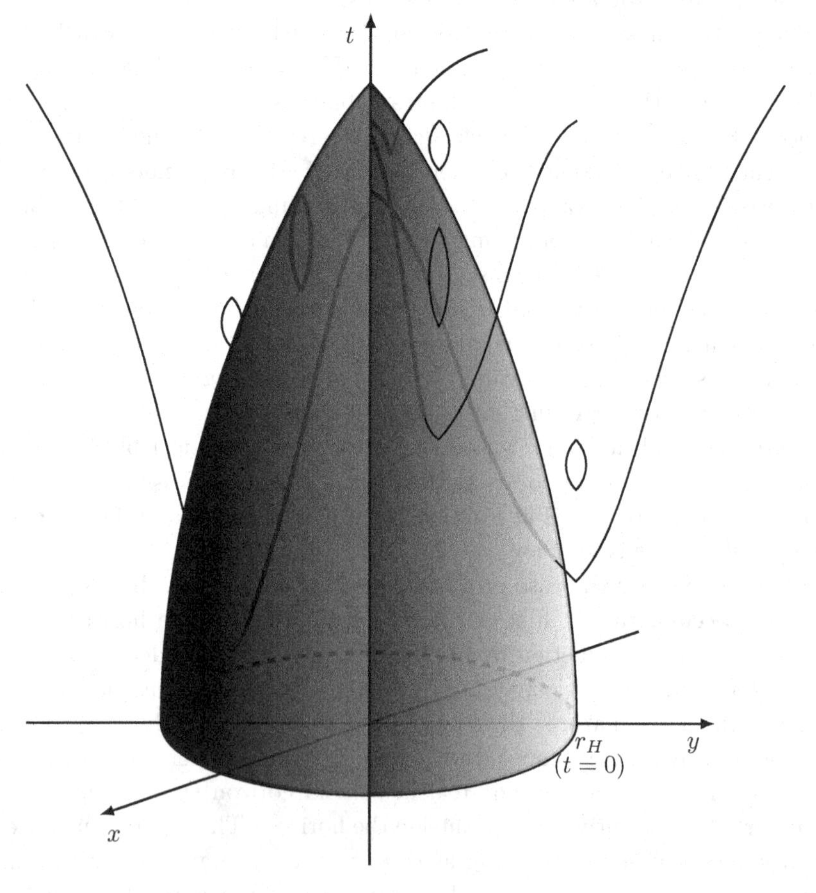

Fig. 2.1 A very naive picture of an evaporating black hole. Loops are virtual pairs, whereas long lines are real escaping (or trapped). The horizon radius r_H decreases as a function of t because of evaporation.

In any case, the role of the collapse for the emission must be emphasized. The point is that the final state of the collapse will result in general at most in a Kerr-Newman solution, which is a stationary state, of which the stationary Schwarzschild solution is a particular case. In such a situation, a mixing of negative and positive energy states is unexpected. The mechanism of superradiance [Zel'dovich (1972); Starobinsky (1973); Unruh (1974)] has been previously shown to allow a loss of angular momentum and charge from the black hole, as clarified firstly by Zel'dovich [Zel'dovich (1972)], and works well also for stationary states. But the emission of mass requires a mixing of positive and negative energy states, which, correspondingly, needs a dynamical non stationary spacetime configuration. This is offered by the collapse, which is then fundamental in the whole process. Nevertheless, despite the collapse process may be arbitrarily complicated, the possible final configurations are quite elementary, being parameterized by at most three parameters: the mass, the gauge charge and the angular momentum. As a consequence one expects that the particle emission should be independent from the details of the collapse.

2.1 Particle creation by black holes: The computation

The process we have in mind is expected to be different from the superradiance mechanism, so that it is sufficient to consider the simplest possible configuration parameterized only by the mass M, whose final state is represented by the static Schwarzschild solution:

$$ds^2 = -\left(1 - \frac{2M}{r}\right) dt^2 + \frac{dr^2}{1 - \frac{2M}{r}} + r^2(d\theta^2 + \sin^2\theta\, d\phi^2). \tag{2.5}$$

The Schwarzschild coordinates cover only the external solution, whereas the global solution is given in Kruskal-Szekeres coordinates and is schematized in the conformal Carter-Penrose diagram shown in Chapter 1. However, this solution describes an eternal black hole. The conformal diagram describing the global solution corresponding to a realistic black hole should contain also the collapsing star, with the formation of the black hole. If one considers the simplest case of a spherical collapse, then, as a consequence of the Birkhoff theorem the external solution is always described by the Schwarzschild metric, so that of the Carter-Penrose diagram it will survive the part corresponding to the outside of the star and the future infinity, as pictured in the figure.

The internal solution is completely different, since it does not contain the past event horizon, neither the physical singularity nor the further asymptotical flat region. All what remains is a time-like curve representing the centre of the star, which is the vertical line in the corresponding conformal diagram reproducing the original picture and reported in Fig. 2.2.

Now let us consider a Hermitian massless Klein-Gordon field operator ϕ. First, it can be written as

$$\phi = \sum_i (f_i a_i + \bar{f}_i a_i^\dagger), \tag{2.6}$$

where the sum is formal (t.i. it may contain both sums and integrals), and the f_i form an orthonormal complete set of solutions over $\mathcal{I}^-$ of the massless Klein-Gordon equation on the Schwarzschild background. This

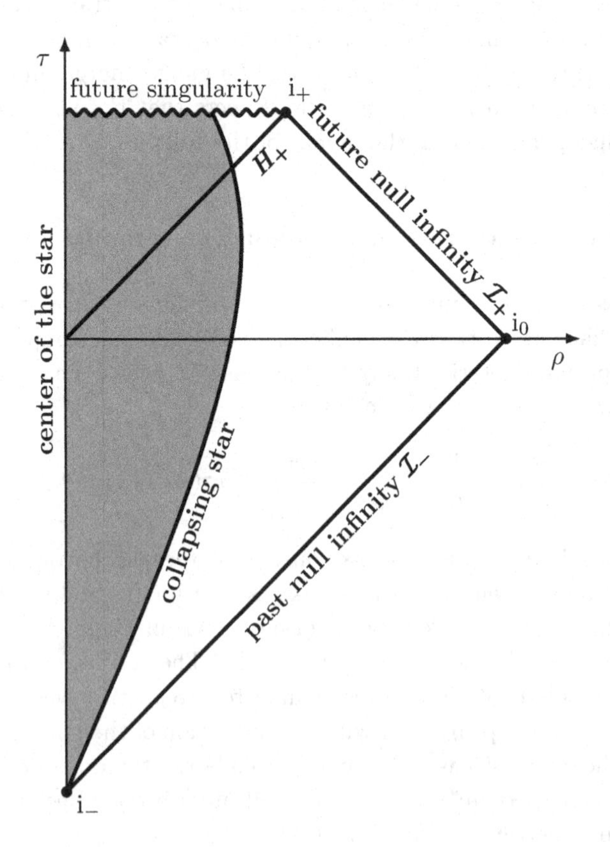

Fig. 2.2 Carter-Penrose diagram of a collapsing star.

means that

$$\frac{i}{2}\int_{\mathcal{I}^-}(f_i\bar{f}_{j;a} - f_j\bar{f}_{i;a})d\Sigma^a = \delta_{ij}. \tag{2.7}$$

Then, a_i and $a_i^\dagger$ are the annihilators and creators of of positive frequency states with respect to the affine parameter over $\mathcal{I}^-$. At this point a comment is in order. $\mathcal{I}^-$ is part of the asymptotically flat boundary of the spacetime. But the question is just how to define the points at infinity as boundary points, since they are infinitely distant from the neighbour points. It is here that Penrose proposed to give meaning to this concept by thinking at them from a conformal point of view, t.i. as a three dimensional boundary $\mathcal{I}$ of a four dimensional conformally finite spacetime. More precisely, let us consider a non-physical metric $\tilde{g}_{ab}$ related to the physical metric g_{ab} by the conformal scaling

$$\tilde{g}_{ab} = \Omega^2 g_{ab} \tag{2.8}$$

such that the infinity becomes finite and regular. This means that $\mathcal{I}$ correspond to the points where $\Omega = 0$, but regularity requires $\Omega_{;a} = 0$. Then one works with the non-physical spacetime $(\tilde{\mathcal{M}}, \tilde{g}_{ab})$, keeping in mind that all conformally invariant quantities will maintain the same meaning in the physical spacetime $(\mathcal{M}, g_{ab})$ as asymptotic properties. In particular, the boundary results to be a null surface. Indeed, if following Penrose we define the quantity

$$S_{ab} := \frac{1}{6}Rg_{ab} - R_{ab}, \tag{2.9}$$

than it is related to the conformal correspondent $\tilde{S}_{ab}$ by the relation

$$S^a{}_b = \Omega^2\tilde{S}^a{}_b + 2\Omega\Omega^{;a}{}_{;b} - \Omega^{;c}\Omega_{;c}\delta^a_b. \tag{2.10}$$

If only conformal fields can extend to infinity (as, for example, the electromagnetic field) then the scalar curvature vanishes at infinity, $R|_{\mathcal{I}} = 0$, so that taking the trace of (2.10) and using $\Omega_{\mathcal{I}} = 0$, we get $\Omega^{;c}\Omega_{;c} = 0$, which shows that the boundary is lightlike.

Now, the question is if the choice of the positive frequencies is well defined on $\mathcal{I}^-$. This is related to the existence of asymptotic Poincaré like symmetries on the flat asymptotic region. These asymptotic symmetries are described by the Bondi-Metzner-Sachs group (see Chapter 1). This has been studied in particular by Sachs [Sachs (1962)] whose idea was to recover the flat spacetime symmetries for the asymptotically flat spacetimes. Surprisingly, he discovered that the asymptotic symmetries form a larger group that he baptized the Generalized Bondi-Metzner group or GBM group, today known indeed as the Bondi-Metzner-Sachs (BMS) group. This group contains a normal abelian subgroup, whose factor group results to be isomorphic to the isotropic homogeneous Lorentz group. In particular, it contains an infinity of subgroups isomorphic to the orthochronous Poincaré group, and a four-dimensional abelian subgroup which can be identified with the group of translations. The ones leaving invariant the polar angles θ and ϕ are the Bondi-Metzner-Sachs super translations. In the Penrose description [Penrose (1963)], the BMS group has a very simple geometrical interpretation: it is the group of the conformal auto-transformations of $\mathcal{I}^+$ (or $\mathcal{I}^-$), which preserve angles and null angles, the last ones being the angles between tangent vectors such that $n^a = -\Omega^{;a}$ is a linear combination of them. Thus, although the super translations could make ambiguous the definition of a positive energy, the fact that they are just part of a conformal group of auto-transformations ensures that they do not affect the positivity of energy, which is thus well defined.

Note that we are discussing about a massless field, of which $\mathcal{I}^-$ is a good Cauchy surface. Then, the field is completely determined by these data and is then everywhere well defined. On the opposite, the future null infinity is not a good Cauchy surface since it can encode the data regarding the region external to the horizon, the remaining data being encoded in the (future) event horizon H.

With respect to the future surface, then, the field operator can be decomposed as

$$\phi = \sum_i \{p_i b_i + \bar{p}_i b_i^\dagger + q_i c_i + \bar{q}_i c_i^\dagger\}, \tag{2.11}$$

where p_i defines the *purely outgoing modes*, which vanish on the horizon, and q_i are the *non-outgoing* modes with vanishing Cauchy data on the null future $\mathcal{I}^+$. In other words, (p_i, q_i) form an orthonormal complete system on the Cauchy surface $\Sigma = \mathcal{I}^+ \bigcup H$. An observer at future infinity and far from the horizon should interpret $b_i^\dagger$ as the creators of outgoing positive energy particles. Thus, the p_i must contain only positive frequencies w.r.t. the affine parameters of the null geodesic generators of $\mathcal{I}^+$. For the remaining modes q_i it is not necessary to fix a condition of positivity for the frequencies for computing the outgoing spectrum. Thus, one can avoid the problem of understanding if such a condition should be implemented and w.r.t. which time parameter.

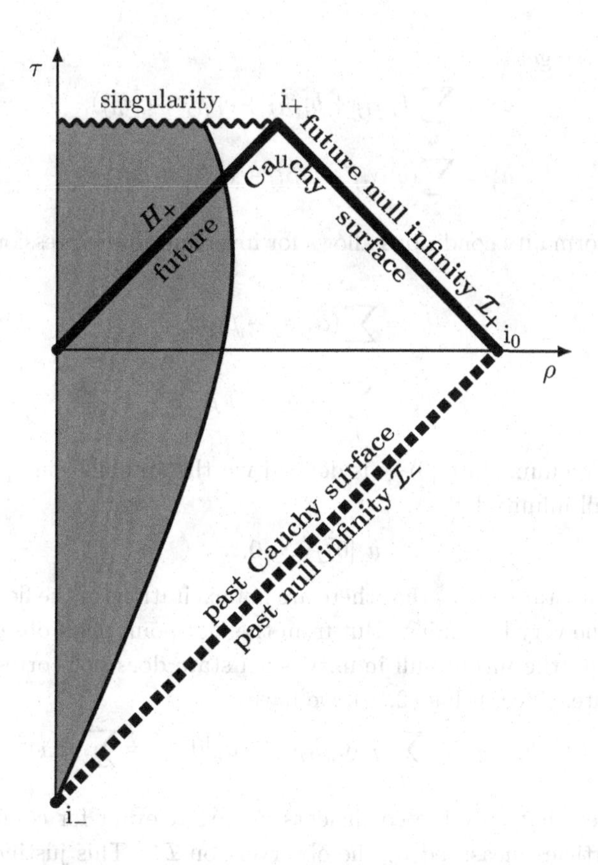

Fig. 2.3 Past and future Cauchy surfaces for the massless Klein-Gordon equation.

The completeness of the $f_i, \bar{f}_{ii}$ system implies that there must exist transformation matrices $\alpha, \beta, \gamma, \eta$ such that

$$p_i = \sum_j (\alpha_{ij} f_j + \beta_{ij} \bar{f}_j), \qquad q_i = \sum_j (\gamma_{ij} f_j + \eta_{ij} \bar{f}_j). \tag{2.12}$$

Thus

$$\phi = \sum_{i,j} (\alpha_{ij} f_j b_i + \beta_{ij} \bar{f}_j b_i + \bar{\alpha}_{ij} \bar{f}_j b_i^\dagger + \bar{\beta}_{ij} f_j b_i^\dagger + \gamma_{ij} f_j c_i + \eta_{ij} \bar{f}_j c_i + \bar{\gamma}_{ij} \bar{f}_j c_i^\dagger + \bar{\eta}_{ij} f_j c_i^\dagger), \tag{2.13}$$

from which we get

$$a_j = \sum_i (b_i \alpha_{ij} + b_i^\dagger \bar{\beta}_{ij} + c_i \gamma_{ij} + c_i^\dagger \bar{\eta}_{ij}), \tag{2.14}$$

$$a_j^\dagger = \sum_i (b_i^\dagger \bar{\alpha}_{ij} + b_i \beta_{ij} + c_i^\dagger \bar{\gamma}_{ij} + c_i \eta_{ij}). \tag{2.15}$$

The orthonormality conditions allows for inverting this expressions in order to get

$$b_i = \sum_j (\bar{\alpha}_{ij} a_j - \bar{\beta}_{ij} a_j^\dagger), \tag{2.16}$$

$$c_i = \sum_j (\bar{\gamma}_{ij} a_j - \bar{\eta}_{ij} a_j^\dagger). \tag{2.17}$$

The initial vacuum state $|0\rangle_{\mathcal{I}^-}$ is defined via the annihilation operators on the past null infinity by

$$a_j |0\rangle_{\mathcal{I}^-} = 0, \tag{2.18}$$

for all j. This state means that there are not excitations of the field (particle states) at the very beginning. But from the viewpoint of an observer in the far future, on the future null infinity, such state does not correspond to a vacuum state, since, using (2.16), we have

$$_{\mathcal{I}^-}\langle 0| b_i^\dagger b_i |0\rangle_{\mathcal{I}^-} = \sum_{jk} {}_{\mathcal{I}^-}\langle 0| \bar{\beta}_{ij} a_j \beta_{ik} a_k^\dagger |0\rangle_{\mathcal{I}^-} = \sum_k |\beta_{ik}|^2. \tag{2.19}$$

Thus, we see that only the coefficients β_{ij} are relevant for computing the created particles measured by the observers on $\mathcal{I}^+$. This justifies the previous claim about the irrelevance of choosing a positivity condition over the Cauchy data on the event horizon. By construction the coefficients β_{ij} are the asymptotic future coefficients. If non vanishing, they determine a stationary final flux of particles, and is thus expected to depend on the final configuration only and not on the details of the history of the collapse. Thus one could expect that during the collapse it is produced only a finite number of particles, which are finally dispersed and replaced by the final stationary flux. It is obvious that such a final picture can be taken seriously only for time intervals such that the lost mass is very small compared to the total black hole mass. As we will see, the flux increases as the mass decreases, so that the final destination should be similar to a true explosion, even though a precise account of this would probably require to take account of quantum gravity corrections.

Now we we are ready to compute the asymptotic coefficient, following the original paper. Thus we are interested only in the relations among the past coefficient functions f_i and the future coefficient functions p_j. To this end, we introduce the Eddington-Finkelstein null coordinates

$$u = t - r - 2M \log \left| \frac{r}{2M} - 1 \right|, \tag{2.20}$$

$$v = t + r + 2M \log \left| \frac{r}{2M} - 1 \right| \tag{2.21}$$

which represent advanced and retarded time coordinates respectively. Because of their meaning, it is convenient to expand the solutions of the wave equation in terms of the Fourier modes w.r.t. the time v for the f_i, and w.r.t. the time u for the p_i (see Chapter 1). Thus

$$p_{\omega lm} = \frac{1}{\sqrt{2\pi\omega}} \frac{1}{r} P_{\omega lm}(r) e^{i\omega u} Y_{lm}(\theta, \phi), \tag{2.22}$$

$$f_{\omega' lm} = \frac{1}{\sqrt{2\pi\omega'}} \frac{1}{r} F_{\omega' lm}(r) e^{i\omega' v} Y_{lm}(\theta, \phi), \tag{2.23}$$

where Y_{lm} are the usual scalar spherical harmonics. The relations we are looking for (the continuum version of the first of (2.12)) are

$$p_{\omega lm} = \sum_{l'm'} \int_0^\infty (\alpha_{\omega lm;\omega' l'm'} f_{\omega' l'm'} + \beta_{\omega lm;\omega' l'm'} \bar{f}_{\omega' l'm'}) d\omega'. \tag{2.24}$$

In order to simplify the details, let us assume that the collapse of the star is spherical. Then, the spherical symmetry is a symmetry at any instant and the angular momentum is conserved. Therefore, the above coefficients are diagonal w.r.t. the indices $\{(lm), (l'm')\}$, and the relation becomes

$$p_{\omega(lm)} = \int_0^\infty (\alpha_{\omega\omega'(lm)} f_{\omega'(lm)} + \beta_{\omega\omega'(lm)} \bar{f}_{\omega'(lm)}) d\omega'. \tag{2.25}$$

Here the angular momentum labels are put in parenthesis to indicate that they can (and will) be dropped because, by diagonality, are just further labels. The strategy for computing the coefficients $\beta_{\omega\omega'}$ is to consider p_ω as propagating backward in time starting from $\mathcal{I}^+$, with vanishing Cauchy data on the event horizon H. Part of this backward evolution will result in a scattering on the star, deviating the particles toward $\mathcal{I}^-$ with the same frequency as the starting future one. This is because we are assuming that the star is very massive w.r.t. the energy scale of this process and any back reaction on the star, caused by the scattering, is totally negligible. Thus, such process contributes to the coefficient $\alpha_{\omega\omega'}$ with a diagonal contribution $\delta_{\omega\omega'}$. But a second fraction $p_\omega^{(2)}$ of p_ω will describe a backward flux, which enters the collapsing body. In its latest times of collapse, when the horizon is near to be formed, the modes entering the star will be subjected to a

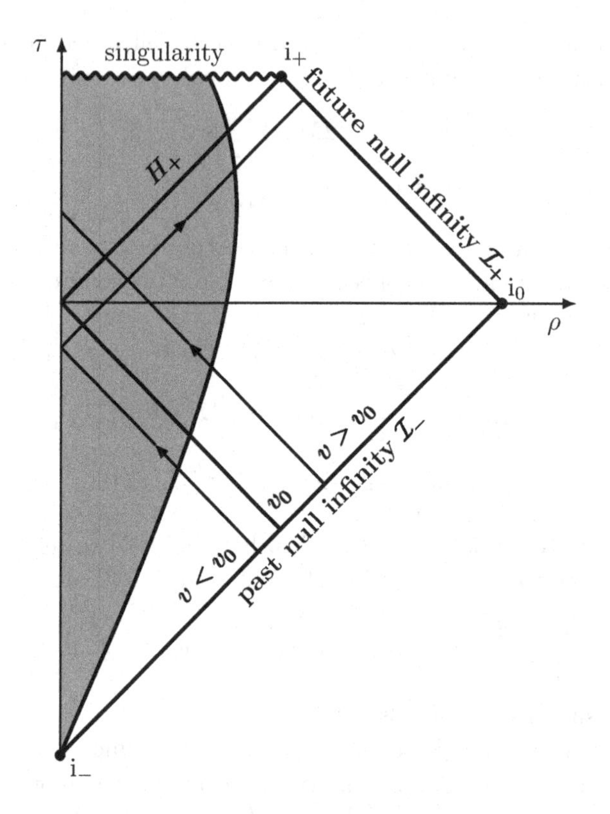

Fig. 2.4 Early times emitted light rays ($v < v_0$) reach the future null infinity.

very strong blueshift. Here the argument of Hawking is that because of such blueshift, the frequency of the propagating mode is strongly blue shifted. In such situation the dispersion becomes very weak and one can describe the propagation of the Klein-Gordon massless field, inside the collapsing body, in terms of pure geometrical optics, since the star becomes very transparent for very high frequencies. The coordinate v is a Killing parameter on the slice $\mathcal{I}^-$, which parameterizes the null geodesics entering the horizon when $v > v_0$, v_0 being the instant at which it starts the formation of the event horizon, see Fig. 2.4.

Thus, the part of backward p entering the collapsing body and reaching the past null infinity, will correspond to times earlier than v_0. Particles incoming (forward in time) from $\mathcal{I}^-$ at $v > v_0$ will fall into the horizon and will never reach the future null infinity. On the other hand, independently

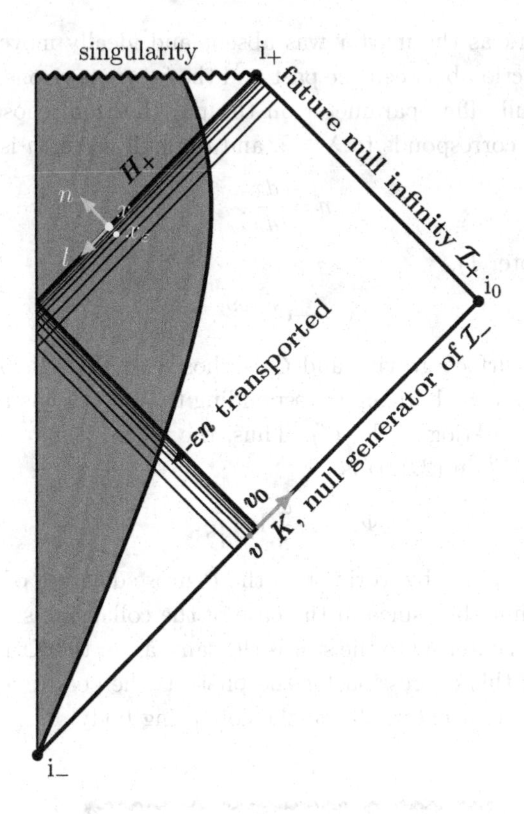

Fig. 2.5 Determining the front waves.

on how much $v < v_0$ is close to v_0, since the retarded time u goes to infinity at H, there will be an infinity of surfaces of constant phase accumulating on the horizon beyond the fixed one (corresponding to $u = u_0$). Then, between the fixed v and v_0 there is an infinite number of cycles that can propagate up to the future null infinity. This allows for getting an estimation of $p_\omega^{(2)}$ as follows. We will refer to Fig. 2.5.

On an arbitrary point x of the horizon fix a null vector l tangent to a null generatrix of the horizon, directed backward w.r.t. v, and the orthogonal null vector n oriented radially inward, normalized so that $n \cdot l = -1$. For a positive small number ε, the $-\varepsilon n$ will shift x a little bit to x_ε on a null surface external to the horizon. Now, let us transport l and n along the null generatrix passing through x; correspondingly, the point x_ε will move on the surface u =constant, on which the phase of $p_\omega^{(2)}$ is constant. For a while let us work as the matter was absent and ideally move the point x until the intersection between the past and the future horizons, see Fig. 2.6. Let λ be the null affine parameter on the past horizon, chosen such that the intersection corresponds to $\lambda = 0$, and the null vector n is just

$$n = \frac{dx}{d\lambda}. \tag{2.26}$$

Then (see Chapter 1),

$$\lambda = -Ce^{-\kappa u}, \tag{2.27}$$

where κ is the surface gravity and C is chosen so that (2.26) is fulfilled. Then, a shift from the horizon corresponding to $\lambda = -\varepsilon$, has retarded time coordinate $u_\varepsilon = -\frac{1}{\kappa}(\log \varepsilon - \log C)$. Thus, the phase of $p_\omega^{(2)}$ on the surface $u = u_\varepsilon$ ($e^{i\omega u} =: e^{i\Phi}$ in (2.22)) is

$$\Phi_\varepsilon = -\frac{\omega}{\kappa} \log \frac{\varepsilon}{C}. \tag{2.28}$$

This has been obtained by working on the Penrose diagram of the Kruskal-Szekeres spacetime, but since in the case of the collapsing star, the future part of the region external to the star is the same as for the Kruskal-Szekeres spacetime, then this expression for the phase is the good one also for the realistic case, at least externally to the collapsing body.

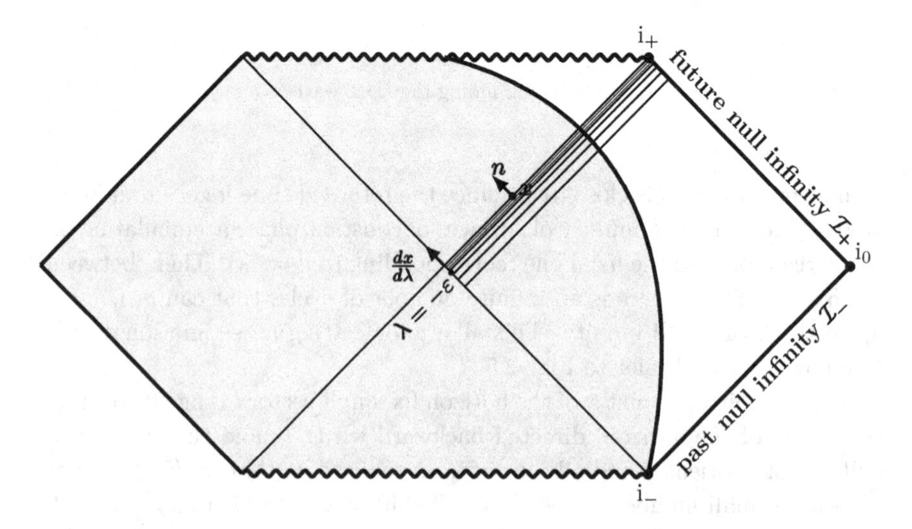

Fig. 2.6 Collapsing star with added a "ghost" eternal Schwarzschild Carter-Penrose diagram.

When the vector $-\varepsilon n$ is parallely transported along the null surface, continuing inside the star, assuming the geometrical optics approximation to be valid, until the end point of the horizon and then back to $\mathcal{I}^-$, it will finally lie on $\mathcal{I}^-$ parallel to the null Killing generator K of the past null infinity, see Fig. 2.5. In particular, then, $DK = n$ for some constant D there. Note that along the transport, the vector $-\varepsilon n$ continues to connect the null geodetic passing through x and prolonged down to $\mathcal{I}^-$ (in $v = v_0$) to the surface of constant phase for $p^{(2)}_\omega$. Since v is a Killing coordinate, see Chapter 1, then the small displacement $v - v_0$ from v_0 corresponding to $-\varepsilon n$, must coincide with the vector $(v - v_0)K$, from which we get $-\varepsilon D = v - v_0$, and hence

$$\Phi_\varepsilon(v) = -\frac{\omega}{\kappa}\log\frac{v_0 - v}{CD}. \tag{2.29}$$

This expression is valid for small ε, t.i. for small $v_0 - v$; otherwise the geometric optics approximation fails and the surface of constant phase may be strongly deformed inside matter. We then arrive at a good approximation for the wave function $p^{(2)}_\omega$ valid for instants $v > v_1$, with $0 < v_0 - v_1 \ll 1$:

$$p^{(2)}_\omega \simeq \begin{cases} 0 & v > v_0, \\ \frac{1}{\sqrt{2\pi\omega}}\frac{1}{r}P^-_\omega(2M)\exp\left(-i\frac{\omega}{\kappa}\log\frac{v_0 - v}{CD}\right) & 0 < v_0 - v \ll 1, \end{cases} \tag{2.30}$$

where $P^-_\omega(2M)$ is the value of the analytic prolongation of P_ω on the past horizon. At this point, we can compute the asymptotic expression for the coefficients of the matrices α and β, simply by taking the Fourier transform of the asymptotic expansion for $p^{(2)}_\omega$. Indeed, looking at (2.23) and (2.22), we se that (after forgetting the Y_{lm} terms)

$$\int_{-\infty}^{\infty} dv e^{-i\omega' v} p^{(2)}_\omega = \alpha_{\omega\omega'}\frac{1}{\sqrt{2\pi\omega'}}\frac{1}{r}2\pi F_{\omega'}(r). \tag{2.31}$$

If we now suppose $\omega' \gg \omega$, in the l.h.s. of this formula there will contribute the terms for which the frequency ω is strongly blue-shifted to ω', which happens near $v = v_0$. Thus, we can compute this integral by substituting $p^{(2)}_\omega$ with its asymptotic expression. We have to compute the integral

$$\int_{-\infty}^{v_0} dv\, e^{-i\omega' v}(v_0 - v)^{-i\frac{\omega}{\kappa}} = e^{-i\omega' v_0}\int_0^\infty ds e^{i\omega' s} s^{-i\frac{\omega}{\kappa}}. \tag{2.32}$$

Since we are interested in $\omega' > 0$, we can compute this integral by deforming the path of integration to the positive imaginary axis. Thus the integral becomes

$$e^{-i\omega' v_0}(i)^{1-i\frac{\omega}{\kappa}}\int_0^\infty ds\, e^{-\omega' s} s^{-i\frac{\omega}{\kappa}} = e^{-i\omega' v_0}(-i\omega')^{-1+i\frac{\omega}{\kappa}}\Gamma\left(1 - i\frac{\omega}{\kappa}\right), \tag{2.33}$$

where Γ is the usual Euler gamma function. In conclusion, for large values of ω' (w.r.t. ω) we get

$$\alpha^{(2)}_{\omega\omega'} \simeq \frac{1}{2\pi}\sqrt{\frac{\omega'}{\omega}}\frac{P^{-}_{\omega}(2M)}{F_{\omega'}(r)}(CD)^{i\frac{\omega}{\kappa}}e^{-i\omega' v_0}(-i\omega')^{-1+i\frac{\omega}{\kappa}}\Gamma\left(1-i\frac{\omega}{\kappa}\right). \quad (2.34)$$

In order to compute $\beta_{\omega\omega'}$, we see tat one must integrate versus $e^{+i\omega' v}$ in place of $e^{-i\omega' v}$. From this it easily follows that $\beta^{(2)}_{\omega\omega'} = -i\alpha^{(2)}_{\omega,-\omega'}$. This means that we have then to prolongate the solution varying ω' to $-\omega'$. Now, notice that, since $p^{(2)}_{\omega}$ vanishes for $v > v_0$, then its Fourier transform is analytic in the upper half plane.[1] Then, if we try to compute β by prolonging from ω' to $-\omega'$, we see that a logarithmic singularity is met at $\omega' = 0$. If we were able to do an exact computation, then, obviously, we would obtain that the singularity is at $\omega = \omega'$ in place of $\omega' = 0$. But, in our approximation, it is clear that, for large ω', considering the singularity as it was in $\omega' = 0$ is a good approximation. In doing the prolongation, we can deform the path with a counter clock small semi circle around the singularity, to avoid it. This deformation shows that almost nothing changes apart from a nontrivial monodromy correction coming from the term

$$(-i\omega')^{-1+i\frac{\omega}{\kappa}}, \quad (2.36)$$

and which give rise to the main result:

$$|\alpha^{(2)}_{\omega\omega'}| = e^{\frac{\pi\omega}{\kappa}}|\beta^{(2)}_{\omega\omega'}|. \quad (2.37)$$

The total number of emitted particles reaching $\mathcal{I}^+$ is

$$d\omega\int_0^{\infty}|\beta_{\omega\omega'}|^2 d\omega' = \infty, \quad (2.38)$$

since $|\beta_{\omega\omega'}|^2 \sim \frac{1}{\omega'}$ for large ω'. This is because the particles are created in a finite steady rate of emission for an infinite time. This can be understood by analyzing the process in terms of a complete set of wave packets:

$$p_{jn} = \int_{j\varepsilon}^{\varepsilon+j\varepsilon} e^{-2\pi i\frac{\omega}{\varepsilon}}p_{\omega}d\omega, \quad (2.39)$$

[1] Indeed,

$$\int_{-\infty}^{v_0} dte^{-i\omega t}f(t) \quad (2.35)$$

must be convergent essentially in $-\infty$ (since the other extreme of integration is finite).

for j, n integers and j, ε positive. They represent wave packets which, for very small ε, are peaked around a large retarded time $u = 2\pi n/\varepsilon$, with frequency $j\varepsilon$ and width $1/\varepsilon$. Expressed in terms of the f modes these are

$$p_{jn} = \int_0^\infty d\omega(\alpha_{jn,\omega} f_\omega + \beta_{jn,\omega} \bar{f}_\omega), \tag{2.40}$$

with

$$\alpha_{jn;\omega} = \frac{1}{\sqrt{\varepsilon}} \int_{j\varepsilon}^{\varepsilon+j\varepsilon} e^{-2\pi i \frac{\omega}{\varepsilon}} \alpha_{\omega'\omega} d\omega', \tag{2.41}$$

$$\beta_{jn;\omega} = \frac{1}{\sqrt{\varepsilon}} \int_{j\varepsilon}^{\varepsilon+j\varepsilon} e^{-2\pi i \frac{\omega}{\varepsilon}} \beta_{\omega'\omega} d\omega'. \tag{2.42}$$

Large values of n correspond to waves reaching the future null infinity at a very late retarded time $u = 2\pi n/\varepsilon$. For these modes it is consistent to adopt the asymptotic expansions for $\alpha_{\omega\omega'}$ and $\beta_{\omega\omega'}$ in order to do computations. Indeed, by using such expressions we get, for example,

$$\begin{aligned} |\alpha_{jn\omega}| =& \left| \frac{1}{\sqrt{\varepsilon}} \int_{j\varepsilon}^{j\varepsilon+\varepsilon} \frac{1}{2\pi} \frac{P_{\omega'}^-(2M)}{F_\omega^-(r)} (CD)^{i\frac{\omega'}{\kappa}} e^{-i\omega v_0} \left(\frac{\omega}{\omega'}\right)^{\frac{1}{2}} \right. \\ & \left. \times \Gamma(1 - i\omega'/\kappa)(-i\omega)^{-1+i\frac{\omega'}{\kappa}} e^{-2\pi i n\omega'\varepsilon^{-1}} \, d\omega' \right| \\ =& \frac{1}{\pi} \frac{P_{j\varepsilon}^-(2M)}{F_\omega^-(r)} \frac{1}{\sqrt{\omega j\varepsilon}} \left| \Gamma\left(1 - i\frac{j\varepsilon}{\kappa}\right) \right| \left| \frac{\sin\left[\frac{\varepsilon}{2}\left(\kappa^{-1}\log\omega - 2\pi n\varepsilon^{-1}\right)\right]}{\sqrt{\varepsilon}\left(\kappa^{-1}\log\omega - 2\pi n\varepsilon^{-1}\right)} \right|. \end{aligned} \tag{2.43}$$

From the last factor, we see that for large n, the main contributions come from $\omega \sim \exp(2\pi n\kappa/\varepsilon)$ so, for very high frequencies. Thus, these contributions do not depend on the details of the collapse.

One can then compute the number of particles, created in the wave packet p_{jn}, which reach $\mathcal{I}^+$:

$$N_{jn} = \int_0^\infty |\beta_{jn\omega}|^2 d\omega. \tag{2.44}$$

To do this, let us consider the wave packet p_{jn} propagating backward from $\mathcal{I}^+$. A fraction Γ_{jn} will enter the collapsing body, whereas a fraction $1-\Gamma_{jn}$ will be scattered away. The entering fraction is

$$\Gamma_{jn} = \int_0^\infty (|\alpha_{jn\omega}^{(2)}|^2 - |\beta_{jn\omega}^{(2)}|^2) d\omega, \tag{2.45}$$

the negative sign being due to the fact that, as discussed above, the $|\beta_{jn\omega}^{(2)}|^2$ created particles enters the horizon with negative frequency. From (2.37) we have

$$|\alpha_{jn\omega}^{(2)}| = e^{\frac{\pi j n}{\kappa}} |\beta_{jn\omega}^{(2)}|, \tag{2.46}$$

which used in the above expression gives

$$\Gamma_{jn} = (e^{\frac{2\pi\omega}{\kappa}} - 1) \int_0^\infty |\beta^{(2)}_{jn\omega}|^2 \, d\omega, \tag{2.47}$$

and, then,

$$N_{jn} = \frac{\Gamma_{jn}}{e^{\frac{2\pi\omega}{\kappa}} - 1}. \tag{2.48}$$

The point is that, for late retarded times, the fraction of particles entering the body, is almost the same fraction of particles that would have crossed the past horizon if an eternal black hole replaced the collapsing body (see Fig. 2.7). Then, by symmetry, such factor Γ_{jn} is the same as the fraction entering the future event horizon of a wave packet started from $\mathcal{I}^-$ and partially absorbed by the black hole. Thus, there is the same process of emission and absorption as for a black body. Thus, a black hole should really appear to a far observer as a black body at a temperature

$$T_{BH} = \frac{\kappa}{2\pi}. \tag{2.49}$$

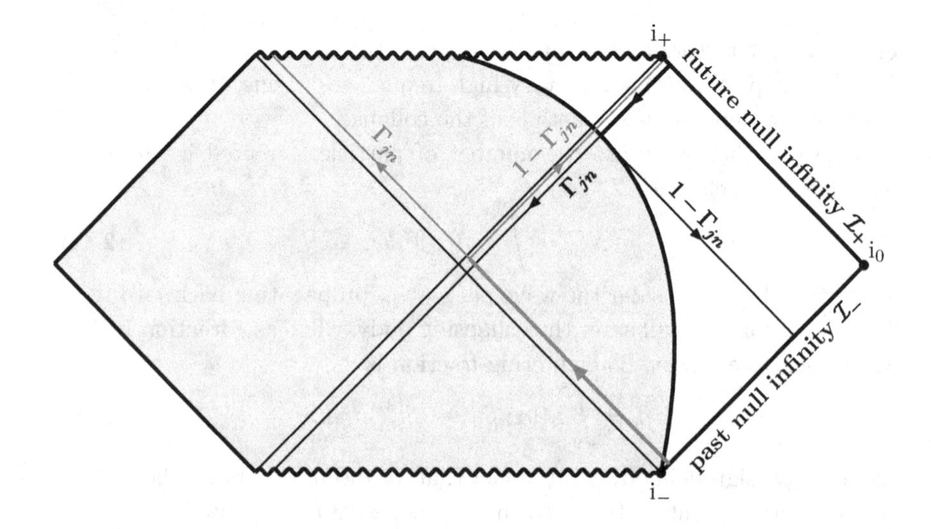

Fig. 2.7 Absorption and emission behavior of the horizons.

2.2 Particle creation by black holes: Dependence on the details

We claimed that the final picture of a stationary configuration with a constant steady flux emitted toward infinity at a constant temperature proportional to the surface gravity at the future event horizon is independent from the details of the collapse. Indeed, we have seen that only the very late part of the collapse, very near to the formation of the horizon, gives rise to the steady state emission, so that only the asymptotic expression of the wave packets are relevant. On the other side, such very late lapse of time, where all is governed essentially by the surface gravity, contains an infinite number of wavefronts, whereas just a finite number of photons, emitted in earlier times, will be sensible to the details of the collapse and, so, dependent on how the collapse exactly takes place.

However, we have considered just the special case of a massless scalar field in the presence of a spherically collapsing body. Thus, more ingredients could change the final result: a non symmetrical collapse, the spin of the field, or its mass, the angular momentum of the collapsing body, or its electrical charge, the effects of the back reaction. Let us report here the comments in [Hawking (1975)].

2.2.1 *Nonsymmetrical collapse*

We consider a massive body, which collapses under the action of gravitational forces, finally giving rise to a Schwarzschild black hole. We will assume that, despite of the symmetry of the final configuration, the implosion of the star takes place without any particular symmetry. This means that for late times the spacetime is described by an almost-stationary state. This is equivalent to say that on $\mathcal{I}_+$ it does exist a system of Bondi coordinates with respect to which all data in spherical modes with positive frequencies. This privileged Bondi coordinates may not exist on $\mathcal{I}_-$, since on it there is not necessarily any quasi-stationarity condition, nor spatial symmetries. Nevertheless, on $\mathcal{I}_-$ we can chose an arbitrary system of Bondi coordinates, no matter which one.

Now, let us consider a null geodesic generator of the event horizon. We can prolongate it backward in the past until it intersects $\mathcal{I}_-$ in a point y on a null geodesic generator $\tilde{\gamma}$ of $\mathcal{I}_-$. In y we can choose two light-like vectors, $\boldsymbol{\hat{n}}$ and $\boldsymbol{\tilde{n}}$, the first one tangent to γ and the latter tangent to $\tilde{\gamma}$. Finally, we transport this pair of vectors along γ forward in time until a point x where the metric is almost the Schwarzschild one. In order to help intuition we refer to the Carter-Penrose diagram depicted in Fig. 2.8. However, such a picture is only symbolic, in the sense that there are no symmetries far from the almost Schwarzschild region so that a meaningful diagram should be four dimensional.

Let us consider a solution $p_{\omega lm}$ which from $\mathcal{I}_+$ propagate backward in time, partially entering the collapsing body, in a fraction $q_{\omega lm}$ back to $\mathcal{I}_-$.

In y, $-\varepsilon\hat{\boldsymbol{n}}$ will connect y to a point of phase

$$-\phi = -i\frac{\omega}{\kappa}\log\frac{\varepsilon}{E} \tag{2.50}$$

for $q_{\omega lm}$ for some constant E (which will depend on the generator $\tilde{\gamma}$). If we transport $\tilde{\boldsymbol{n}}$ along γ, through the collapsing body and then in the late future until reaching a point x on H, where the metric is almost the Schwarzschild

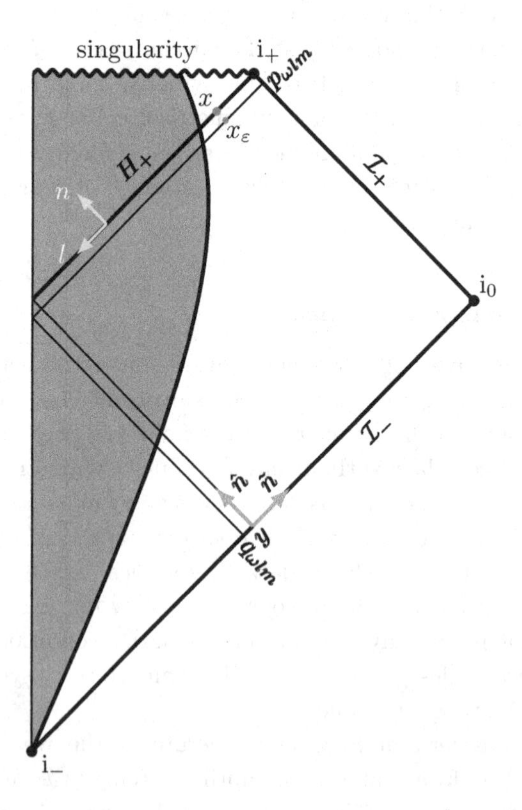

Fig. 2.8 Indicative diagram of a nonsymmetric collapse ending in a static spherical configuration.

one, then, there it will be a combination of the null vectors $\boldsymbol{l}$ and $\boldsymbol{n}$, and will connect x with a surface with the same constant phase $i\phi$ for $p_{\omega lm}$. By reasoning like in the spherical collapse, we then conclude that, with respect to the coordinate v on $\mathcal{I}_-$, the phase of the solution on the null generator $\tilde{\gamma}$ will be

$$i\phi = -i\frac{\omega}{\kappa}\log\frac{v_0 - v}{K}, \tag{2.51}$$

and the solution will take the form

$$q_{\omega lm} \sim Ae^{-i\frac{\omega}{\kappa}\log\frac{v_0 - v}{K}}, \tag{2.52}$$

where A, K and v_0 depend from the given null generator $\tilde{\gamma}$. In the arbitrarily chosen Bondi coordinates on $\mathcal{I}_-$, the harmonic components will have complicate expressions depending on the lack of symmetries. Nevertheless, the dependence on the frequencies, given by the above asymptotic behavior, is the same as for the spherical collapse, so that one can employ exactly the same arguments to deduce that in the final stage the black hole will emit particles at a temperature

$$T = \frac{\kappa}{2\pi}. \tag{2.53}$$

2.2.2 *The spin of the fields*

It is possible to see that the same kind of thermal emission takes place at the same temperature $T = \kappa/2\pi$, governed by the surface gravity κ, for any kind of massless particles, regardless of their spin. In the case of photons or gravitons, the expression for the field on $\mathcal{I}_-$ generated by positive frequency modes on $\mathcal{I}_+$ will have the form

$$q_\omega \simeq \begin{cases} 0 & v > v_0, \\ \frac{\Gamma(s)}{\sqrt{2\pi\omega}}\frac{1}{r}P_\omega^-(2M)\exp\left(-i\frac{\omega}{\kappa}\log\frac{v_0 - v}{CD}\right) & 0 < v_0 - v \ll 1, \end{cases} \tag{2.54}$$

where here $\Gamma(s)$ is a blue shift factor dependent on the spin, that is on their vectorial or tensorial behavior. Since the scalar product on the fields now involve the metric tensor, these factors are exactly the ones canceling out the contribution of the metric, so that the scalar products behave indeed as scalars. Thus, the asymptotic expressions of the coefficient α and β are just the same as for the scalar field, which means that also photons and gravitons are radiated away thermally.

For massless fermions, like could be neutrinos, the components with negative frequency contribute to the probability flux of the collapsing body with an opposite sign as compared to the bosonic case. From this it simply follows that the emission law is governed by the factor

$$N_{jn} = \frac{\Gamma_{jn}}{e^{\frac{2\pi\omega}{\kappa}} + 1}, \tag{2.55}$$

which is in perfect accordance with th Fermi-Dirac statistics, as one should expect for a thermal emission of fermions.

2.2.3 *Massive fields*

Another issue to be considered is if and haw a mass could change the emission process. In this case, the conformal diagrams of Carter and Penrose are not suitable, since massive particles not entering the horizon will not reach the future null infinity, but at most the "point" i_+. In place of the conformal infinity, in this case it would be more suitable to introduce the projective infinities and use them in order to define the ingoing and outgoing states. However, one can consider the simplification where the final asymptotic state is a Schwarzschild or a Kerr one, and, after separating the variables, define the positive frequencies by means of the Killing vector field of the temporal translations, for both the final and the initial states.

A second issue in the case of massive particles is that if the collapse takes place after an infinite time in which the body has kept finite size, then in the past there could exist bound states. This cannot happen in the far future since then any particle is destined to escape to infinity or fall into the horizon (indeed, the Dirac equation on a Schwarzschild or a Kerr black hole solution, and the same for other linearized massive fields, there is no mass-gap and, indeed, an eventual bound state would be imbedded in the continuum spectrum). So, one may wonder if the presence of eventual bound states in the past may change the situation. However, the major contribution of the emission comes essentially from the moment when the body falls into the forming horizon. Thus, the previous story of the collapse is essentially irrelevant, and one is free to imagine that the final state of the collapse comes from a body which started with an infinitely extended size. Since in this situation no past bound states are allowed, we thus conclude that bound states have no influence on the final processes.

With these observations in mind, we see that there is not any essential difference with respect to the massless case, so that we can proceed in showing that the rate of emission will be again of the form

$$N_{jn} = \frac{\Gamma_{jn}}{e^{\frac{2\pi\omega}{\kappa}} \mp 1}, \tag{2.56}$$

where the sign will be $-$ for bosons and $+$ for fermions. It is worth to mention that in this formula ω includes the mass of the particles, so that in general, for astrophysical black holes, the probability of emitting massive particles is very small.

2.2.4 *Angular momentum*

Now, let us consider the case when the collapsing body brings an angular momentum leading to a Kerr black hole in the final state. In this case the equation of motion of scalar fields, photons and gravitons, as well as for neutrinos, can be analyzed by separating variables. In particular, on $\mathcal{I}_-$ we can pick an arbitrary system of Bondi coordinates, and decompose the field ψ in terms of entering modes $\{f_{\omega lm}\}$, with respect to the retarded time and the angular coordinates on $\mathcal{I}_-$. On $\mathcal{I}_+$ we have a privileged set of Bondi coordinates, determined by the asymptotic state, with respect to which we can define the outgoing modes $\{p_{\omega lm}\}$. Moving it back in the past, one can compute the fraction $q_{\omega lm}$ entering the body and reaching the past $\mathcal{I}_-$. Since the horizon H is rotating with angular velocity Ω with respect to the conformal null infinity, the frequency near the horizon will be $\omega - m\Omega$ in place of ω. Thus, one will get for the steady rate of emitted particles

$$N_{jnlm} = \frac{\Gamma_{jnlm}}{e^{2\pi(\omega - m\Omega)/\kappa} \mp 1}. \tag{2.57}$$

From this expression we see that particles with positive m have a higher probability to be emitted, which means that the radiations will give rise to a diminution of the angular momentum of the black hole. However, we also see that for m large enough and positive the denominator changes sign. However, this does not imply an inversion of the flux of particles, since also the sign Γ_{njlm} changes for bosons. This phenomenon is called *superradiance* and can be explained as follows. For a stationary black hole, the variation of mass dM is related to a change in the horizon area dA, angular momentum dj and electric charge dq by

$$dM = \frac{\kappa}{8\pi} dA + \Omega dj + V dq, \tag{2.58}$$

where Ω is the angular velocity of the horizon and V its potential. Let us consider a wave packet incident to the horizon with frequency ω, angular momentum m and charge e. If a given fraction ν of it enters the horizon, so that $dM = \nu\omega$, we get

$$\frac{\kappa}{8\pi} dA = dM(1 - m\Omega/\omega - eV/\omega). \tag{2.59}$$

Since bosons satisfy the weak energy condition (so that the local energy density is positive), in a classical process the area of the horizon can only increase and, for $e = 0$ and $m\Omega > \omega$, necessarily we have $dM > 0$. This corresponds to an inversion of the sign of the ingoing probability flux or, equivalently, to an inversion of the sign of Γ_{jnlm} since the reflection flux is higher than the incident one. This does not happens for fermions, which are not constrained to satisfy the weak energy condition, and have positive definite scalar product.

2.2.5 *Electric charge*

Let us consider a charged black hole. In the presence of neutral fields nothing of special happens apart from the fact that the surface gravity is lower than for uncharged black holes so that the emission takes place at a lower temperature. For a charged massless scalar field the equation of motion is

$$0 = g^{\mu\nu}(\nabla_\mu - ieA_\mu)(\nabla_\nu - ieA_\nu)\phi. \tag{2.60}$$

The propagating wave vector for a wave p_ω is thus

$$i\boldsymbol{k} = \nabla \log p_\omega - ie\boldsymbol{A} \tag{2.61}$$

and in the WKB limit its propagation equation is

$$k^\mu \nabla_\mu k_\nu = -eF_{\nu\mu}k^\mu, \tag{2.62}$$

where $\boldsymbol{F} = d\boldsymbol{A}$ is the field strength. This means that in general infinitesimal vectors $\boldsymbol{z} = -\varepsilon\boldsymbol{n}$ connect points that have a gauge invariant phase difference of $ik_\mu z^\mu$. If we consider propagation along $\boldsymbol{z}$

$$k^\mu \nabla_\mu z_\nu = -eF_{\nu\mu}z^\mu, \tag{2.63}$$

we see that $\boldsymbol{z}$ will connect surfaces with constant gauge invariant phase. Now, in the final stage, where the solution is stationary, one can fix a gauge such that $\boldsymbol{A}$ is stationary and vanishes on $\mathcal{I}_+$. With these conditions the equation of motion becomes separable, with solutions having dependence $e^{i\omega u}$ from the retarded time. Let us fix a point x on the horizon in the far future, t.i. in the asymptotically stationary region. In x let us chose a null vector $\boldsymbol{l}$, tangent to a null geodesic generator, and a null normal vector $\boldsymbol{n}$ radially inward directed. The vector $\boldsymbol{z} = -\varepsilon\boldsymbol{n}$ connects x to a surface of phase

$$\Phi = -\frac{\omega}{\kappa}\log\frac{\varepsilon}{C} \tag{2.64}$$

for some constant C. However, the gauge invariant phase is

$$\phi_A = -\frac{\omega - eV}{\kappa} \log \frac{\varepsilon}{C} \tag{2.65}$$

with $V = k^\mu A_\mu$, and it converges to Φ only at $\mathcal{I}_+$. Now, let us propagate $\boldsymbol{l}$ back in time, with

$$l^\mu \nabla_\mu l_\nu = -eF_{\nu\mu} l^\mu, \tag{2.66}$$

until it meets in y a null geodesic generator γ of $\mathcal{I}_-$. Next, let us propagate $\boldsymbol{z}$ along the integral curve of $\boldsymbol{l}$,

$$l^\mu \nabla_\mu z_\nu = -eF_{\nu\mu} z^\mu, \tag{2.67}$$

backward to y. Then, $\boldsymbol{z}$ will connect y to surfaces with constant gauge invariant phase. Near $\mathcal{I}_-$ we can change gauge in order to make $\tilde{A}$ vanishing on $\mathcal{I}_-$. In this gauge the fraction of p_ω entering the body, along the generators of $\mathcal{I}_-$ will have phase

$$\Phi = -\frac{\omega - eV}{\kappa} \log \frac{v_0 - v}{K}, \tag{2.68}$$

where K is constant on each null generator of the conformal past infinity. Thus, we conclude that the emission will take place with frequency $\omega - eV$ in place of ω:

$$N_{inlm} = \frac{\Gamma_{jnlm}}{e^{(\omega - eV)/\kappa} \mp 1}. \tag{2.69}$$

Notice that for bosons we have again the same subtlety on the signs, but again we can explain it by means of superradiance. In particular, particles with the same sign of the charge as for the black hole charge (which has the sign of V) are emitted with higher probability and then the steady flow of particle will imply a discharge of the black hole.

2.2.6 *Back reaction*

The final comments regard the effects that the radiation has on the black hole itself. All the discussions considered up to now, show that all depends on the form of the metric in the very final stage of the collapse, when the horizon is forming an infinite flow of particles, created just out near the horizon, run away toward $\mathcal{I}_+$ with a steady rate. This looks in contrast with any local description, since the collapsing body should need to know exactly when the horizon is forming, while the event horizon depends also on the whole future story, being it a global observer independent concept. It is the apparent horizon to be defined locally.

Let us consider an observer who is freely falling into the black hole in a certain point of the horizon. He will be locally inertial and is then able to realize a local inertial frame with diameter of order M. In this local patch he can describe the field, say a scalar field, by means of a complete set of solutions $\{f_\omega\}$ of the equations of motion, satisfying the condition

$$\frac{i}{2}\int_\Sigma (f_\omega \nabla_\mu \bar{f}_{\omega'} - \bar{f}_{\omega'} \nabla_\mu f_\omega) d\Sigma^\mu = \delta(\omega - \omega') \tag{2.70}$$

on a Cauchy surface Σ, and which have dependence $e^{i\omega t}$ on the local time. This system will correctly define positive and negative frequencies for $\omega > M$, but not for $\omega < M$, since M is the size of the local frame. Differently from the previously constructed solutions $\{p_\omega\}$, the $\{f_\omega\}$ are continuous in the whole local path and, then, on the horizon. This means that their prolongation back to the conformal past null infinity are also continuous, there is no any discontinuity at any point v_0, and, as a consequence, no particles production. It is just the discontinuity at v_0 of the $\{p_\omega\}$ that give rise to particles creation, thanks to the tail $1/\omega'$ in its Fourier transform, for large values of the negative frequencies ω'. The $\{f_\omega\}$, for $\omega > M$, on $\mathcal{I}_-$ will have Fourier transforms whose negative frequency components will have small frequency, an then to them there will correspond very few particles having $\omega > M$. On the other hand, particles with $\omega < M$ cannot be seen by the free falling observer, since there would need detectors larger than the locally inertial patch.

This reasoning confirms that the emission of particles is not a local effect, but a global one. A free falling observer will not see an infinite number of particles flowing away from the horizon (he will not even see the horizon itself), he at most can see an indetermination in the energy which will be of the order

$$\Delta E \sim B^2 \sim M^{-4}, \tag{2.71}$$

B being the curvature. Nevertheless, both the local inertial observer and the asymptotic observer on $\mathcal{I}_+$ must experiment a back reaction on the black hole: because of the evaporation the total mass (and charge and angular momentum) must decrease balancing the steady flow of particles. The non locality of the process means that we cannot expect to be able to describe the back reaction by means of a local energy-momentum tensor. Rather, the free falling observer should take account of the outgoing energy (due to the escaping particles) by means of the fluctuations of order M^{-4}, which is expected to make ambiguous the position of the horizon. To the decreasing mass it should correspond an outgoing flux of energy. Indeed, if one cannot speak about a local energy momentum tensor, the global flux of energy through a macroscopic surface is perfectly defined (in the same way as the energy density pseudo-tensor of the gravitational field is ambiguous but its flux through surfaces is not since two such tensors give the same flux).

So, let us consider the energy momentum tensor density of the scalar field ψ

$$T_{\mu\nu} = \partial_\mu \psi \partial_\nu \psi - \frac{1}{2} g_{\mu\nu} \partial_\rho \psi \partial_\sigma \psi g^{\rho\sigma}. \qquad (2.72)$$

In the quantum case one has to subtract infinities in order to get a well defined result, for example by choosing a normal ordering for oscillator modes, with respect to some vacuum state. However, a nice thing is that for a quasi stationary region, all operators $T_{\mu\nu}$ renormalized in such the way that $\nabla_\mu T^{\mu\nu} = 0$, which are stationary (so that their Lie derivative $\mathcal{L}_{\boldsymbol{K}} T_{\mu\nu} = 0$, where $\boldsymbol{k}$ is the Killing vector field associated to time translations) and which coincide on $\mathcal{I}_+$ give rise to the same flux through any surface of radius r external to the horizon. Near $\mathcal{I}_+$ it is convenient to renormalize the field by choosing the normal ordering associated to the positive frequencies defined by the Killing vector $\boldsymbol{K}$ associated to the time translation for the quasi stationary state. This is divergent on the horizon, but it gives a finite flux, it is sufficient to take the limit $r \to r_H$ of the flux through the spheres of radius $r > r_H$. If one would get something that is finite on the horizon, one has to violate the weak energy condition, with the price that negative energy densities are not locally observable.

In order to evaluate the ordered operator one fixes a set $\{q_i\}$ of solutions describing waves passing through the event horizon and representing positive frequencies with respect to the time parameter defined by $\boldsymbol{K}$ along the generators of the horizon in the final quasi stationary state. This condition is not determined during the time dependent collapsing phase, but this should not affect the wave packets in the very late time. If one forms wave packets $\{q_{jn}\}$, then, going back in time, a fraction Γ_{jn} of them will enter the potential barrier around the black hole, going back on $\mathcal{I}_-$ with the same frequency it has on the horizon. This generates a $\delta(\omega - \omega')$ in the coefficients $\alpha_{jn\omega'}$. The remaining fraction $1 - \Gamma_{jn}$ is reflected back by the barrier, enters the collapsing body and goes out until reaches $\mathcal{I}_-$. Here it will have exactly the same form as for $p^{(2)}_{jn\omega'}$ so that, for large ω', one has

$$|\alpha^{(2)}_{jn\omega'}|^2 = e^{\frac{\pi\omega}{\kappa}} |\beta^{(2)}_{jn\omega'}|^2. \qquad (2.73)$$

Thus, one could be tempted to conclude that the rate of number of particles passing through the horizon in a picked packet for late times is (for boson)

$$N = \frac{1 - \Gamma_{jn}}{e^{2\frac{\pi\omega}{\kappa}} - 1}. \tag{2.74}$$

Now, for a fixed ω, because of the potential barrier, the coefficients Γ_{jn} tends to zero when the angular momentum number l increases. Thus, any mode of the wave packets with large l will bring a factor $(e^{2\frac{\pi\omega}{\kappa}} - 1)^{-1}$ of particles, for a total infinite number of particles passing the horizon. But this would be inconsistent with what we said above: these should be just fluctuations of order M^{-4}. This may due to the fact that $\{p_{jn}\}$ and $\{q_{jn}\}$ are bases of solutions just out of the horizon and not on the horizon. This is why one has to consider the flux out from the horizon and then consider the limit on the horizon. To this aim it is convenient to introduce the wave packets $\chi_{jn} = p_{jn}^{(2)} + q_{jn}^{(2)}$ that are the fractions passing through the collapsing body, and $y_{jn} = p_{jn}^{(1)} + q_{jn}^{(1)}$ which are the part propagating backward to $\mathcal{I}_-$ along the quasi stationary final part. In the initial states y_{jn} do not contain particles, whereas in χ_{jn} there are $(e^{2\frac{\pi\omega}{\kappa}} - 1)^{-1}$. These will appear leaving the body just out of the horizon, propagating radially outward. The mean flux of energy among the retarded times u_1 and u_2 will be

$$\mathcal{E} = \frac{1}{u_2 - u_1} \int_{u_1}^{u_2} \langle 0_- | T_{\mu\nu} | 0_- \rangle k^\mu d\Sigma^\nu, \tag{2.75}$$

and it is directed outward and equals the flux of energy of a warm body. The $\{y_{jn}\}$ do not contain negative frequencies on $\mathcal{I}_-$ an then do not contribute to such a flux. On the other hand,

$$\chi_{jn} = \int_0^\infty \{\zeta_{jn\omega'} f_{\omega'} + \xi_{jn\omega'} \bar{f}_{\omega'}\} d\omega', \tag{2.76}$$

and near $\mathcal{I}_+$ $\chi_{jn} = (\Gamma_{jn})^{\frac{1}{2}} p_{jn}$. Thus one gets

$$\mathcal{E} = \frac{1}{u_2 - u_1} Re \left[\sum_{jn} \sum_{j''n''} \int_0^\infty \int_{u_1}^{u_2} \omega\omega'' \Gamma_{jn}^{\frac{1}{2}} p_{jn} \bar{\xi}_{jn\omega'} (\bar{\Gamma}_{j''n''}^{\frac{1}{2}} \bar{p}_{j''n''} \xi_{j''n''\omega'} \right.$$
$$\left. - \Gamma_{j''n''}^{\frac{1}{2}} p_{j''n''} \zeta_{j''n''\omega'}) d\omega' du \right], \tag{2.77}$$

where ω and ω'' are the frequencies of the wave packets p_{jn} and $p_{j''n''}$ respectively. If $u_2 - u_1 \to \infty$, the second term in the integrand vanishes, whereas the first term gives a non vanishing contribution only when

$(j'', n'') = (j, n)$. The same arguments we repeated more then one time then gives (for bosons)

$$\int_0^\infty |\beta_{jn\omega'}|^2 d\omega' = \frac{1}{e^{2\pi\frac{\omega}{\kappa}} - 1} \tag{2.78}$$

so that

$$\mathcal{E} = \int_0^\infty \frac{\Gamma_\omega \omega}{e^{2\pi\frac{\omega}{\kappa}} - 1}\, d\omega, \tag{2.79}$$

where $\Gamma_\omega = \lim_{n\to\infty} \Gamma_{jn}$ is the fraction of incident particles which would be absorbed by the black hole. This is exactly the emission rate calculated initially from the viewpoint of the observers on $\mathcal{I}_+$. Each energy-momentum tensor that on $\mathcal{I}_+$ agrees with the one normally ordered there, and which is asymptotically stationary and conserved, would give rise to the same flux of outgoing energy.

The conclusion is that the area of the black hole decreases and it cannot be considered stationary in the final stage. However, if the black hole mass is much larger than the Planck mass, then the evolution can be considered as in an adiabatic regime, since the evolution of the area remain very slow with respect to the time a photon employs for crossing the horizon. So it is reasonable to assume that the quasi stationary evolution is a good approximation far from the Planck scales. After the Planck scale is reached, one should need to introduce new suitable concepts of some version of a quantum gravity. In any case, no much energy remains at this point and the simplest evolution one can expect is simply that the remaining small black hole simply disappears (or gives rise to a naked singularity). It is difficult to imagine that baryons or leptons could reappear since all energy has been radiated away thermally.

2.3 A very short list of further readings

We provide a very short (too short, indeed) list of further readings, with reference to some early seminal papers on Hawking radiation [Gibbons (1975); Wald (1975); Parker (1975); Page (1976b,a, 1977); Gerlach (1976); Davies and Fulling (1977); Davies *et al.* (1976)]. See also [Page (2005)]. Several other papers concerning the subject are in the bibliography and are quoted in the following chapters. Furthermore, we list the following references on the stimulated Hawking effect [Wald (1976); Bekenstein and Meisels (1977); Audretsch and Muller (1992)] (see also [Frolov and Novikov (1998)]), because, even if it is not so important in the case of astrophysical black holes, stimulated radiation plays an important role in the case of analogous black holes (see the second part of the book).

Chapter 3

Thermality of Hawking radiation: From Hartle-Hawking to Israel and Unruh

In this chapter we explore various aspects of thermality of Hawking radiation. We first take into account a masterpiece by Hartle and Hawking on black hole radiation [Hartle and Hawking (1976)]. Therein, thermality of black holes is shown to occur by using path integral techniques. A fundamental role is played by the fact that amplitudes for particle emission are linked with amplitudes for particle absorption. Both processes are actually involved in pair creation near the black hole horizon. On the grounds of analyticity properties of the propagator, which enters the probability calculations, and in which is implicit the selection of the quantum state for quantum fields on a black hole background, it is found that probability for particle absorption and for particle emission are related in such a way that particles perceived by a static observer at fixed radius r far from the horizon share a blackbody spectrum at the Hawking temperature. There appear both technical and physical relevant arguments, as we show below.

Furthermore, the periodicity of the propagator is also taken into account, and we shall consider also another important contribution contained in [Gibbons and Perry (1978)]. Periodicity in imaginary time is in relation with the thermal quantum state, describing a black hole in thermal equilibrium, i.e. the Hartle-Hawking state. This state is also called Hartle-Hawking-Israel state because of another seminal paper, due this time to Israel [Israel (1976)]. We shall introduce also Israel very interesting picture.

It is worth mentioning a further seminal paper by Unruh [Unruh (1976)], which has given rise to a whole huge literature concerning the so-called Unruh effect. The last section of this chapter is devoted to a short summary of this very important contribution.

3.1 Hartle-Hawking approach to black hole radiance

We consider only the situation where an eternal black hole geometry is given. For details about the collapse process, we refer to [Hartle and Hawking (1976)]. Substantially, what happens is that, after the collapse has completed, within a sufficiently long time many properties of the black hole background can be replaced with the corresponding ones of the eternal black hole (which is, of course, an idealization). Moreover, we discuss only the case of scalar particles on a Schwarzschild black hole, referring the reader to the original publication for the other cases.

The pair creation process and its characteristics are the goal of the approach we are discussing. As a necessary premise, we recall a fundamental property of quantum field theory: particle worldlines, in a relativistic theory, are non-necessarily always forward in time, as one would expect naively. Indeed, it is possible to meet both worldlines which start backward in time and end forward in time, which correspond to a pair creation process, and also worldlines which start forward in time and end backward in time, which correspond to a pair annihilation process. Indeed, we recall that, according to Feynman-Stueckelberg interpretation [Stueckelberg (1941); Feynman (1949)], a particle propagating backward in time has to be considered as an antiparticle propagating forward in time. This fact is also crucial for a full interpretation of the process at hand, as remarked also in [Gibbons and Hawking (1977b); Schulman (1981)]. Indeed, in such a case a worldline starting from the black hole singularity in the black hole region of the Penrose diagram and ending at future null infinity in region I of the Penrose diagram is considered. The worldline part in region I, starting from near the horizon and ending at future null infinity, represents a particle escaping from the black hole (forward in time), whereas the remaining part, mostly propagating in the black hole sector, corresponds to an antiparticle (backward in time). The dramatic change in the nature forward-backward in time occurs near the horizon, as an effect of the strong gravitational field of the black hole. The gravitational field plays the role of external classical field which allows such a change, involving energy conservation (as expected, due to the static nature of the geometry) and violation of momentum conservation (due to the presence of the external field itself).

We depart from the original path of thought in [Hartle and Hawking (1976)] and we give a slightly different presentation, which appears to be less unusual from a quantum field theory point of view. Our basic point is that one is interested to measure the flux of particles at a surface $\Sigma_{r'}$

which corresponds to the surface r' =const in the region I of the Penrose diagram, which is the measure that a static observe could perform. Then one has to consider the standard integral

$$\int_{\Sigma_{r'}} d\sigma^{\mu}(x')\phi^{*}(x')\overleftrightarrow{\partial^{x'}}_{\mu}\phi(x'), \tag{3.1}$$

which is a standard trick in scattering theory. As a matter of fact, we only need to consider

$$n^{\mu}(x')J_{\mu}(x'), \tag{3.2}$$

i.e. the current density in the direction of the unit normal n^{μ} to the surface $\Sigma_{r'}$. Indeed, note that $\phi^{*}(x')\overleftrightarrow{\partial}_{\mu'}\phi(x') \propto J_{\mu}$, where J^{μ} is the standard current associated with the field. The propagator comes into play when one imposes that the field at x' is related to the field at x, where x is a surface Σ_r at r =const in the black hole region in the Penrose diagram. Note that such a surface is spacelike, as in the black hole region t and r exchange their roles with respect to the external region. In particular, we have that Green functions are involved in identities associated with (a generalization of) Stokes theorem (see e.g. [Barton (1963); DeWitt (2003)]). We are interested in the following identity, involving the well-known separation in the field $\phi(x)$ of a positive frequency part $\phi^{+}(x)$ and of a negative frequency one $\phi^{-}(x)$, and of the properties of the Feynman propagator $K(x,x')$, which for $x^{0'} > x^{0}$ allows us to write

$$\phi^{+}(x') = \int_{\Sigma_r} d\sigma^{\mu}(x)K(x,x')\overleftrightarrow{\partial}_{\mu}\phi^{+}(x), \tag{3.3}$$

(see e.g. [Bjorken and Drell (1964, 1965); Greiner and Reinhardt (2013)], and of course [Feynman (1949)]). It relates the values of ϕ^{+} on a Cauchy surface to which x belongs, to the value of ϕ^{+} in a different point x' with $x^{0'} > x^{0}$. Formally, the spacelike surface Σ_r in the black hole sector can plays the role of Cauchy surface. We obtain for (3.3)

$$\int_{\Sigma_r} d\sigma^{\lambda}(x)\int_{\Sigma_r} d\sigma^{\rho}(y)\phi^{+*}(x)\overleftrightarrow{\partial^{x}}_{\lambda}K^{*}(x,x')n^{\mu}(x')\overleftrightarrow{\partial^{x'}}_{\mu}K(y,x')\overleftrightarrow{\partial^{y}}_{\rho}\phi^{+}(y). \tag{3.4}$$

Let us consider particle detectors at fixed large radius $r' = R$ in region I. These detectors measure particles which are positive frequency (i.e. positive norm) in time t' in the same region, and we know that separation of variables allows to write the scalar particle wave packets as follows for positive frequency:

$$f(t',r',\Omega') = \sum_{l=0}^{\infty}\sum_{m=-l}^{l}\int_{\mathbb{R}_>}\frac{dE}{2\pi}\, c(l,m,E)\exp(-iEt')f_{Elm}(r')Y_{lm}(\Omega')$$
$$=: \sum_{l=0}^{\infty}\sum_{m=-l}^{l}\int_{\mathbb{R}_>}\frac{dE}{2\pi}\, c(l,m,E)\exp(-iEt')F_{Elm}(r',\Omega'), \tag{3.5}$$

where Y_{lm} are the usual harmonic spherical functions. As to the initial state, to be considered at Σ_r, for consistency with the assumption of positive frequency (norm) just made, particles are described by a packet

$$h(t,r,\Omega) = \sum_{l=0}^{\infty}\sum_{m=-l}^{l}\int_{\mathbb{R}_>}\frac{dE}{2\pi}\, q(l,m,E)\exp(-iEt)h_{Elm}(r)Y_{lm}(\Omega)$$
$$=: \sum_{l=0}^{\infty}\sum_{m=-l}^{l}\int_{\mathbb{R}_>}\frac{dE}{2\pi}\, q(l,m,E)\exp(-iEt)H_{Elm}(r,\Omega), \tag{3.6}$$

where t, r, Ω belong to a r =constant surface in the black hole region, and chosen in such a way that for $r < r_+$ one obtains positive norm states *for the Hartle-Hawking vacuum* (see also the following sections and the following chapter; cf. [Massar and Parentani (2000)]). Please, note the aforementioned subtlety: the Feynman propagator for the Hartle-Hawking vacuum does *not* propagate forward in time the standard timelike Killing modes, but the positive norm modes of the Hartle-Hawking vacuum. Furthermore, one can exploit also the following structure for K:

$$K(x,x') = K(r,r',\Omega,\Omega',t-t'), \tag{3.7}$$

and then obtain for (3.2)

$$\int_{-\infty}^{\infty} dt \int_{-\infty}^{\infty} d\bar{t} \int d\Omega \int d\bar{\Omega} \sum_{l,m}\sum_{\bar{l}\bar{m}} \int_{\mathbb{R}_>}\frac{dE}{2\pi}\int_{\mathbb{R}_>}\frac{d\bar{E}}{2\pi}$$
$$q^*(l,m,E)q(\bar{l},\bar{m},\bar{E})e^{iEt}e^{-i\bar{E}\bar{t}}Y^*_{lm}(\Omega)Y_{\bar{l}\bar{m}}(\bar{\Omega})$$
$$h^*_{Elm}(r)\overleftrightarrow{\partial}_r K^*(r,r',\Omega,\Omega',t-t')\overleftrightarrow{\partial}_{r'}\left(K(\bar{r},r',\bar{\Omega},\Omega',\bar{t}-t')\overleftrightarrow{\partial}_{\bar{r}}h_{\bar{E}\bar{l}\bar{m}}(\bar{r})\right)_{\bar{r}=r}. \tag{3.8}$$

Note that, as Σ_r occurs twice, in the last expression one has to take into account the problem of how to avoid an ambiguity (our choice is to introduce a surface $\Sigma_{\bar{r}}$ and put $\bar{r} = r$ at the end of the calculations).

In order to display the essentials of the calculation, we first consider the 2D case (no angular part). We are interested in simplifying the time dependence of the above expression, so we consider

$$I := \int_{-\infty}^{\infty} dt \int_{-\infty}^{\infty} d\bar{t} \int_{\mathbb{R}_>} \frac{dE}{2\pi} \int_{\mathbb{R}_>} \frac{d\bar{E}}{2\pi}\, q^*(E)q(\bar{E})e^{iEt}e^{-i\bar{E}\bar{t}}$$
$$\times h_E^*(r)\overleftrightarrow{\partial}_r K^*(r,r',t-t')K(\bar{r},r',\bar{t}-t')\overleftrightarrow{\partial}_{\bar{r}} h_{\bar{E}}(\bar{r}). \tag{3.9}$$

Let us set

$$K_E^*(r,r',t-t') := h_E^*(r)\overleftrightarrow{\partial}_r K^*(r,r',t-t') \tag{3.10}$$

and consider

$$II := \int_{-\infty}^{\infty} dt e^{iEt} K_E^*(r,r',t-t'). \tag{3.11}$$

We have

$$II = e^{iEt'} \int_{-\infty}^{\infty} dt e^{iE(t-t')} K_E^*(r,r',t-t'), \tag{3.12}$$

and the redefinition $\tau(t) = t - t'$ leads to

$$II = e^{iEt'} \int_{-\infty}^{\infty} d\tau e^{iE\tau} K_E^*(r,r',\tau), \tag{3.13}$$

Analogously, we get

$$\begin{aligned} III: &= \int_{-\infty}^{\infty} d\bar{t} e^{-i\bar{E}\bar{t}} K_{\bar{E}}(r,r',\bar{t}-t') \\ &= e^{-i\bar{E}t'} \int_{-\infty}^{\infty} d\bar{\tau} e^{-i\bar{E}\bar{\tau}} K_{\bar{E}}(r,r',\bar{\tau}), \end{aligned} \tag{3.14}$$

where $\bar{\tau} = \bar{t} - t'$. and

$$K_{\bar{E}}(\bar{r},r',\bar{t}-t') = K(\bar{r},r',\bar{t}-t')\overleftrightarrow{\partial}_{\bar{r}} h_{\bar{E}}(\bar{r}). \tag{3.15}$$

Let us introduce the Fourier transform of the propagator

$$\epsilon_E(r,r') := \int_{-\infty}^{\infty} dw \exp(-iEw) K_E(r,r',w). \tag{3.16}$$

Then we find

$$I = \int_{\mathbb{R}_>} \frac{dE}{2\pi} \int_{\mathbb{R}_>} \frac{d\bar{E}}{2\pi}\, q^*(E)q(\bar{E})\epsilon_E^*(r,r')\epsilon_{\bar{E}}(r,r')e^{i(E-\bar{E})t'}. \tag{3.17}$$

Finally, we have to insert it in (3.1) thus getting

$$\int_{\mathbb{R}_>} \frac{dE}{2\pi}\, |q(E)|^2 \epsilon_E^*(r,r')\overleftrightarrow{\partial}_{r'} \epsilon_E(r,r'). \tag{3.18}$$

Notice that the above integral exhausts the flux calculation only in a 2D case, where no angular contributions appear. In the case of a full 4D calculation, one must take into account that, due to separation of variables, further simplifications occur in the flux calculation, still this does not change the fact that there exists the relation between the absorption probability and the emission probability which is explained in the following.

The 4D calculation proceeds exactly in the same way. If we introduce the functions

$$K_{Elm}(r,r',\Omega',w) := \int d\Omega Y_{lm}(\Omega) K(r,r',\Omega,\Omega',t-t') \overleftrightarrow{\partial}_r h_{Elm}(r) \tag{3.19}$$

and

$$\epsilon_{Elm}(r,r',\Omega') := \int_{-\infty}^{\infty} dw e^{-iEw} K_{Elm}(r,r',\Omega',w), \tag{3.20}$$

and similar expressions for $\epsilon^*_{Elm}(r,r',\Omega')$, then (3.8) becomes

$$\int_{\mathbb{R}_>} \frac{dE}{2\pi} \int_{\mathbb{R}_>} \frac{d\bar{E}}{2\pi} \sum_{lm} \sum_{\bar{l}\bar{m}} q^*(l,m,E) q(\bar{l},\bar{m},\bar{E})$$
$$\epsilon^*_{Elm}(r,r',\Omega') \overleftrightarrow{\partial}_{r'} \epsilon_{\bar{E}\bar{l}\bar{m}}(r,r',\Omega') e^{i(E-\bar{E})t'} \tag{3.21}$$

and (3.1) takes the form

$$\sum_{lm} \sum_{\bar{l}\bar{m}} \int_{\mathbb{R}_>} \frac{dE}{2\pi} q^*(l,m,E) q(\bar{l},\bar{m},E) \epsilon^*_{Elm}(r,r',\Omega') \overleftrightarrow{\partial}_{r'} \epsilon_{E\bar{l}\bar{m}}(r,r',\Omega'). \tag{3.22}$$

Notice that, by computing a flux through the r' =const surface, we get directly the probability rate (and not the amplitude). We point out that, coming back to equation (3.8), we can also rewrite it as

$$\sum_{l,m} \sum_{\bar{l}\bar{m}} \int_{\mathbb{R}_>} \frac{d\omega}{2\pi} \int_{\mathbb{R}_>} \frac{d\bar{\omega}}{2\pi} \int d\Omega \int d\bar{\Omega} Y_{lm}(\Omega) Y^*_{\bar{l}\bar{m}}(\bar{\Omega}) q^*(l,m,\omega) q(\bar{l},\bar{m},\bar{\omega})$$
$$h^*_{\omega lm}(r) \overleftrightarrow{\partial}_r \epsilon^*_\omega(r,r',\Omega,\Omega') \overleftrightarrow{\partial}_{r'} \epsilon_{\bar{\omega}}(\bar{r},r',\bar{\Omega},\Omega') \overleftrightarrow{\partial}_{\bar{r}} h_{\bar{\omega}\bar{l}\bar{m}}(\bar{r}) \Big|_{\bar{r}=r}, \tag{3.23}$$

where the Fourier transform of the full propagator appears:

$$\epsilon_\omega(r,r',\Omega,\Omega') := \int_{-\infty}^{\infty} dw \exp(-i\omega w) K(r,r',\Omega,\Omega',w). \tag{3.24}$$

We can now come back to the original calculation. The basic amplitude to be calculated, according to Hartle and Hawking, involves the Fourier transform of the propagator (3.24) (see also [Gibbons and Hawking (1977b);

Schulman (1981)]). Equation (3.24) and, in particular, its analyticity properties play a fundamental role in determining the thermal character of the Hawking radiation which is associated with the pair creation process at hand. We need two ingredients. On one side, we have to complexify the manifold. In standard Kruskal coordinates, by defining $t_R := Re(t)$ and $t_I := Im(t)$, we have in the black hole sector

$$U = |U| \exp(-it_I \kappa), \tag{3.25}$$

$$V = |V| \exp(it_I \kappa); \tag{3.26}$$

As to analyticity properties of the propagator K, for a complete discussion involving path-integral techniques, we refer to the original paper [Hartle and Hawking (1976)]. Herein we limit ourselves to point out that there are two main characteristics: on one hand, by passing to complexified Kruskal coordinates U, V, one may show that the propagator is periodic in imaginary time with period equal to β_H. The analyticity of the propagator on the complexified horizon in the upper half U plane and in the lower half V plane emerges, which is at the root of the aforementioned periodicity. There is a strip of width $\beta_H/2$ where the propagator is analytic. Just above and just below the aforementioned strip, there appear singularities associated with null geodesics [Hartle and Hawking (1976)].

Analyticity properties of K allow to relate equation (3.24) to the one obtained by shifting $t \mapsto t - i\frac{\pi}{\kappa}$. Then, we obtain (we maintain the original notation here)

$$\begin{aligned} \epsilon_{E_j}(r, r', \Omega, \Omega') &= \exp\left(-\pi\frac{E_j}{\kappa}\right) \int_{-\infty}^{\infty} dt \exp(-iE_j w) K\left(r, r', \Omega, \Omega', t - i\frac{\pi}{\kappa}\right) \\ &=: \exp\left(-\pi\frac{E_j}{\kappa}\right) \tilde{\epsilon}_{E_j}(r, r', \Omega, \Omega'). \end{aligned} \tag{3.27}$$

In order to allow a correct interpretation of the amplitude in which $\tilde{\epsilon}_{E_j}$ appears, we need to observe that the shift $t \mapsto t - i\frac{\pi}{\kappa}$ amounts to $U \mapsto -U, V \mapsto -V$, i.e. to a reflection of the Kruskal coordinates corresponding to time-reversal. If we consider the surface Σ_+ which corresponds, in the white hole sector, to the reflection of surface Σ_r in the black hole sector, and the amplitude for a particle starting at x' on Σ_+ to reach x near the future horizon, then we find that such an amplitude corresponds to a process where a particle emerging from the past horizon is absorbed at the future horizon. This absorption process is obtained by time-reversal from the previous one. As the probability corresponding to the process is invariant under time reversal, and time reversal connects that amplitude to the one of absorption, one obtains

$$P_{emission} = \exp(-\beta_H E) P_{absorption}, \tag{3.28}$$

where we have put $E_j \mapsto E$ and $\beta_H = \frac{2\pi}{\kappa}$. As a consequence, we have the probability of emission, compared with the probability of absorption, satisfies the same relation as in the case of common blackbodies. Thus, the black hole behaves as a blackbody at the Hawking temperature.

3.2 Gibbons-Perry analysis for thermality

In independent calculations, the analyticity properties of the propagator appear to be in a straightforward way interpreted as associated with a thermal equilibrium in [Gibbons and Perry (1978)], where thermal field theory is considered and, in particular, thermal green functions are introduced. Again, in the Schwarzschild case, it is realized that, in the Euclidean section, the two-point function corresponding to Hartle and Hawking boundary conditions (i.e. equilibrium at the Hawking temperature), is periodic in imaginary time. This has a direct resemblance with usual thermal field theory formalism at a given temperature T, with the difference that, in this case, the temperature is fixed by the geometry of the black hole, and it is the Hawking temperature. This thermality is also related to the well known KMS condition which has to be satisfied by quantum fields at thermal equilibrium.

Let us introduce the Wightman functions in ordinary quantum field theory:

$$G^+(x,x') = <0|\phi(x)\phi(x')|0>, \tag{3.29}$$

$$G^-(x,x') = <0|\phi(x')\phi(x)|0>, \tag{3.30}$$

where $|0>$ stays for the vacuum state. In flat spacetime, the thermal Wightman functions are

$$G_T^+(x,x') = \sum_n P_n < f_n|\phi(x)\phi(x')|f_n >, \tag{3.31}$$

$$G_T^-(x,x') = \sum_n P_n < f_n|\phi(x')\phi(x)|f_n >, \tag{3.32}$$

where

$$P_n = \frac{1}{Z}\exp(-\beta E_n), \tag{3.33}$$

and, if H is the Hamiltonian operator associated with the field theory at hand,

$$H|f_n >= E_n|f_n > . \tag{3.34}$$

As $\phi(t+t',\mathbf{x}) = \exp(iHt')\phi(t,\mathbf{x})\exp(-iHt')$ holds true, it follows

$$G_T^+(t,\mathbf{x};t',\mathbf{x}') = G_T^-(t,\mathbf{x};t'-i\beta,\mathbf{x}'), \tag{3.35}$$

$$G_T^-(t,\mathbf{x};t',\mathbf{x}') = G_T^+(t-i\beta,\mathbf{x};t',\mathbf{x}'). \tag{3.36}$$

Then, by considering $t = -i\tau$, one finds that the Feynman thermal function

$$G_T^F(t,\mathbf{x};t',\mathbf{x}') = \sum_n P_n < f_n|T(\phi(x)\phi(x'))|f_n >, \tag{3.37}$$

is periodic in the Euclidean time τ with period β. The same periodicity property in the Euclidean time is shared by the Feynman Green function on the Hartle-Hawking-Israel state, as shown in [Hartle and Hawking (1976); Gibbons and Perry (1978)].

The argument developed in [Gibbons and Perry (1978)] is twofold. See also [Gibbons and Perry (1976); Fulling and Ruijsenaars (1987)]. As usual, a spherical box around the black hole, with Dirichlet boundary conditions on the box walls, is introduced. From one hand, one constructs the thermal state at the Hawking temperature for a scalar field as seen from the static observer in the external region of the black hole, obtaining the Green function

$$G_{TS}(x,x') := \begin{cases} i\sum_n \frac{\psi_n(x)\psi_n(x')}{E_n}\left[(1+n)e^{-iE_n(t-t')} + ne^{iE_n(t-t')}\right] & t > t', \\ i\sum_n \frac{\psi_n(x)\psi_n(x')}{E_n}\left[(1+n)e^{iE_n(t-t')} + ne^{-iE_n(t-t')}\right] & t < t', \end{cases} \tag{3.38}$$

where $n := (e^{\beta E_n} - 1)^{-1}$ is the mean number of particles in the Bose-Einstein distribution and $\psi_n(t,\mathbf{x}) =< t,\mathbf{x}|f_n >$.

On the other hand, one proceeds as follows. Let us consider a basis of solutions for the wave equation in Kruskal UV coordinates, and require that they are positive frequency on the past horizon in the lower half U-plane. In order to obtain the Hartle-Hawking state $|HH>$, we also require that analyticity occurs also in the lower half V-plane on the future horizon. Then, the Green function

$$G_{HH}(t,\mathbf{x};t',\mathbf{x}') :=< HH|T(\phi(x)\phi(x'))|HH >, \tag{3.39}$$

is analytic in Kruskal coordinates, except for points connected by null geodesics, and in particular it is analytic in the upper half U-plane as $V = 0$, and also in the lower half V-plane as $U = 0$, and then shares the same analyticity properties as deduced in the Hartle-Hawking approach. Furthermore, from the definition of U, V one realizes that in terms of complex values of $t = t_R + it_I$ (with real t_R, t_I) a periodicity in imaginary time with period $\beta = 2\pi/\kappa$ occurs, and one infers that periodicity and analyticity properties of G_{HH} above are the same as for G_{TS}.

3.3 Thermofield dynamics and Hartle-Hawking-Israel state

The work we are going to describe represents a very interesting bridge between Thermofield Dynamics and black hole physics, which has been introduced by W. Israel just after the introduction of the paradigm of black hole evaporation and of the framework of Thermofield Dynamics, It is in some sense quite surprising that a bridge (indeed, a somewhat puzzling one) between the two frameworks can be found, as they seem to be quite disconnected each other, and, moreover, that not only Thermofield Dynamics can be applied to black holes, but also that black holes involve a non-trivial interpretation for a part of the formalism of Thermofield Dynamics.

Since the discover of the Hawking effect, a very fine construction of the thermal state living on a finite temperature black hole manifold characterized by a bifurcate Killing horizon was introduced by Israel [Israel (1976)] on the grounds of Unruh analysis [Unruh (1976)] and of the thermofield approach to thermal physics introduced by Takahashi and Umezawa [Takahasi and Umezawa (1975); Umezawa *et al.* (1982); Umezawa (1993)]. Israel discovered that the HH state (hence called also Hartle-Hawking-Israel state) corresponds to the thermal vacuum of thermofield approach with the temperature equal to the black-hole temperature, and that the would-be fictitious states of the thermofield approach correspond to the states in the left wedge of the extended solution (if one is living in the right wedge). In what follows, our focus is to the case of black hole backgrounds with a single temperature, and then β is to be meant in the black hole case as the inverse black hole temperature. We deal with a scalar field, so the bosonic case is discussed. We point out that the following analysis holds true also for fermions in a generic thermal state with inverse temperature β.

3.3.1 *Thermofield Dynamics*

Thermofield Dynamics represents a formalism for describing quantum field theory at finite temperature. The key idea is to obtain a formalism where the ensemble average of any operator at thermal equilibrium is replaced by an expectation value of the same operator with respect to a so-called thermal vacuum:

$$\langle A\rangle_\beta = \frac{1}{Z(\beta)} Tr(\exp(-\beta H)A) \mapsto < 0(\beta)|A|0(\beta) > . \tag{3.40}$$

In the previous equation, $Z(\beta)$ is the partition function and H is the Hamiltonian operator for the system at hand, and $|0(\beta) >$ is the thermal vacuum. It is easy to show that it is not possible to obtain that $|0(\beta) >$ stays in the same Hilbert space $\mathcal{H}$ as the bosonic theory at hand, but, in order to implement the above mapping, it is necessary to duplicate the degrees of freedom by introducing a fictitious copy $\tilde{\mathcal{H}}$ of the original Hilbert space, with associated creation and annihilation operators. The thermal state $|0(\beta) >$ can be arranged for in the tensor product $\mathcal{H} \otimes \tilde{\mathcal{H}}$. Let us introduce in the vacuum states $|0 >$ and $|\tilde{0} >$, with corresponding annihilator operators $a_l, \tilde{a}_l$ and $b_l, \tilde{b}_l$ for particles and antiparticles respectively. One can also introduce thermal state operators, according to the standard constructions in thermofield dynamics [Umezawa *et al.* (1982); Umezawa (1993); Das (1997); Khanna *et al.* (2009)], and introduce the thermal state $|O(\beta) >$ and thermal state annihilation operators $a_l(\beta), \tilde{a}_l(\beta), b_l(\beta), \tilde{b}_l(\beta)$, which are such that

$$a_l(\beta)|O(\beta) >= \tilde{a}_l(\beta)|O(\beta) >= b_l(\beta)|O(\beta) >= \tilde{b}_l(\beta)|O(\beta) >= 0. \tag{3.41}$$

We are mainly interested in the following Bogoliubov relations:

$$a_l = c_l a_l(\beta) - s_l \tilde{a}_l^\dagger(\beta), \tag{3.42}$$

$$a_l^\dagger = c_l a_l^\dagger(\beta) - s_l \tilde{a}_l(\beta), \tag{3.43}$$

and the analogous ones for b-operators (which correspond to operators for antiparticles, i.e. for negative frequency states; cf. [Umezawa *et al.* (1982)]), with

$$c_l := \frac{1}{\sqrt{1-\exp(-\beta|\omega|)}}, \tag{3.44}$$

$$s_l := \frac{\exp(-\frac{1}{2}\beta|\omega|)}{\sqrt{1-\exp(-\beta|\omega|)}}. \tag{3.45}$$

As mentioned above, the formalism [Takahasi and Umezawa (1975)] is introduced in order to express a statistical mean of an observable A in terms of a expectation value on a "thermal vacuum state":

$$< A >= Tr(\rho A) \equiv < O(\beta)|A|O(\beta) > . \tag{3.46}$$

The equivalence in (3.46) is implementable only in an extended Hilbert space. Given a physical system described by a Hamiltonian H in a Hilbert space $\mathcal{H}$, one has to introduce a fictitious Hilbert space $\tilde{\mathcal{H}}$ and an Hamiltonian $\tilde{H}$ describing an identical physical system. Total Hilbert space and total Hamiltonian are

$$\begin{aligned} \mathcal{H}_{tot} &= \mathcal{H} \otimes \tilde{\mathcal{H}} \\ H_{tot} &= H - \tilde{H}. \end{aligned} \tag{3.47}$$

A physical (fictitious) n–particle state whose energy is E_n is denoted by $|n>$ ($|\tilde{n}>$) and $|n, \tilde{n}>=|n>|\tilde{n}>$ is the n–particle tensor product state. Vacuum state in the extended Hilbert state is

$$|O(\beta)>= \frac{1}{\sqrt{Z(\beta)}} \sum_n e^{-\beta E_n/2} \, |n, \tilde{n}>; \tag{3.48}$$

$Z(\beta)$ is a normalization.

In the calculation of (3.46) one has to sum over all the fictitious states $\tilde{n}$; the result is

$$< A >=< O(\beta)|A|O(\beta) >= \frac{1}{Z(\beta)} \sum_n e^{-\beta E_n} < n|A|n > . \tag{3.49}$$

3.3.2 *Israel contribution*

Between Thermofield Dynamics and black hole thermodynamics, as discussed by Israel [Israel (1976)], the following identifications are possible

$$\begin{aligned} |O(\kappa)> &\equiv |\text{HH} - \text{vacuum} > = |O(\beta)> \\ < O(\kappa)|A|O(\kappa)> &= Tr_{II}(\rho A) = \frac{1}{Z(\kappa)} \sum_n e^{-\frac{2\pi}{\kappa} E_n} < n|A|n > \\ \mathcal{H}_{II} &= \tilde{\mathcal{H}} \end{aligned} \tag{3.50}$$

κ is the surface gravity, and $|n>$ are n–particle (Boulware) states. Physical Hilbert space available to a static observer (region I) is related to the physical Hilbert space of Thermofield Dynamics; statistical means in (3.50) are relative to the region I, and the fictitious space is identified with the space of states in region II (time reversed of region I). The main point consists in realizing that the relation between Minkowski vacuum and Rindler vacuum introduced in [Unruh (1976)] can be also implemented in the case of eternal black holes, where the so-called Hartle-Hawking-Israel state replaces Minkowski vacuum and Boulware vacuum replaces Rindler vacuum. The HH state is obtained by requiring that the wave function is analytic both in the lower U plane and in the lower V plane. We shall follow [Isham (1977)]. See also [Fulling (1977); Fabbri and Navarro-Salas (2005); Wald (1995)].

Let us e.g. solve the massless Klein-Gordon equation

$$\Box\phi = 0 \tag{3.51}$$

in the external region of the wedge I of the Penrose diagram for the Schwarzschild solution. Separation of variables leads to solutions of the form

$$\phi = \exp(-i\omega t) Y_{lm}(\Omega) f_{\omega lm}(r_*), \tag{3.52}$$

where r_* is the tortoise coordinate. The radial part of the wavefunction $f_{\omega lm}(r_*)$ satisfies

$$\left[\frac{1}{r^2}\partial_{r_*} r^2 \partial_{r_*} - \left(1 - \frac{r_+}{r}\right)\frac{l(l+1)}{r^2} + \omega^2\right] f_{\omega lm}(r_*) = 0, \tag{3.53}$$

where $r_+ = 2M$ is the black hole radius. A basis can be chosen such that, near the horizon $r = r_+$, the wave function behaves as

$$\overrightarrow{f}_{\omega lm}(r_*) \sim e^{i\omega r_*} + Re^{-i\omega r_*}, \tag{3.54}$$

$$\overleftarrow{f}_{\omega lm}(r_*) \sim Te^{-i\omega r_*}, \tag{3.55}$$

where R, T are the reflection coefficient and the transmission coefficient respectively [Fulling (1977); Isham (1977)].

We can introduce the following basis for region I (R-region)

$${}^{R}_{+}\overrightarrow{\phi}_{\omega} = \frac{1}{2\pi\omega}\exp(-i\omega t) Y_{lm}(\Omega)\overrightarrow{f}_{\omega lm}(r_*)\theta_R, \tag{3.56}$$

$${}^{R}_{+}\overleftarrow{\phi}_{\omega} = \frac{1}{2\pi\omega}\exp(-i\omega t) Y_{lm}(\Omega)\overleftarrow{f}_{\omega lm}(r_*)\theta_R, \tag{3.57}$$

where θ_R indicates that they are supported only in the R-region, and also the 'twin solutions' in the region II (L-region):

$${}^{L}_{+}\overrightarrow{\phi}_{\omega} = \frac{1}{2\pi\omega}\exp(i\omega t) Y_{lm}(\Omega)\overrightarrow{f}^{*}_{\omega lm}(r_*)\theta_L, \tag{3.58}$$

$${}^{L}_{+}\overleftarrow{\phi}_{\omega} = \frac{1}{2\pi\omega}\exp(i\omega t) Y_{lm}(\Omega)\overleftarrow{f}^{*}_{\omega lm}(r_*)\theta_L, \tag{3.59}$$

where θ_L indicates that they are supported only in the L-region. The solutions

$${}_{+}\overrightarrow{\phi}_{\omega} = \frac{1}{\sqrt{2\sinh(\frac{\pi\omega}{\kappa})}}\left(\exp\left(\frac{\pi\omega}{2\kappa}\right){}^{R}_{+}\overrightarrow{\phi}_{\omega} + \exp\left(-\frac{\pi\omega}{2\kappa}\right){}^{L}_{+}\overrightarrow{\phi}^{*}_{\omega}\right), \tag{3.60}$$

$${}_{+}\overrightarrow{\phi}_{-\omega} = \frac{1}{\sqrt{2\sinh(\frac{\pi\omega}{\kappa})}}\left(\exp\left(\frac{\pi\omega}{2\kappa}\right){}^{R}_{+}\overrightarrow{\phi}^{*}_{\omega} + \exp\left(-\frac{\pi\omega}{2\kappa}\right){}^{L}_{+}\overrightarrow{\phi}_{\omega}\right), \tag{3.61}$$

define a set of functions which are analytical and bounded in the lower half complex U-plane on $V = 0$ and in the lower half complex V-plane on $U = 0$, and the same is true for

$$ {}_{+}\overleftarrow{\phi}_{\omega} = \frac{1}{\sqrt{2\sinh(\frac{\pi\omega}{\kappa})}} \left(\exp\left(\frac{\pi\omega}{2\kappa}\right) {}^{R}_{+}\overleftarrow{\phi}_{\omega} + \exp\left(-\frac{\pi\omega}{2\kappa}\right) {}^{L}_{+}\overleftarrow{\phi}^{*}_{\omega} \right), \tag{3.62} $$

$$ {}_{+}\overleftarrow{\phi}_{-\omega} = \frac{1}{\sqrt{2\sinh(\frac{\pi\omega}{\kappa})}} \left(\exp\left(\frac{\pi\omega}{2\kappa}\right) {}^{R}_{+}\overleftarrow{\phi}^{*}_{\omega} + \exp\left(-\frac{\pi\omega}{2}\right) {}^{L}_{+}\overleftarrow{\phi}_{\omega} \right). \tag{3.63} $$

Regularity at the horizon is better appreciated by taking into account fully the distributional nature of the aforementioned superpositions of Schwarzschild modes [Sciama *et al.* (1981); Brout *et al.* (1995a)]: indeed, one finds that near the horizon they can be approximated as Rindler modes

$$ {}_{+}\overrightarrow{\phi}_{\omega} \propto (U - i\epsilon)^{i\frac{\omega}{\kappa}}, \tag{3.64} $$

and that analogous identities hold true. This has a relation also with the construction of so-called straddling modes for the Unruh vacuum (see Chapter 4).

The field ϕ can be expressed both in terms of the original Schwarzschild modes and in terms of the Kruskal ones just obtained. In particular, we obtain [Israel (1976)]

$$ |HH> = \exp(-iG)|B> = \exp\left[\sum_{\omega}\left(\frac{1}{2}\log\left(1 - \exp\left(-\frac{\pi}{2\kappa}\omega\right)\right)\right)\right] \times \exp\left[\exp\left(-\frac{\pi}{\kappa}\omega\right)\sum_{j}\left(a^{R\dagger}_{\omega j}a^{L\dagger}_{\omega j}\right)\right]|B>. \tag{3.65} $$

We have also

$$ |HH> = \frac{1}{Z(\kappa)}\prod_{\omega j}\left(\sum_{n_{\omega j}=0}^{\infty}\exp\left(-n_{\omega j}\frac{\pi\omega}{\kappa}\right)\right)|n_{\omega j}>^{L}|n_{\omega j}>^{R}. \tag{3.66} $$

States $|n_{\omega j}>^{L}, |n_{\omega j}>^{R}$ are Boulware states with $n_{\omega j}$ particles. Killing observers in the R-region cannot measure L-states, so a trace over the latter ones is required. As a consequence, for the expectation value of an operator (observable) A^{R}, constructed by means of Boulware vacuum creation and annihilation operators, one obtains

$$ < HH|A^{R}|HH >= \frac{1}{Z(\kappa)}\sum_{n}\exp(-\beta^{*}_{H}E_{n}) < n|A^{R}|n >, \tag{3.67} $$

where $E_{n} = \frac{1}{K}\sum_{\omega j} n_{\omega j}\omega$, $\beta^{*}_{H} = \beta_{H}K$ (K stays for the norm of the timelike Killing vector), $Z = \sum_{n}\exp(-\beta^{*}_{H}E_{n})$. The relation with thermofield dynamics is then implemented, with the identification of the fictitious states with the states in the unaccessible L-region. Further aspects of the relation between Thermofield Dynamics and geometry are discussed in [Laflamme (1989)]. See also [Jacobson (1994)] and [Martellini *et al.* (1978); Moschella and Schaeffer (2009a,b)].

3.4 Unruh's cornerstone

The seminal paper by Unruh [Unruh (1976)], with the guiding idea to simplify calculations still maintaining physical meaning of the equations, becomes a fundamental one. There appear various achievements, between which the most important for our framework are the definition of the vacuum state (called Unruh state) for a star collapsing to a black hole, and the analysis of the Hawking process in light of the Rindler-Minkowski situation for a scalar field. The second picture gave origin to the so called Unruh effect, and we note that in [Wald (1995)] there is a distinction between the Hawking effect (which is involved only with the collapse picture according to [Wald (1995)]), and the Unruh effect in curved spacetime (which involves also what other authors call Hawking effect for eternal black holes). The relation between quantization in Rindler coordinates and standard quantization in Minkowski coordinates is discussed, and results are analogous to the ones discussed in the previous section.

One main trick consists in realizing that normal modes for a black hole can be defined on the past horizon by requiring that they are eigenfunctions of the Lie Derivative with respect to $\xi := \partial_U$:

$$\xi \phi_\omega = -i\omega \phi_\omega, \tag{3.68}$$

and solutions of the form $\phi_\omega = \exp(-i\omega U) Y_{lm}(\theta, \varphi)/\sqrt{(2\pi|\omega|)}$ is for $\omega > 0$ a complete set of positive frequency modes. This approach replaces the collapse process with boundary conditions on the past horizon.

A further fundamental contribution is associated with particle detectors in Rindler case. We don't delve into this topic.

We cannot even imagine to provide a serious bibliography on the Unruh effect, we limit ourselves to quote two papers [Ginzburg and Frolov (1987); Takagi (1986)].

Chapter 4

The tunneling approach

Since the original work by Hawking, the fundamental mechanism leading to Hawking radiation is represented by quantum mechanical tunneling of a particle through the black hole horizon [Hawking (1975)]. In particular, in the geometrical optics approximation, which is justified by the fact that near the horizon particle energies are severely boosted, because of an hard blueshift, a classically forbidden tunneling of a particle from inside the horizon to outside can occur, according to still different mechanisms. Together with other approaches, so-called tunneling methods appeared since the initial period in black hole evaporation calculations. Herein, we mean to delve into this very intuitive framework, in which the original suggestion on the pair creation mechanism is implemented. Of course, the present approach is strictly related to the Hartle-Hawking approach we have considered in Chapter 3. Still, we stress that the tunneling method is very important in its own, because it represents a simple tool to show that the Hawking effect is actually present if a non-degenerate horizon is at hand in a stationary black hole metric. All the main features can be reproduced, at a very lower cost, both from a conceptual point of view and also from a computational one. Furthermore, it also provides us an useful tool for generalizing the Hawking effect to situations where backreaction effects are taken into account, giving rise to non-strictly thermal features.

It is quite impossible to cover exhaustively the existing literature on this topic, and we limit ourselves to quote mainly some seminal papers and a review. We also limit our considerations to the Schwarzschild case.

4.1 Damour-Ruffini

The paper by Damour and Ruffini represents the first example of tunneling approach to the Hawking radiation. What follows is substantially contained in the original work [Damour and Ruffini (1976)], but also refers to the analysis in [Visser (2003)]. The main idea consists in the following steps: a) introducing a chart (coordinate system) which is regular on the black hole horizon; b) solving the Klein-Gordon equation for a scalar field in the external region; c) by appealing to analyticity properties for particle/antiparticle states, inherited by the axioms holding in the ordinary Minkowski spacetime, writing down an outgoing mode which lives both inside and outside the horizon (straddling mode henceforth); d) normalize the straddling mode. Then one obtains the Hawking effect.

The step (a) is implemented by choosing the so-called advanced Eddington-Finkelstein coordinates:

$$ds^2 = -(1 - r_+/r)dv^2 + 2dvdr + r^2 d\Omega^2. \tag{4.1}$$

Herein, $r_+ = 2M$. It holds $v = t + r_*$, where r_* is the usual tortoise coordinate. ∂_v is a timelike Killing vector in the external region. As a matter of facts, we can limit ourselves to considering a 2D version of the above metric, where the transverse contribution is neglected. This amounts to an evaluation of a purely s-wave contribution to the Hawking radiation. Moreover, we consider only a massless scalar field. Still, this is enough for safely reproducing the essential features of the radiation process. As to (b), the 2D metric

$$ds^2_{2D} = -(1 - r_+/r)dv^2 + 2dvdr \tag{4.2}$$

leads to the following Klein-Gordon equation for a scalar field:

$$\Box\phi = 0 \Longleftrightarrow \partial_r(2\partial_v\phi + (1 - r_+/r)\partial_r\phi) = 0. \tag{4.3}$$

We first look for solutions in the external region $r > r_+$. A variable separation ansatz $\phi(v, r) = h(v)f(r)$ leads to solutions of the following form:

$$\phi(v, r) = C_1 e^{2i\omega r}(r - r_+)^{2i\omega r_+} e^{-i\omega v} + C_2 e^{-i\omega v}, \tag{4.4}$$

so that we can get a purely ingoing solution

$$\phi_{in} = N_{in} e^{-i\omega v}, \tag{4.5}$$

and a purely outgoing one

$$\phi_{out} = N_{out} e^{2i\omega r}(r - r_+)^{2i\omega r_+} e^{-i\omega v}. \tag{4.6}$$

Note that $\phi_{out} = N_{out}e^{-i\omega u}$, where $u = t - r_*$. Note also that the ingoing mode is well defined everywhere (also inside the black hole horizon). A very different status has the outgoing mode, whose continuation inside the horizon is nontrivial, due to the fact that $r = r_+$ represents a logarithmic branch point. In step (c), the assumption is the following one [Damour and Ruffini (1976)]: *the wave function $\phi(x)$ describing a particle state (positive norm) can be analytically continued to complex points of the form $z = x+iy$ if $y \in V_-$* (V_- is the past light cone); *the wave function $\phi(x)$ describing an antiparticle state (negative norm) can be analytically continued to complex points of the form $z = x + iy$ if $y \in V_+$* (V_+ is the future light cone).

In the case at hand, positive norm is equivalent to positive frequency ω. The vector field ∂_r (components $(0,1)$) is lightlike everywhere, and past-directed [Damour and Ruffini (1976)]. As a consequence, the shift $r \mapsto r-i\epsilon$ (with $0 < \epsilon \ll 1$) leads to a unique continuation to an antiparticle state. Analyticity in the lower half plane (in r) occurs. In particular, one can appeal to the behavior of the distribution $(x \pm i\epsilon)^\lambda$, with real x ([Gel'fand and Shilov (1964)], p.59):

$$(x + i\epsilon)^\lambda = \begin{cases} |x|^\lambda e^{i\pi\lambda} & x < 0, \\ x^\lambda & x > 0, \end{cases} \tag{4.7}$$

$$(x - i\epsilon)^\lambda = \begin{cases} |x|^\lambda e^{-i\pi\lambda} & x < 0, \\ x^\lambda & x > 0, \end{cases} \tag{4.8}$$

which, in the present case, gives us for $(r - r_+ - i\epsilon)^{2i\omega r_+}$

$$(r - r_+ - i\epsilon)^{2i\omega r_+} = \begin{cases} (r_+ - r)^{2i\omega r_+} e^{2\pi\omega r_+} & r < r_+, \\ (r - r_+)^{2i\omega r_+} & r > r_+. \end{cases} \tag{4.9}$$

Thanks to the above formulas, we can provide the following identification:

$$\begin{aligned} \phi_{straddle} &= N_{straddle}\phi_{out}(v, r - r_+ - i\epsilon) \\ &= N_{straddle}(\phi_{out}(v, r - r_+)\theta(r - r_+) \\ &\quad + e^{2\pi\omega r_+}\phi_{out}(v, r_+ - r)\theta(r_+ - r)). \end{aligned} \tag{4.10}$$

The splitting of the straddling mode is into a wave outgoing from the horizon and a wave falling onto the singularity. Note that the mode ϕ_{out} has to be normalized only in the external region, but the straddling mode has to be normalized in the whole region $r > 0$. As in the original paper, we can assume that the state ϕ_{out} is already normalized. Then, by taking into account the antiparticle nature of $\phi_{straddle}$, from

$$(\phi^{\omega_1}_{straddle}, \phi^{\omega_2}_{straddle}) = -\delta(\omega_1 - \omega_2), \tag{4.11}$$

where $(\cdot, \cdot)$ is the usual Klein-Gordon inner product, we obtain

$$|N_{straddle}|^2 = \frac{1}{e^{4\pi\omega r_+} - 1}. \tag{4.12}$$

We can also say that the straddling mode is thermally populated by outgoing (escaping) modes [Visser (2003)]. The outgoing flux of particles to infinity per unit time and unit frequency is given by $|N_{straddle}|^2/(2\pi)$. The pair-creation process takes place at the horizon, where the following interpretation holds: a particle wave of strength $|N_{straddle}|^2$ is outgoing from the horizon and reaches infinity with positive energy flux (in 4D, one should take backscattering into account). A negative norm wave with positive energy flux is outgoing in the past reaching the singularity; equivalently, an antiparticle wave with negative energy flux is ingoing in the future into the singularity.

The straddling mode can be associated with modes defining the Unruh vacuum [Massar and Parentani (2000); Visser (2003)]. It is the vacuum state for a free-falling observer (which is allowed to check both the exterior region and the interior one). A Bogolubov transformation between the usual basis with respect to the Killing vector ∂_t can also be implemented, with the same results as above. The dynamical process associated with the gravitational collapse is the reason why the vacuum fluctuation is actually broken in a particle-antiparticle pair [Parentani and Brout (1992)].

4.2 Hamilton-Jacobi tunneling method

The basic idea of the method is very simple, and consists in adopting the WKB approximation and computing the tunneling probability for a straddling mode. Subtleties occur if singular coordinate systems on the horizon are adopted, as pointed out in [Vanzo (2011); Vanzo *et al.* (2011)]. Former calculations appear in [Srinivasan and Padmanabhan (1999)], and further developments are contained in [Shankaranarayanan *et al.* (2002, 2001)]. A thorough analysis and review is contained in [Vanzo *et al.* (2011)], to which we refer the reader for a more complete list of references. Relevant papers for this section are [Srinivasan and Padmanabhan (1999); Shankaranarayanan *et al.* (2002, 2001); Kerner and Mann (2006); Vanzo *et al.* (2011); Akhmedov *et al.* (2006)].

As a basic ingredient of the approach, we have the classical action S of particles (massless or not), to be computed along trajectories which pass through the horizon. Of course, the calculation is semiclassical, and lowest order results are enough to realize that the Hawking effect is in act. The semiclassical emission rate is given by

$$\Gamma_{emission} = \exp\left(-2\mathrm{Im}S\right). \tag{4.13}$$

The right-hand side is easily realized to correspond to the standard form for the rate of emission associated with tunneling through a potential barrier in the WKB approximation. Still, the barrier is non-standard, being present a single turning point against the usual couple of turning points for standard barriers. In other terms, whereas in standard barriers one faces with a (finite extent) region where classical motion is forbidden, but where there is still a non-zero quantum mechanical probability for a transition to a different region where classical motion is again allowed, in the present case, apart for a peculiar but only qualitative picture involving backreaction due to Parikh and to be discussed in the following section, there is no such a situation. The horizon plays the role of a unattainable limiting region for signals in the inner region of the black hole, much more than a real potential barrier. Moreover, a very non-trivial transition between a spacetime region with a time-dependent metric (black-hole region) to a static region (exterior of the black hole) is being occurring, so it is the case to remark that 'standard interpretations' are not so well grounded or, at the very least, free of misinterpretations. See also [Moretti and Pinamonti (2012)].

A further remark is that thermality will occur with inverse temperature β if [Hartle and Hawking (1976)]

$$\Gamma_{emission} = \exp\left(-\beta\omega\right)\Gamma_{absorption}. \tag{4.14}$$

As to the relation with the WKB approximation for the wave equations, we point out that both the Klein-Gordon equation and the Dirac equation have the same semiclassical limit at the lowest order approximation, which is represented by the Hamilton-Jacobi (HJ henceforth) equation.

The tunneling approach can also be related to the original path of thought of the seminal paper by Hartle and Hawking. Indeed, if with x, x' we indicate $x' := (\vec{x}', t')$ and $x := (\vec{x}, t)$, the semiclassical kernel for a particle to propagate from x to x' in the saddle point approximation is

$$K(x', x) \propto e^{i/\hbar S(x', x)}, \tag{4.15}$$

where S is the classical action functional, and

$$S(x', x) = \int_{\gamma(x', x)} dS, \tag{4.16}$$

with $\gamma(x', x)$ is the classical path between x' and x. See [Srinivasan and Padmanabhan (1999)].

4.2.1 *Canonical invariance*

Some authors stress that (4.13) has the drawback to be not invariant under canonical transformations, as is remarked first in [Chowdhury (2008)] and then in [Akhmedov *et al.* (2006, 2007)]. Such an invariance is achieved by the expression $\Gamma_{emission} = \exp\left(-\text{Im} \oint pdq\right)$, where $\oint$ henceforth indicates that both the contribution of the outgoing path $\int_{\rightarrow}$ and the contribution of the ingoing one $\int_{\leftarrow}$ are taken into account. See also [Vanzo *et al.* (2011)]. Moreover, one has to take into account the temporal contribution to $\exp\left(-\text{Im} \oint pdq\right)$ in order to obtain a consistent implementation of the tunneling picture [Akhmedov *et al.* (2008); Akhmedova *et al.* (2008, 2009)]. Explicitly, one has $\Gamma = \exp(-\text{Im}\left(\omega \Delta t_{out} + \omega \Delta t_{in} - (\int p^{out} dq - \int p^{in} dq)\right))$, where Δt_{out} refers to the temporal contribution for outgoing particles, and analogously for Δt_{in} [Akhmedov *et al.* (2008)]. See also [de Gill *et al.* (2010); Tao *et al.* (2017)] for further discussion.

It is interesting to point out that, by defining the so-called Poincaré-Cartan one-form in the extended phase space, limiting our attention to a single particle Hamiltonian suitable for our purpose: [Arnold (1989); de Gosson (2016)]

$$\alpha_H := \sum_{i=1}^{3} p_i dq^i - H(\vec{q}, \vec{p}, t) dt, \tag{4.17}$$

the action functional becomes

$$S(x', x) = \int_{\gamma(x', x)} \alpha_H, \tag{4.18}$$

and that one can obtain an invariant expression simply by considering

$$\Gamma_{emission} = \exp\left(-2\text{Im} \oint \alpha_H\right), \tag{4.19}$$

which takes automatically into account the temporal part, is invariant under canonical transformations and also relativistically, as it can be easily realized. In particular, we can add $p^0 := -H, q^0 := t$ and obtain $\alpha_H = p_\mu dq^\mu$. Note also that, by fixing the initial point x' one has [Arnold (1989); de Gosson (2016)]

$$\alpha_H = dS. \tag{4.20}$$

4.2.2 *The null geodesic method*

The null geodesic method was introduced in [Kerner and Mann (2006)]. The naive idea is to use null geodesics and null regular coordinates in order to avoid subtleties and spurious contributions arising by using singular coordinates. We shall follow the improved version appearing in [Vanzo (2011); Vanzo *et al.* (2011)]. By following [Vanzo *et al.* (2011)], it is interesting to write down the action as follows:

$$S = \int dS = \int_{\gamma} (\partial_{x^{\mu}} S) dx^{\mu}, \tag{4.21}$$

where dS is the one-form corresponding to the differential of S, and an integration along an oriented, null path is understood, and this is at the root of the so called null geodesic method [Vanzo *et al.* (2011)]. In such a way, dS is written in terms of the differential of coordinates times the conjugate momenta $p_i = (\partial_{x^{\mu}} S)$ for $\mu = 0, 1, 2, 3$ (a change of sign in the 0-component can occur with respect to this definition). See below for an example. The choice to follow null geodesics in regular coordinate systems has some evident benefits: the horizon appears in a very simple way in these coordinates, and once more no spurious behavior due to singularities is introduced. Furthermore, if $\int_{\rightarrow}$ indicates an integral along an outgoing null geodesic curve and $\int_{\leftarrow}$ indicates an integral along an ingoing null geodesic curve, we get [Vanzo (2011); Vanzo *et al.* (2011)]

$$\text{Im} \oint dS = \text{Im}\left(\int_{\rightarrow} dS - \int_{\leftarrow} dS \right) = \frac{\pi \omega}{\kappa} \tag{4.22}$$

where κ is the surface gravity. In particular, one can show that [Vanzo (2011); Vanzo *et al.* (2011)]

$$\Gamma_{emission} = \exp\left(-2\left(\text{Im}\left(\int_{\rightarrow} dS - \int_{\leftarrow} dS \right) \right) \right) = \exp\left(-\frac{2\pi}{\kappa} \omega \right). \tag{4.23}$$

For the proof, see the following subsection.

Once more, we stress that regular coordinate systems are to be privileged with respect to singular ones, because they have the bonus of simplifying the procedure, and do not require 'special' prescriptions which are simply associated with the use of singular coordinate systems (the latter case is the first developed: see e.g. [Srinivasan and Padmanabhan (1999)]). Subtleties occur if singular coordinate systems on the horizon are adopted, as pointed out e.g. in [Akhmedov *et al.* (2006, 2007, 2008); Mitra (2007); Akhmedova *et al.* (2008, 2009); Vanzo (2011); Vanzo *et al.* (2011)]. In particular, in [Mitra (2007)] the imaginary part of a integration constant usually neglected can still play an important role to restore the 'factor two' missing in previous literature.

In order to be more explicit, let us revisit the original calculation of [Damour and Ruffini (1976)], discussed in the previous section, in light of the present method [Vanzo (2011); Vanzo *et al.* (2011)]. The HJ equation inherited by the Klein-Gordon equation is

$$(\partial_r S)(2\partial_v S + (1 - r_+/r)\partial_r S) = 0. \tag{4.24}$$

We are only interested to outgoing modes (the ingoing ones do not provide any contribution to the imaginary part of the effective action, being regular on the horizon in the given coordinate system [Vanzo *et al.* (2011)]). Due to the fact that ∂_v is a timelike Killing vector, we can put $\omega := -\partial_v S =$const, and then for a straddling mode we get

$$p_r \equiv k(r) := (\partial_r S) = 2\omega \frac{1}{(1 - r_+/r)}. \tag{4.25}$$

Then we obtain a contribution to the imaginary part of the effective action only by the integration in r, which is our paradigmatic example of transmission integral (cf. the standard WKB calculations of transmission amplitudes in ordinary quantum mechanics):

$$\text{Im}S = \text{Im} \int_{r_1}^{r_2} dr \; (\partial_r S) = \text{Im} \int_{r_1}^{r_2} dr \; k(r), \tag{4.26}$$

where $r_1 < r_+ < r_2$. Indeed, we have to regularize the simple pole appearing in

$$\int_{r_1}^{r_2} dr \; k(r) = \int_{r_1}^{r_2} dr \; \frac{2\omega}{(1 - r_+/r)} \tag{4.27}$$

by means of the shift $r - r_+ \mapsto r - r_+ - i\epsilon$, in the limit as $\epsilon \to 0^+$. This amounts to a definition of the aforementioned integral, which is otherwise ill-defined. We obtain

$$\int_{r_1}^{r_2} dr \; k(r) = \int_{r_1}^{r_+ - \epsilon} dr \; k(r) + \int_{C_\epsilon} dz \; k(z) + \int_{r_+ + \epsilon}^{r_2} dr \; k(r), \tag{4.28}$$

where C_ϵ is an anti-clockwise semicircle centered in r_+ and in the lower half r-plane. The only contribution to the imaginary part of the action S is due to $\int_{C_\epsilon} dz \; k(z)$. A simple application of the fractional residue theorem ([Gamelin (2001)], p. 209) leads to

$$\lim_{\epsilon \to 0^+} \int_{C_\epsilon} dz \; k(z) = i\pi \text{Res}[k(z), r_+], \tag{4.29}$$

which in the present case gives

$$\text{Im}S = 2\pi r_+ \omega. \tag{4.30}$$

Then, the correct result is obtained:

$$\Gamma_{emission} = \exp\left(-4\pi r_+ \omega\right). \tag{4.31}$$

Absorption occurs instead with probability equal to 1, which is both the classical result and also compatible with the detailed balance argument.

Note that, in comparison with the calculation in the previous section, the WKB approximation of the wave function which is related to the present approach gives rise to the same straddling mode, apart for the normalization. This can be realized as follows: the phase factor in the WKB expression of the wave function involves an arbitrary initial point r_0, in the sense that

$$\phi(r,v) = \exp(-i\omega v) \exp\left(i \int_{r_0}^{r} dx\ k(x)\right) \tag{4.32}$$

is a way to express the wave function and r_0 can be chosen outside as well as inside the horizon. If it is chosen outside, there is no problem in defining ϕ_{out}, but, if we want to define $\phi_{straddle}$, we must impose some analytic continuation condition on the integral:

$$\int_{r_0}^{r} dx\ k(x) = \int_{r_0}^{r_+ + \epsilon} dx\ k(x) + \int_{-C_\epsilon} dz\ k(z) + \int_{r_+ - \epsilon}^{r} dx\ k(x), \tag{4.33}$$

where $-C_\epsilon$ stays for the clockwise version of C_ϵ. The integral $\int_{-C_\epsilon} dz\ k(z)$ provides $\exp(\beta\omega/2)$ as a relative weight between the two parts (inner and outer) of the mode at hand. This fact corroborates also the $i\epsilon$ regularization of the integral, which simply corresponds to the unique analytic continuation of a escaping mode ϕ_{out} to the inner side of the horizon, as seen.

It is also interesting to note that, by using the standard textbook Kruskal coordinates U, V, the same conclusion can be reached.

We stress that $\Gamma_{emission}$ is associated with the conditional probability rate $\tilde{P}_\omega(1|0)$ to create a pair in the state labeled by ω, given the probability rate to get zero particles $\tilde{P}_\omega(0)$: the absolute probability rate is obtained as usual:

$$\tilde{P}_\omega(1) = \tilde{P}_\omega(1|0)\tilde{P}_\omega(0). \tag{4.34}$$

We shall come back on this topic in subsection 4.2.4, where a deeper discussion is found.

4.2.3 *The analyticity argument*

We follow [Vanzo (2011)] and the null geodesic method in order to compute the ratio between the emission rate and the absorption rate. Let us consider standard Kruskal coordinates for Schwarzschild spacetime:

$U = -e^{-\kappa u}$, $V = e^{\kappa v}$, where $u = t - r_*$, $v = t + r_*$, r_* is the tortoise coordinate. U, V can be extended to cover the maximally extended Schwarzschild spacetime, which represents the so-called eternal version of the black hole, as well known. Four distinct regions are covered: (I) $U < 0, V > 0$, (F) $U > 0, V > 0$, (P) $U < 0, V < 0$, (II) $U > 0, V < 0$. We can also define the following complexification: $\tilde{U} = Ue^{i\lambda}$, $\tilde{V} = Ve^{-i\lambda}$, with $\lambda \in [0, \pi]$. Then for the differential of the action we obtain

$$\begin{aligned} dS &= (\partial_{\tilde{U}} S) d\tilde{U} + (\partial_{\tilde{V}} S) d\tilde{V} \\ &= (\partial_U S) dU + (\partial_V S) dV + [U(\partial_U S) - V(\partial_V S)] i d\lambda. \end{aligned} \tag{4.35}$$

Let the straddling path be $a \to b \to c$, where a is in the region (F) and b, c are in the region (I) (see Fig. 4.1).

By analytically continuing the point a to complex values (at constant θ, ϕ), the integral on the straddling path is equivalent to the integral on the path $a' \to b \to c$ plus the integral along the semicircle from a to a' obtained by keeping U, V constant and varying λ from 0 to π. a' is in the region (P). The integral of the action along the novel path $a' \to b \to c$ allows us to obtain the absorption probability rate $\Gamma_{absorption}$: the path represents a particle which straddles the past horizon, by passing from region (P) to region (I); by time reversal invariance, the rate of this process is the same as the rate for passing from region (I) to region (F) through the future horizon [Vanzo (2011)]. It is easily shown that it holds

$$U(\partial_U S) - V(\partial_V S) = -\frac{1}{\kappa}\partial_t S = \frac{\omega}{\kappa}. \tag{4.36}$$

As a consequence, we obtain [Vanzo (2011); Vanzo *et al.* (2011)]

$$\text{Im} \int_{a \to b \to c} dS = \text{Im} \int_{a' \to b \to c} dS + \frac{\pi \omega}{\kappa}, \tag{4.37}$$

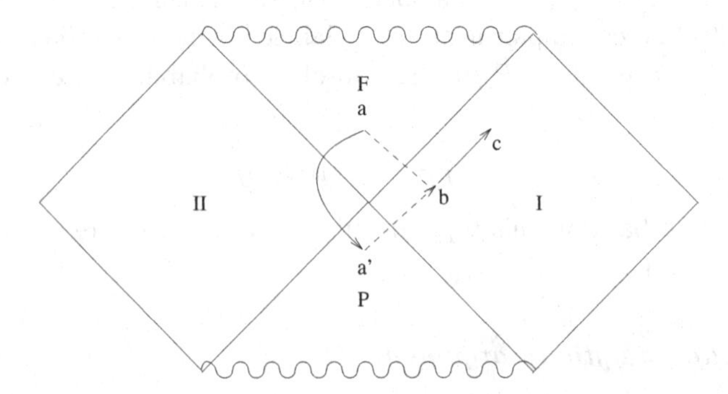

Fig. 4.1 Penrose diagram where the relevant paths discussed in the text are displayed. As observed in [Vanzo (2011)], the sketched paths are not planar (a continuation to complex geometry is understood).

from which detailed balance follows. This analytic argument corroborates the common $i\epsilon$ (Feynman) regularization for the pole on the horizon. Being based on the analyticity on the Kruskal extension, involving also the past horizon, it is appropriate to the Hartle-Hawking vacuum (thermal equilibrium).

We stress that the null geodesic method, as discussed in [Vanzo (2011); Vanzo *et al.* (2011)], is covariant and invariant under canonical transformations, and equivalent to the HJ method.

4.2.4 *A trick à la Nikishov*

The following trick , inspired by [Nikishov (1969); Damour (1975)], is based on the idea that the probability rate of pair creation near the black hole horizon $\Gamma_{emission}$ can be interpreted as the conditional probability rate for the creation of a couple of particles from the vacuum (to be intended as the state of absence of particles). This interpretation is non-standard, but is based on the fact that the WKB approximation is able to provide us the correct behavior of the wave function of the straddling mode, up to the normalization. This suggests that $\Gamma_{emission}$ is just more the square of the relative weight between the outer part and the inner part of the straddling mode than the pair-creation rate itself. This argument is to be compared with the argument in [Sannan (1988)], which is relative to the original picture by Damour and Ruffini. Part of this discussion appeared also in [Belgiorno *et al.* (2017a)].

For the following general picture, we refer to [Nikishov (1969); Damour (1975)]. The approach we are going to follow can be reconduced to the mainstream calculations for instability in quantum field theory in presence of external fields, where a point of main interest is represented by the imaginary part of the effective action W is the signal of particle production. Indeed, the permanence of the vacuum has probability < 1: particle creation occurs with probability (per unit time)

$$P_{0_{in}\to 0_{out}} = \exp\left(-2\mathrm{Im}W\right). \tag{4.38}$$

One can notice the resemblance with the formula defining $\Gamma_{emission}$, but the relation between $\mathrm{Im}W$ and $\Gamma_{emission}$ is not so straightforward. Still, it exists and is found below.

Basically, the following idea is pursued. We proceed as in [Damour (1975)] for the general picture.

Let us introduce, for a diagonal scattering process,

$$n_i^{IN} = R_i n_i^{OUT} + T_i p_i^{OUT}, \tag{4.39}$$

where n_i stays for a negative energy mode and p_i for a positive energy one. In case an inner product different from the standard one for bosonic and fermionic fields occurs, 'positive energy' should be replaced by 'positive norm' (and analogously for negative energy). p_i^{IN}, n_i^{IN} form a scattering basis for the IN states, and p_i^{OUT}, n_i^{OUT} form a scattering basis for the OUT states. T_i is the transmission coefficient and R_i is the reflection one. It is evident that above we have written a Bogoliubov transformation between IN and OUT states, so the following identification is also true:

$$R_i = \alpha_i, \tag{4.40}$$

$$T_i = \beta_i. \tag{4.41}$$

Moreover, one defines as in [Damour (1975)]

$$\eta_i := |T_i|^2, \tag{4.42}$$

which, as shown below, coincides with the mean number per unit time and unit volume of created particles. One has

$$|R_i|^2 = 1 \mp \eta_i, \tag{4.43}$$

where, here and in the sequel, the upper sign holds for fermions and the lower one for bosons. By interpreting *à la* Stueckelberg the scattering process, one can also obtain

$$n_i^{OUT} = R_i^{-1} n_i^{IN} - R_i^{-1} T_i p_i^{OUT}, \tag{4.44}$$

which is interpreted as the scattering of a negative mode incident from the future and which is in part refracted in the past and in part reflected in the future. The new reflection amplitude $-R_i^{-1}T_i$ is such that the reflection coefficient

$$|R_i^{-1}T_i|^2 = \frac{\eta_i}{1 \mp \eta_i} = \tilde{P}_i(1|0) \tag{4.45}$$

can be interpreted as the conditional probability rate $\tilde{P}_i(1|0)$ for the creation of the pair n_i^{OUT}, p_i^{OUT}, starting from absence of particles in that state. The conditional probability rate for n couples is

$$\tilde{P}_i(n|0) = (\tilde{P}_i(1|0))^n. \tag{4.46}$$

Of course, in the fermionic case only $n = 1$ is allowed. The probability rate for n couples is

$$\tilde{P}_i(n) = \tilde{P}_i(n|0)\tilde{P}_i(0). \tag{4.47}$$

$\tilde{P}_i(0)$ represents the probability rate that no particles are created in the given state i. It can be calculated as follows:

$$\sum_n \tilde{P}_i(n) = 1 = \tilde{P}_i(0) \sum_n \tilde{P}_i(n|0), \tag{4.48}$$

and then

$$\tilde{P}_i(0) = \frac{1}{\sum_n \tilde{P}_i(n|0)} = \left(1 \pm \tilde{P}_i(1|0)\right)^{\mp 1} = (1 \mp \eta_i)^{\pm 1}. \tag{4.49}$$

The persistence of the vacuum is given by

$$P_{0_{in} \to 0_{out}} = \exp(-2\text{Im}W) = \prod_i \tilde{P}_i(0). \tag{4.50}$$

Then we can infer

$$2\text{Im}W = -\log\left(\prod_i \tilde{P}_i(0)\right) = \mp \sum_i \log(1 \mp \eta_i) = \sum_i \sum_{k=1}^{\infty} (\pm)^{k+1} \frac{1}{k} \eta_i^k. \tag{4.51}$$

As to the mean number of created couples, we have

$$< n_i >= \sum_n n\tilde{P}_i(n) = \tilde{P}_i(0) \sum_n n(\tilde{P}_i(1|0))^n = \eta_i = |\beta_i|^2. \tag{4.52}$$

As a consequence, in the above formulas we realize that $\eta_i \mapsto < n_i >$ is allowed.

Let us apply the above picture to our specific case. We interpret $\Gamma_{emission}$ as follows:

$$\Gamma_{emission} = \tilde{P}_\omega(1|0), \tag{4.53}$$

where ω identifies the quantum state. In the present case, we get for bosons

$$\tilde{P}_\omega(0) = 1 - \exp(-\beta\omega), \tag{4.54}$$

with $\beta = 4\pi r_+$. As a consequence, one gets

$$\tilde{P}_\omega(n) = (1 - \exp(-\beta\omega)) \exp(-\beta\omega n). \tag{4.55}$$

It is then easy to show that the mean number of created pairs in the state with energy ω is

$$< n_\omega >= \sum_{n=0}^{\infty} n \exp(-\beta\omega n) = \frac{1}{\exp(\beta\omega) - 1}, \tag{4.56}$$

which is the correct result. This argument is substantially equivalent to the one of [Kim and Hwang (2011)]. Note that a thermal particle distribution is obtained without resorting to detailed balance arguments.

It is also worth noticing that

$$\mathrm{Im}W = -\frac{1}{2}\int d\omega\ \log(1-\exp(-\beta\omega)), \tag{4.57}$$

which is the expected result [Stephens (1989); Kim and Hwang (2011)]. The calculation in the fermionic case is analogous, and is based on the fact that the WKB approximation for the Dirac equation coincides with the HJ equation. The only change is the statistics. Extensions to the cases where one takes into account also the backscattering (which is mandatory in 4D) are discussed in [Sannan (1988)].

We point out again the substantial difference between the expressions for $\Gamma_{emission}$ and $\exp(-2\mathrm{Im}W)$ appearing in (4.38). As shown above, $\Gamma_{emission}$ is the conditional probability (4.53) for the emission of a pair labeled by ω, whereas $\exp(-2\mathrm{Im}W)$ is the probability that, for any field mode with label ω, there is not a quantum instability in the field at hand, and then it involves a sum over all values of ω

4.2.5 *The 4D case: Role of the transverse coordinates*

In order to simplify our discussion, we have restricted our analysis so far to a 2D metric, i.e. we have neglected the part of the metric involving spherical coordinates θ, φ. What happens if we restore the aforementioned part of the metric? As a matter of facts, from the point of view of the calculation of $\Gamma_{emission}$, nothing substantially changes. Indeed, as shown e.g. in [Visser (2003)], the 2D calculation reproduces all the essential features of the intrinsic mechanism of pair-creation. Near the horizon, where the phenomenon takes place, further contributions associated with the mass, and the angular momentum of the field are substantially negligible, so that the aforementioned results still hold true. Of course, this does not mean that no effect at all is to be expected. Indeed, a very important contribution arising from the angular momentum barrier is the backscattering, which still affects the spectrum of the particles actually able to reach the static observer at infinity. The grey body factor which affects the black hole radiation spectrum in 4D is typically a 4D effect which cannot be reproduced in a 2D reduction of the process. It is true that the spectrum main contribution arises from s-waves which do not feel the barrier. Still, its effects are non-negligible in general.

4.3 Parikh-Wilczek approach

In the case of the tunneling approach by Parikh-Wilczek [Parikh and Wilczek (2000)], a further ingredient is added to the picture of Hawking radiation. Indeed, backreaction, differently from the HJ approach, is taken into account through the influence of the radiation emission on the black hole mass. Conservation of energy takes a very relevant role. In the Parihk-Wilczek approach to the Hawking effect tunneling through the horizon of a particle can happen because of a quite unexpected mechanism, where the tunneling particle sets up the barrier by energy conservation, as nicely described by Parikh [Parikh (2004)]. Indeed, as the particle emission takes place, the black hole radius shrinks, because the black hole mass shrinks due to the energy subtracted by the escaping particle. In this sense, the particle itself secretly defines its own barrier. Also in this method, coordinates well-defined on the horizon are to be preferable because they are not involved in subtle problems of analytical continuation. The spectrum results to be non-strictly thermal because of the backreaction which takes place explicitly in the method, and which is fundamental for its feasibility [Parikh and Wilczek (2000); Parikh (2004)].

The approach is based on the following ideas. One starts from the exact semiclassical description of a spherical shell interacting with a black hole background. A constrained quantization of such a system is developed, and then a WKB approximation is set up [Kraus and Wilczek (1995)]. In a semiclassical picture, the spherical shell represents an s-wave emission from the background (particle emission), and, in the process, energy is conserved. The action for the system is

$$S = \int dt\ [p_C \dot{\hat{r}} - M_+], \tag{4.58}$$

where p_C is the conjugate momentum to the radius variable $\hat{r}$ and M_+ is the total mass of the system. It holds $M_+ = M - \omega$, where M is the background mass (which is kept constant in the process in [Kraus and Wilczek (1995)], whereas in [Parikh and Wilczek (2000)] is kept constant the total energy; see also below), and ω is the shell energy. M_+ coincides with the Hamiltonian of the system. Then, apart from an inessential constant, one gets a form for the action where the contribution of the particle is evident:

$$S = \int dt\ [p_C \dot{\hat{r}} + \omega]. \tag{4.59}$$

In the WKB approximation, $\hat{r}(t)$ extremizing the classical action are the null geodesics of the metric

$$ds^2 = -dt^2 + \left(dr + \sqrt{\frac{2(M-\omega)}{r}}dt\right)^2, \tag{4.60}$$

which can be realized to be formally coincident with Schwarzschild metric in Painlevè-Gullstrand coordinates. In the passage from [Kraus and Wilczek (1995)] to [Parikh and Wilczek (2000)] one has to make the shift $M + \omega \mapsto M - \omega$ [Parikh and Wilczek (2000)]. Then two further basic steps are to be implemented: a) the semiclassical approach to the imaginary part of the effective action is considered, where again it holds $\Gamma_{emission} = \exp(-2\text{Im}S)$, and b) a particular trick for the calculation of the transmission integral is adopted. Given the discussion in the previous section, as to (a) we limit ourselves to recall that the transmission integral through the horizon for a particle is involving only p_r; then only (b) has to be discussed. The main point is the following trick:

$$\text{Im}S = \text{Im}\int_{r_1}^{r_2} dr\, p_r = \text{Im}\int_{r_1}^{r_2} dr \int_0^{p_r} dp'_r, \tag{4.61}$$

where $r_1 < r_+ < r_2$. Then, Hamilton equations allow to write

$$\dot{r} = \left.\frac{dH}{dp_r}\right|_r \Longrightarrow dp_r = \frac{dH}{\dot{r}}, \tag{4.62}$$

and, by taking into account that $H = M - \omega$, we get

$$\text{Im}S = \text{Im}\int_{r_1}^{r_2} dr \int_0^{\omega} \frac{d\omega'}{\dot{r}}. \tag{4.63}$$

We are only interested in the outgoing null geodesics (no contribution to the imaginary part of the action arises in the given metric for the ingoing null geodesics):

$$\dot{r} = 1 - \sqrt{\frac{2(M-\omega)}{r}}. \tag{4.64}$$

Then we obtain

$$\text{Im}\int_0^{\omega} d\omega' \int_{r_1}^{r_2} dr \frac{1}{1 - \sqrt{\frac{2(M-\omega')}{r}}} = 4\pi\omega\left(M - \frac{\omega}{2}\right), \tag{4.65}$$

where again the contribution to the imaginary part of the effective action arises from the simple pole at $r = 2M$ (which is evidently the horizon of the unperturbed geometry) and is calculated as in the previous section. The novelty is also evident: there is a further contribution, associated with energy conservation (backreaction), which makes not possible to obtain a strictly ω-independent temperature (see e.g. [Parikh (2004); Corda (2011)]).

Chapter 5

The anomaly route to Hawking radiation

In this chapter, we describe the somewhat unexpected relation between Hawking radiation and quantum anomalies. The history of this subject begins at the early days of Hawking's discovery, indeed in a seminal paper [Christensen and Fulling (1977)] it was shown that the conformal anomaly can be used in order to infer information about the renormalized quantum stress-energy tensor through its covariant conservation equation. In particular, in 2D, the quantum stress-energy tensor is completely determined by the anomaly, up to boundary conditions to be imposed in order to find out integration constants. This route inspired several works and developments, still it seems to be not at the roots of the phenomenon, because even in 2D we must impose suitable boundary conditions and it is specialized only to the case of conformally invariant couplings (which, of course, do not correspond to the general case).

Only in recent years anomalies have come back into the interest of experts in the field. Two seminal works by Robinson and Wilczek, and by Iso, Umetsu and Wilczek, shed new light on the subject [Robinson and Wilczek (2005); Iso *et al.* (2006b)]. In the former it was shown how the consistent gravitational anomaly can be used in order to find out the Hawking flux, in the latter the covariant gravitational and gauge anomalies are taken into account for generating the Hawking flux.

5.1 Christensen-Fulling way (1977)

The recipes are very few: a) covariant conservation of the quantum stress-energy tensor, b) classically conformally invariant action for the quantum fields and quantum trace anomaly, c) boundary conditions for determining integration constants. As to (c), the requirement of a nonsingular behavior (absence of divergences for a free-falling observer) at the horizon leads to the Unruh state stress-energy tensor in case of a real collapse situation, and to the Hartle-Hawking state stress-energy tensor in case of an eternal black hole. In the 4D case, the trace anomaly is not enough for determining all the components of the quantum stress-energy tensor, still it is very useful for gaining information. In the original paper [Christensen and Fulling (1977)], the 2D case is shown to lead to a thermal state under the aforementioned regularity requirement, and also the converse is shown to be true: starting from a thermal state at infinity, it is impossible to generate a regular state unless a nonvanishing trace anomaly is present.

The paper [Christensen and Fulling (1977)] is a seminal paper which has laid the grounds of calculations for the stress-energy tensor of Hawking radiation. We do not delve into the modern developments, and refer to [Frolov and Novikov (1998)].

In what follows, and also in the following section, we indicate with $T_{\mu\nu}$ the renormalized quantum stress energy tensor. As such, an expectation value on the associated quantum state should be indicated, but, as usual, we do not mark suitably the quantum stress energy tensor to distinguish it from its classical counterpart. This is an abuse of language, but simplifies the notation. We also recall that Wald set up an axiomatic approach for dealing with the renormalized quantum stress energy tensor [Wald (1977, 1978)], aimed at specifying properties that the renormalized stress energy tensor should satisfy in order to get a sensible picture for quantum backreaction in curved space.

We start from the covariant conservation equation for a conformally invariant scalar field:

$$\nabla_\mu T^\mu_\nu = 0. \tag{5.1}$$

The stress energy tensor in 4D is assumed to be spherosymmetric and static, as is the metric of the Schwarzschild black hole. Then the following components are possibly nonvanishing: $T^t_t, T^t_r, T^r_t, T^r_r, T^\theta_\theta, T^\phi_\phi$, where one has $T^\theta_\theta = T^\phi_\phi$. All components are moreover functions only of r.

Furthermore, T^t_t is obtained from the trace equation:

$$T^t_t = T^\mu_\mu - T^r_r - 2T^\theta_\theta. \tag{5.2}$$

The conservation equation (5.1) leads to the following equation for T^r_t:

$$\left(\partial_r + \frac{2}{r}\right) T^r_t = 0, \tag{5.3}$$

whose solution is

$$T_t^r = \frac{K1}{r^2}, \tag{5.4}$$

where $K1$ is an integration constant. For T_θ^θ and T_ϕ^ϕ one obtains the spherosymmetric constraint

$$T_\theta^\theta = T_\phi^\phi. \tag{5.5}$$

The equation for T_r^r is

$$\partial_r T_r^r + \left(\frac{r_+}{2r^2\left(1 - \frac{r_+}{r}\right)} + \frac{2}{r} \right) T_r^r - \frac{1}{r}\left(T_\theta^\theta + T_\phi^\phi + \frac{r_+}{2r^2\left(1 - \frac{r_+}{r}\right)} T_t^t \right) = 0. \tag{5.6}$$

Then we obtain

$$T_r^r = \frac{1}{2r^2\left(1 - \frac{r_+}{r}\right)} \left[K2 + \int_{r_+}^{r} dx \frac{1}{2} \left(r_+ T_\mu^\mu(x) + (4x - 6r_+) T_\theta^\theta(x) \right) \right], \tag{5.7}$$

where $K2$ is an integration constant. The complete solution is expressed in terms of two integration constants $K1, K2$ and of two functions $T_\mu^\mu(r), T_\theta^\theta(r)$. The trace anomaly is known, so there remains only an undetermined function $T_\theta^\theta(r)$. In order to match the original paper, we put $K1 = -4K/r_+^2$ and $K2 = 4(Q - K)/r_+^2$, where the integration constants are K, Q.

Christensen and Fulling show that it is possible to obtain

$$T_\nu^\mu = \sum_{i=1}^{4} T_{(i)\ \nu}^{\mu}, \tag{5.8}$$

where each term satisfies (5.1). In particular,

$$T_{(1)\ \nu}^{\mu} = \text{diag}\left(\frac{-1}{r^2\left(1 - \frac{r_+}{r}\right)} H(r) + \frac{1}{2} T_\mu^\mu(r), \frac{1}{r^2\left(1 - \frac{r_+}{r}\right)} H(r), \frac{1}{4} T_\mu^\mu(r), \frac{1}{4} T_\mu^\mu(r) \right) \tag{5.9}$$

is the only term having non-zero trace;

$$T_{(2)\ \nu}^{\mu} = \frac{4K}{r_+^2 r^2} \frac{1}{\left(1 - \frac{r_+}{r}\right)} \begin{pmatrix} 1 & -1 & 0 & 0 \\ 1 & -1 & 0 & 0 \\ 0 & 0 & 0 & 0 \\ 0 & 0 & 0 & 0 \end{pmatrix} \tag{5.10}$$

is the only term involving fluxes (non-diagonal entries);

$$T^{\mu}_{(3)\ \nu} = \mathrm{diag}\left(\frac{-1}{r^2\left(1-\frac{r_+}{r}\right)}G(r) - 2\Theta(r), \frac{1}{r^2\left(1-\frac{r_+}{r}\right)}G(r), \Theta(r), \Theta(r)\right) \tag{5.11}$$

has a traceless part whose $\theta\theta$-component is non-zero;

$$T^{\mu}_{(4)\ \nu} = \frac{4Q}{r_+^2 r^2}\frac{1}{\left(1-\frac{r_+}{r}\right)}\mathrm{diag}\,(-1,1,0,0) \tag{5.12}$$

is the only non-regular term on the future horizon [Frolov and Novikov (1998)]. The functions $H(r), G(r), \Theta(r)$ are

$$H(r) = \frac{1}{2}\int_{r_+}^{r} dx\left(x - \frac{r_+}{2}\right)T^{\mu}_{\mu}(x), \tag{5.13}$$

$$G(r) = 2\int_{r_+}^{r} dx\left(x - \frac{3r_+}{2}\right) \tag{5.14}$$

$$\Theta(r) = T^{\theta}_{\theta}(r) - \frac{1}{4}T^{\mu}_{\mu}(r). \tag{5.15}$$

Even if the anomaly is known (see e.g. [Birrell and Davies (1984)]), there is an unknown function $\Theta(r)$ which cannot be calculated in the given framework. Christensen and Fulling provide insights for the behavior of that function, using the expected fluxes at infinity for the three vacuum states (Boulware state, Unruh state and Hartle-Hawking state). We limit ourselves to specify the values of the integration constants for the equilibrium state (Hartle-Hawking): $K = 0 = Q$. This ensures regularity on the horizon of the stress-energy tensor.

In the 2D case the trace anomaly, (5.1) is sufficient to determine, for a static vacuum case, the stress-energy tensor, once the trace anomaly is known. There appear two integration constants, which are fixed by suitable boundary conditions. The stress-energy tensor can be written as follows:

$$T^{\mu}_{\nu} = \sum_{i=1}^{3} T^{\mu}_{(i)\ \nu}, \tag{5.16}$$

where all terms satisfy (5.1) and, moreover, $T^{\mu}_{(1)\ \nu}$ is the only diagonal non-traceless term, $T^{\mu}_{(2)\ \nu}$ contains non-diagonal contributions, and $T^{\mu}_{(3)\ \nu}$ is a diagonal traceless term. The trace anomaly is

$$T^{\mu}_{\mu} = -\frac{R}{24\pi}, \tag{5.17}$$

which, for the metric

$$ds^2_{(2)} = \left(1-\frac{r_+}{r}\right)dt^2 - \left(1-\frac{r_+}{r}\right)^{-1}dr^2, \tag{5.18}$$

where $r_+ = 2M$, becomes $T^\mu_\mu = -\frac{r_+}{12\pi r^3}$. (5.1) amounts to

$$\partial_r T^r_t = 0, \tag{5.19}$$

$$\partial_r \left[\left(1 - \frac{r_+}{r} \right) T^r_r \right] - \frac{r_+}{2r^2} T^\mu_\mu = 0, \tag{5.20}$$

which, together with

$$T^t_t = T^\mu_\mu - T^r_r, \tag{5.21}$$

allow to determine all the components of T^μ_ν. Let us define

$$H(r) = \frac{r_+}{2} \int_{r_+}^{r} dx \frac{1}{x^2} T^\mu_\mu(x) = -\frac{1}{96\pi r_+^2 r^4}(r^2 + r_+^2)(r + r_+)(r - r_+); \tag{5.22}$$

then we get

$$T^\mu_{(1)\ \nu} = \text{diag} \left(-\frac{1}{\left(1 - \frac{r_+}{r}\right)} H(r) + T^\mu_\mu(r), \frac{1}{\left(1 - \frac{r_+}{r}\right)} H(r) \right), \tag{5.23}$$

$$T^\mu_{(2)\ \nu} = \frac{4K}{r_+^2} \frac{1}{\left(1 - \frac{r_+}{r}\right)} \begin{pmatrix} 1 & -1 \\ 1 & -1 \end{pmatrix}, \tag{5.24}$$

and

$$T^\mu_{(3)\ \nu} = \frac{4Q}{r_+^2} \frac{1}{\left(1 - \frac{r_+}{r}\right)} \text{diag}\,(-1, 1)\,. \tag{5.25}$$

K, Q are two integration constants, to be determined.

For the Boulware vacuum, i.e. absence of particle at infinity, one has to require that $K = 0$ and $4Q = -r_+^2 H(\infty) = \frac{1}{96\pi}$, i.e. $Q = 1/(384\pi)$. The stress-energy tensor for this state diverges at $r = r_+$ [Christensen and Fulling (1977); Birrell and Davies (1984); Frolov and Novikov (1998)].

For the Hartle-Hawking state , regularity on the horizon requires $K = 0 = Q$. Then one obtains a thermal stress-energy tensor for a boson gas with temperature $T_H = 1/(4\pi r_+)$.

5.2 Robinson-Wilczek (2005) and Iso-Umetsu-Wilczek (2006)

The methodological key of this approach consists in focusing on what is dangerous in the mode population of the Boulware state, and to cut away near the horizon (which is always a part of the physical region to be considered) the propagating modes which would generate the divergence [Robinson and Wilczek (2005)]. These modes are identified with the ones which propagate light-like, parallely to the horizon, and very near to it. The aforementioned cutting away induces a chiral anomaly contribution very near the horizon, because a truncation of the parallely propagating modes in a region including the horizon determines a chirality in the field. The anomaly is repaired by the presence of a flux of particles at the Hawking temperature [Robinson and Wilczek (2005)]. Together with the methodological key we find also the working hypothesis that the effective action near the horizon, due to the variable separation occurring for spherosymmetric and axially symmetric black holes and to the very rapid vanishing at the horizon of the contribution from the transversal degrees of freedom, involves only the s-wave part whose contribution to the Hawking radiation is essentially 2D [Robinson and Wilczek (2005)]. This actually represents from one hand a very relevant simplification of the calculations, because the effective action to be considered is simply 2D, whichever d-dimensional starting black hole manifold one considers; on the other hand, no backscattering can be accounted for in this approximation.

The d-dimensional metric, in the static case, is

$$ds^2 = f(r)dt^2 - \frac{1}{f(r)}dr^2 - r^2 d\omega_{d-2}^2, \tag{5.26}$$

where $d\Omega_{d-2}^2$ involves the transversal degrees of freedom (a $d-2$-dimensional sphere). We suppose that $f(r)$ is analytic at the horizon $r = r_+$ and that $f'(r_+) \neq 0$, i.e. $\kappa = |f'(r_+)|/2 > 0$.

For simplicity, but without substantial loss of generality, we specialize to the physically interesting case $d = 4$. Let us consider a neutral scalar field minimally coupled, whose action is

$$S = \frac{1}{2}\int d^4x\sqrt{-g}\left(g^{\mu\nu}(\nabla_\mu\phi)(\nabla_n u\phi) - m^2\phi^2 - V(\phi)\right), \tag{5.27}$$

where $V(\phi)$ is a self-interaction potential (e.g. ϕ^4). Variable separation of the classical field equation is allowed:

$$\phi(x) = \sum_{l,m}\phi_{l,m}(t,r)Y_{l,m}(\Omega), \tag{5.28}$$

so that one obtains, by using the standard tortoise coordinate r_*,

$$\partial_t^2\phi_{l,m} - \partial_{r_*}^2\phi_{l,m} - f(r_*)\left(m^2 + \frac{l(l+1)}{r^2} + V(\phi_{l,m})\right)\phi_{l,m} = 0. \tag{5.29}$$

$f(r_*)$ vanishes very rapidly when $r \to r_+$, so as a first approximation one assumes to be enabled to reduce the action to its free t,r-part:

$$S \sim \frac{1}{2}\sum_{l,m}\int dtdxr^2(r_*)\phi_{lm}(\partial_t^2 - \partial_{r_*}^2)\phi_{l,m}. \tag{5.30}$$

It is to be noted that, in order to account for gauge anomalies in the case of the Reissner-Nordsström black hole, it is necessary to leave room for a 'dilatonic' contribution represented by $r^2(r_*)$ in the 2D action [Iso *et al.* (2006b)]. It is worth stressing that the aforementioned reduction of the action associates an infinite number of quantum fields (one for each l, m) in the 2D theory. In this approach of dimensional reduction, one has

$$J^{\mu}_{2D} = \int r^2 d\Omega^2 J^{\mu}_{4D} = 4\pi r^2 J^{\mu}_{4D}, \tag{5.31}$$

$$T^{\mu\nu}_{2D} = \int r^2 d\Omega^2 T^{\mu\nu}_{4D} = 4\pi r^2 T^{\mu\mu}_{4D}. \tag{5.32}$$

This means that, in the d-dimensional case. one has to integrate on a $d-2$-dimensional sphere.

For a Reissner-Nordsström black hole, we have $A_t = -Q/r, A_r = 0$, and also $f(r) = (r - r_+)(r - r_-)/r^2$, with $r_{\pm} = M \pm \sqrt{M^2 - Q^2}$, where M, Q are the mass and the charge of the black hole respectively. Moreover, we consider a complex scalar field ϕ.

Following [Iso *et al.* (2006b,a)], let us now consider the following subdivision of the radial interval:

$$[r_+, \infty) = [r_+, r_+ + \epsilon] \cup (r_+ + \epsilon, \infty), \tag{5.33}$$

in such a way that, for $\epsilon \ll 1$, in the interval $[r_+, r_+ + \epsilon]$ field modes propagating along the lightlike direction parallel to the horizon are cut away, leaving a chiral anomaly in the theory. In the first paper [Robinson and Wilczek (2005)] the assumption about the anomaly region was slightly different: a region of size 2ϵ around the horizon $r = r_+$ was symmetrically disposed, involving also the inner region of the black hole manifold. This is not strictly necessary, what is really important is to involve the horizon into the manifold (note that, strictly speaking, the assumed system of coordinates does not include the horizon).

In the discussion of the anomaly contribution, consistent and covariant anomalies play a relevant role. We recall that the consistent form of the anomaly is the one which satisfies Wess-Zumino consistency condition, whereas the covariant form restores the covariance which is lost in the consistent form. The two forms of the current are not equivalent, but they correspond to the same anomaly cancellation condition [Bertlmann (1996)], which is what matter for the present framework. A further important ingredient is represented by the so-called anomalous Ward identity [Bertlmann (1996); Fujikawa and Suzuki (2004)]. As in [Iso *et al.* (2006b)], we start by describing gauge anomalies. The consistent form is

$$\nabla_\mu J^\mu = \pm \frac{e^2}{4\pi\sqrt{-g}} \epsilon^{\mu\nu} \partial_\mu A_\nu, \tag{5.34}$$

where $\epsilon^{01} = 1$ and where + holds for left-handed fields. The covariant form of the current is

$$\bar{J}^\mu = J^\mu \mp \frac{e^2}{4\pi\sqrt{-g}} A_\lambda \epsilon^{\lambda\mu}, \tag{5.35}$$

which satisfies

$$\nabla_\mu \bar{J}^\mu = \pm \frac{e^2}{4\pi\sqrt{-g}} F_{\lambda\mu} \epsilon^{\lambda\mu}. \tag{5.36}$$

It holds

$$\partial_\mu J^\mu = \partial_r J^r = \frac{e^2}{4\pi} (\partial_r A_t) H, \tag{5.37}$$

where one takes into account that in a static situation J^μ does not depend on t and, moreover, one introduces

$$H \equiv 1 - \theta(r - r_* - \epsilon) = \begin{cases} 1 & r < r_+ + \epsilon, \\ 0 & r > r_+ + \epsilon. \end{cases} \tag{5.38}$$

By construction, the anomaly appears only for $r \in [r_+, r_+ + \epsilon]$. We write

$$J_H^\mu = J_o^\mu \theta(r - r_* - \epsilon) + J_H^\mu H, \tag{5.39}$$

where o labels the outer region $r > r_+ + \epsilon$, H labels the inner region. From $\partial_r J_o^r = 0$ and $\partial_r J_H^r = \frac{e^2}{4\pi}(\partial_r A_t)$, we can infer

$$J_o^r = c_o, \quad r > r_+ + \epsilon, \tag{5.40}$$

$$J_H^r = c_H + \frac{e^2}{4\pi}(A_t(r) - A_t(r_+)), \quad r < r_+ + \epsilon. \tag{5.41}$$

c_o, c_H are two integration constants which have to be fixed in such a way that anomaly cancellation occurs. In particular, c_H is the value of the consistent current on the horizon. In order to see how, we consider the variation of the effective action W under gauge transformations:

$$-\delta_{\lambda_G} W = \int d^2x \sqrt{-g} \lambda_G (\nabla_\mu J^\mu); \tag{5.42}$$

Absence of gauge anomaly requires that $\delta_{\lambda_G} W = 0$. This must happen in the limit as $\epsilon \to 0^+$:

$$0 = -\delta_{\lambda_G} W = \lim_{\epsilon \to 0^+} \int d^2x \sqrt{-g} \lambda_G \left[\partial_r (J_o^r \theta + J_H^r H)\right]$$

$$= \lim_{\epsilon \to 0^+} \int d^2x \sqrt{-g}\lambda_G \left[\frac{e^2}{4\pi}\partial_r(A_t H) + \delta(r - r_+ - \epsilon) \left(J_o^r + \frac{e^2}{4\pi}A_t - J_H^r \right) \right] . \tag{5.43}$$

Formally, we have

$$\delta(r - r_+ - \epsilon) = \exp(-\epsilon\partial_r)\delta(r - r_+), \tag{5.44}$$

and we can realize that, in the limit as $\epsilon \to 0^+$, the term $\partial_r(A_t H)$ gives no contribution, whereas to the remaining one contribute only the values of the currents J_o^r, J_H^r on the horizon, so that

$$0 = -\delta_{\lambda_G} W \Longleftrightarrow c_o = c_H - \frac{e^2}{4\pi}A_t(r_+). \tag{5.45}$$

A gauge invariant condition on the horizon is obtained by requiring that the covariant current vanishes at the horizon [Iso *et al.* (2006b)]:

$$\bar{J}_H^r(r_+) = 0 \Longleftrightarrow c_H = -\frac{e^2}{4\pi}A_t(r_+). \tag{5.46}$$

As a consequence, we get

$$c_o = -\frac{e^2}{2\pi}A_t(r_+) = \frac{e^2 Q}{2\pi r_+}, \tag{5.47}$$

which is the expected value for the charge flux of the Hawking radiation with a chemical potential $eA_t(r_+)$ [Iso *et al.* (2006b)].

We consider also the contribution of the gravitational anomaly associated with right-handed fields [Iso *et al.* (2006b)]. The consistent anomaly is

$$\nabla_\mu T_\nu^\mu = \frac{1}{96\pi\sqrt{-g}}\epsilon^{\beta\delta}\partial_\delta\partial_\alpha\Gamma_{\nu\beta}^\alpha \equiv \mathcal{A}_\nu, \tag{5.48}$$

and the covariant anomaly is

$$\nabla_\mu \bar{T}_\nu^\mu = \frac{1}{96\pi\sqrt{-g}}\epsilon_{\mu\nu}\partial^\mu R \equiv \bar{\mathcal{A}}_\nu. \tag{5.49}$$

The stress-energy tensor is not conserved even classically, due to the presence of a charged field in a charged background and of a dilaton field $\sigma(r) \equiv r_*(r)$. This means that in the anomalous Ward identity one obtains in imposing diffeomorphism-invariance of the effective action, there will appear classical terms which must be present. An anomalous term will also be present, unless is suitably canceled, in analogous way to what happens in the gauge anomaly case. The anomalous Ward identity is [Iso *et al.* (2006b)]

$$\nabla_\mu T^\mu_\nu = F_{\mu\nu} J^\mu + A_\nu \nabla_\mu J^\mu - \frac{1}{\sqrt{-g}} \partial_\nu \sigma \frac{\delta S}{\delta \sigma} + \mathcal{B}_\nu, \tag{5.50}$$

where S is the classical action and $\mathcal{B}_\nu$ is the consistent gravitational anomaly. Explicitly, we have

$$\mathcal{B}_r = 0, \qquad \mathcal{B}_t = \partial_r N^r_t, \tag{5.51}$$

with

$$N^r_t = \frac{1}{192\pi}(f'^2 + f f''); \tag{5.52}$$

for the covariant anomaly we have

$$\bar{\mathcal{B}}_r = 0, \qquad \bar{\mathcal{B}}_t = \partial_r \bar{N}^r_t, \tag{5.53}$$

with

$$\bar{N}^r_t = \frac{1}{192\pi}(-f'^2 + f f''). \tag{5.54}$$

As in the gauge anomaly case, where it was possible to determine only the r-component of the current J^μ, also in this case it is possible to gain knowledge only on T^r_t, and so we choose $\nu = t$ in the anomalous Ward identity. Indeed, we can get rid of the dilaton contribution [Iso *et al.* (2006b)].

We have

$$T^r_t = T_{o}{}^r_t \theta + T_{H}{}^r_t H. \tag{5.55}$$

In the outer region we obtain

$$\partial_r T_{o}{}^r_t = F_{rt} J^r_o = (\partial_r A_t) c_o, \tag{5.56}$$

and then

$$T_{o}{}^r_t = A_t c_o + a_o. \tag{5.57}$$

In the region $[r_+, r_+ + \epsilon]$ we get

$$\begin{aligned} \partial_r T_{H}{}^r_t &= F_{rt} J^r_H + A_t \partial_r J^r_H + \partial_r N^r_t \\ &= F_{rt} \bar{J}^r_H + \partial_r N^r_t = (\partial_r A_t) \bar{J}^r_H + \partial_r N^r_t, \end{aligned} \tag{5.58}$$

and then

$$T_{H}{}^r_t = \int_{r_+}^{r} dx \partial_x \left[c_o A_t + \frac{e^2}{4\pi} A_t^2 + N^r_t \right]. \tag{5.59}$$

As to the effective action, we have

$$-\delta_\xi W = \int d^2x \sqrt{-g} \xi^\nu \nabla_\mu T^\mu_\nu. \tag{5.60}$$

From

$$\nabla_\mu(T_{ot}{}^\mu_t\theta+T_{H}{}^\mu_t H)=(F_{rt}J^r_o\theta+T_{ot}{}^r_t\delta)+\left[\left(c_oA_t+\frac{e^2}{4\pi}A_t^2+N_t^r\right)H-T_{H}{}^r_t\delta\right]$$
$$=(c_o(\partial_rA_t)\theta+T_{ot}{}^r_t\delta)+\left[(\partial_rA_t)\left(c_o+\frac{e^2}{2\pi}A_t\right)H+(\partial_rN_t^r)H-T_{H}{}^r_t\delta\right] \quad (5.61)$$

we find [Iso *et al.* (2006b)]

$$\int d^2x\sqrt{-g}\xi^tT^\mu_\nu=\int d^2x\xi^t\left[c_o(\partial_rA_t)+\partial_r\left[\left(\frac{e^2}{4\pi}A_t^2+N_t^r\right)H\right]\right.$$
$$\left.+\left(T_{ot}{}^r_t-T_{H}{}^r_t+\frac{e^2}{4\pi}A_t^2+N_t^r\right)\delta\right]. \quad (5.62)$$

The latter equation has to be considered in the limit as $\epsilon\to0^+$. The first term is classical, and does not represent a problem, the second one vanishes in the limit (as $H\to0$ in the same limit), whereas the last term is associated with the quantum anomaly and has to vanish. As a consequence

$$T_{ot}{}^r_t(r_+)=T_{H}{}^r_t(r_+)-\frac{e^2}{4\pi}A_t^2(r_+)+N_t^r(r_+)=a_H-2N_t^r(r_+), \quad (5.63)$$

which implies

$$a_o=a_H+\frac{e^2}{4\pi}A_t^2(r_+)-N_t^r(r_+). \quad (5.64)$$

Analogously to what happens in the gauge anomaly case, one requires that the $t-r$ component of the covariant stress-energy tensor vanishes for $r=r_+$:

$$0=\bar{T}^r_t(r_+)=T^r_t(r_+)+\frac{1}{192\pi}(ff''-2(f')^2)(r_+), \quad (5.65)$$

which means

$$a_H=2N_t^r(r_+), \quad (5.66)$$

and then

$$a_o=N_t^r(r_+)+\frac{e^2}{4\pi}A_t^2(r_+)=\frac{e^2Q^2}{4\pi r_+^2}+\frac{\pi T_H^2}{12}, \quad (5.67)$$

where $T_H=1/(4\pi r_+)$ is the black hole temperature. Then one obtains the correct flux of Hawking particles (as far as a 2D theory is considered). The calculations displayed above follow strictly the line of [Iso *et al.* (2006b)]. One can also develop all the calculation with reference only to covariant anomalies as in [Banerjee (2009)], where the case of fermion fields is studied.

It is remarkable as the universality of the Hawking effect emerges from the above calculations: fluxes do not depend on detailed information on the microscopic Lagrangian. The calculation of the other components of the stress-energy tensor involves non-universal terms, and is possible in the case of a neutral black hole (Schwarschild) [Robinson and Wilczek (2005); Robinson (2005)], or, alternatively, by choosing a conformally invariant theory [Iso *et al.* (2006a); Banerjee and Kulkarni (2009)]. The rotating case is studied in [Iso *et al.* (2006a)]. Further discussions and developments are found in [Das *et al.* (2008); Banerjee and Majhi (2009)].

This approach is different from the one discussed in the previous section under many respects. In [Christensen and Fulling (1977)] the conformal anomaly is used in the 2D case together with boundary conditions of regularity of the stress-energy tensor, obtaining the Hawking flux by integrating the equations on the whole spacetime. In the present case, one finds the Hawking flux in a sort of (hyper)local way, simply by requiring the vanishing of the chiral anomaly and of the covariant stress-energy tensor on the horizon. This approach is, by construction, more specifically related to the Hawking effect and to its nature.

There remain a few open problems: thermality is not demonstrated [Iso *et al.* (2006b)], and it is not yet possible to find out grey-body factors for matching physics in 4D, as in the original intentions of the authors. As it stands, it is essentially a 2D method which should have been linked in a stronger way to 4D physics (more in general to a d-dimensional physics). In order to give some more insights about the actual discrepancy between 2D results and 4D ones, we recall e.g. that the stress-energy tensor for the thermal (Hartle-Hawking) state in 2D (which would represent the s-wave contribution in the 4D case) is by no means a good approximation of the full 4D stress-energy tensor on the horizon [Balbinot *et al.* (2001)]. See also [Fabbri and Navarro-Salas (2005)].

Chapter 6

The Euclidean section and Hawking temperature

It is known that, for static black holes with bifurcate horizon, and temperature T_H, it is necessary to impose a periodicity $\beta = \beta_H$ in the imaginary time in order to avoid distributional curvature singularities ($\beta_H = \hbar c/(k_B T_H)$). We reconsider this argument, and show that the Hawking temperature T_H can be found as a consequence of the requirement of extendability of the Euclidean section of the metric up to including the horizon. The extremal case is shown to be special.

Quantum field theory in Euclidean time with periodicity β corresponds to the so-called imaginary time formalism for quantum statistical mechanics in the canonical (or, more generally, grand canonical) ensemble. Since the paper by Hartle and Hawking [Hartle and Hawking (1976)], it was realized that for static black holes it is necessary to impose a periodicity $\beta = \beta_H$ in the imaginary time in order to avoid distributional curvature singularities at the black hole horizon. It was soon realized by Gibbons and Perry [Gibbons and Perry (1978)] that such a periodicity is also associated to the periodicity of thermal green functions in finite temperature quantum field theory. Also, this was implemented in [Gibbons and Hawking (1977a)] in the path integral formalism. We focus on the latter approach.

In the framework of the Euclidean approach to black hole thermodynamics it is postulated that the partition function for scalar quantum matter fields in thermal equilibrium with a black hole is given by [Gibbons and Hawking (1977a)]

$$Z = \int_{P(\beta)} [D\phi] \, exp(-S[\phi]) \tag{6.1}$$

where $S[\phi]$ is the classical action for the free field ϕ

$$S[\phi] = \frac{1}{2} \int d^4x \; \phi(-\Delta + m^2 + \xi R) \; \phi \tag{6.2}$$

and Δ is the Laplacian on functions; m is the mass of the field and ξ is an arbitrary coupling constant of the field with the Ricci scalar R. $P(\beta)$ indicates that periodic boundary conditions on the euclidean time are imposed.

The requirement to avoid curvature singularity at the horizon amounts, as Hawking points out [Hartle and Hawking (1976)], to a fixed choice of the period: $\beta = \beta_H$, where β_H is the inverse Hawking temperature (Hartle-Hawking state). Thus, any equilibrium quantum field theory in the aforementioned formalism can be managed only at the Hawking temperature, if curvature singularities at the horizon have to be avoided. We show that, equivalently, black hole temperature can be fixed by requiring that the $\tau - r$ part of the metric, where τ stays for the Euclidean time, can be extended up to include the black hole radius r_+. We work in natural units.

6.1 Local diffeomorphism and extendability

In the following, we will deal with static spherosymmetric black hole manifolds in the so-called Schwarzschild gauge, where $g_{00} =: g$, and $g_{11} = \frac{1}{g}$. Their metric for the external euclidean region can be written as

$$ds^2 = g(r)d\tau^2 + \frac{1}{g(r)}dr^2 + \alpha(r)^2 d\Omega^2. \tag{6.3}$$

This metric is obtained from the usual pseudo-Riemannian one by means of a simple rotation to imaginary time: $t \mapsto -i\tau$. The function $g(r)$ is characterized by one or more zeroes, and the black hole event horizon is indicated by r_+. The corresponding manifold (in the above coordinates) is diffeomorphic to $\mathcal{M} = S^1 \times (r_+, r_0) \times S^2$, where $r_0 = \infty$ and $\alpha(r) = r$ in the Schwarzschild case. Actually, as it is generally accepted, the thermodynamical manifold in the case of Schwarzschild non–degenerate Killing horizon, encloses also the point $r = r_+$, so that the above metric corresponds only to a incomplete chart partially covering the actual thermodynamical manifold $\mathcal{M}_E$. The latter assumption is grounded on the possibility to extend the physical manifold and to find out non-singular coordinates like Kruskal ones in the Schwarzschild case, in such a way to cover also the horizon. It is to be noticed that the Euclidean section, provided that no conical singularity and no boundary occur, like e.g. in the Schwarzschild case without boundary ($r_0 = \infty$), represents a complete manifold just by including the horizon (which is not true in the pseudo-Riemannian manifold).

In order to study the extendability of the Euclidean section, we need a rescaling and a local diffeomorphism. The following rescaling to dimensionless variables is performed:

$$\tau \in (0, \beta) \mapsto \psi = \frac{2\pi}{\beta}\tau \in (0, 2\pi) \tag{6.4}$$

$$r \in (r_+, r_0) \mapsto \bar{r} = \frac{2\pi}{\beta} r \in \left(\frac{2\pi}{\beta} r_+, \frac{2\pi}{\beta} r_0\right). \tag{6.5}$$

Moreover, the following (local) diffeomorphism on the $\tau - r$ part of the metric is introduced (see also [Frolov *et al.* (1996)]):

$$\begin{aligned} ds^2_{(2)} &= \left(\frac{\beta}{2\pi}\right)^2 \left[g\left(\frac{\beta}{2\pi}\bar{r}\right) d\psi^2 + \frac{1}{g\left(\frac{\beta}{2\pi}\bar{r}\right)} d\bar{r}^2 \right] \\ &= \left(\frac{\beta}{2\pi}\right)^2 H^2(y) \left(y^2 d\psi^2 + dy^2\right), \end{aligned} \tag{6.6}$$

whence

$$g\left(\frac{\beta}{2\pi}\bar{r}\right) =: y^2 H^2(y), \tag{6.7}$$

$$\frac{1}{g\left(\frac{\beta}{2\pi}\bar{r}\right)} =: H^2(y) \left(\frac{dy}{d\bar{r}}\right)^2. \tag{6.8}$$

The above diffeomorphism is such that the $\tau - r$ part of the metric near the horizon can, at least locally, be made conformal to a 2D plane with a suitable choice of β, with conformal factor $\left(\frac{\beta}{2\pi}\right)^2 H^2(y)$. Then, we appeal to Lemma 9.114 in [Besse (1987)], where the possibility to complete the polar coordinates up to include the origin is related to suitable properties of the metric itself.[1] Our argument is the following: as for the metric $y^2 d\psi^2 + dy^2$ the conditions of the aforementioned Lemma are ensured, we require that the limit as $r \to r_+$ of the conformal factor exists, and is positive and finite. On the grounds of the Lemma, we shall show that this amounts to the possibility to get a smooth extension of the metric up to include the horizon. If $r = r_+$ is a simple zero, then a conical singularity is avoided by means of the above procedure, and the correct T_H is found. See Sec. 6.3.

[1] Lemma 9.114 of [Besse (1987)]: *If we identify $\{x \in \mathbb{R}^n, 0 < |x| < \epsilon\}$ with $(0, \epsilon) \times S^{n-1}$ in polar coordinates, the C^∞ metric $dt^2 + \varphi(t)^2 \hat{g}_0$ (where t is the parameter on $(0, \epsilon)$ and $\hat{g}_0$ is a metric on S^{n-1}) extends to a C^∞ Riemannian metric on $\{x \in \mathbb{R}^n, |x| < \epsilon\}$ if and only if $\hat{g}_0$ is λg_{can} where g_{can} is the canonical metric on S^{n-1} and λ some positive constant, and $\frac{1}{\lambda}\varphi$ is the restriction on $(0, \epsilon)$ of a C^∞ odd function on $(-\epsilon, \epsilon)$ with $\frac{1}{\lambda}\varphi'(0) = 1$.*

6.2 Conical singularity and black hole horizon

In this section, we summarize some relevant aspects concerning manifolds with conical singularities. Since the seminal papers by Hawking (see e.g. also [Hawking (1980)]), the Euclidean section of the static black hole manifolds has been pointed out to present a conical singularity, involving distributional curvature contributions at the horizon $r = r_+$, if for the Euclidean time $\tau \in [0, \beta)$ one has $\beta \neq \beta_H$, where, putting aside momentarily natural units, we have

$$\beta_H = \frac{\hbar c}{k_B T_H} \tag{6.9}$$

(k_B is the Boltzmann constant). A conical defect γ is present in the manifold if $\gamma \equiv \beta/\beta_H \neq 1$, where β is the inverse temperature and β_H is the inverse of Hawking temperature. In partition function calculations, the case $\gamma = 1$ is also called "on shell", whereas the case $\gamma \neq 1$ is called "off shell". Relevant references for this topic, albeit devoted to quantum partition function calculations, are [Solodukhin (1995); Fursaev and Solodukhin (1995); Frolov *et al.* (1996)].

In the following we recall some basic definitions and properties, starting from the 2D case, and then considering the 4D one.

6.2.1 *The 2D case*

Let us consider at first the 2D problem of a metric with conical singularity. We start from the following definition [Troyanov (1986)]. Let V_θ be the set

$$V_\theta = \{(r, \phi), r \geq 0, \phi \in (0, \theta)\}, \tag{6.10}$$

where a periodic identification of period θ is understood, and the quotient over equivalent points at different ϕ for $r = 0$ is also given. Let us introduce the following metric g over V_θ

$$ds^2 = r^2 d\phi^2 + dr^2. \tag{6.11}$$

Then (V_θ, g) is called the standard cone of total angle θ. As in [Troyanov (1986)], one can introduce the following quantities:

$$k := 2\pi - \theta, \tag{6.12}$$

which is called the concentrated curvature, and also the weight (or residue) δ,

$$\delta = \frac{\theta}{2\pi} - 1. \tag{6.13}$$

As far as $\theta \neq 2\pi$, at the tip of the cone a distributional curvature contribution, coinciding with k, appears.

Conformal structures can be also introduced [Troyanov (1986)]. Let us first define a function $h : U \subset \mathbb{C} \to \mathbb{R}$, with U open set, to be harmonic with logarithmic singularity of weight δ if $h(z) - \delta \log(|z-p|)$ is harmonic. It is interesting to quote the following proposition [Troyanov (1986)]: *Let S be a 2D surface which is associated with two conformal metrics ds_1, ds_2 ($ds_2^2 = \exp(2h)ds_1^2$). Let ds_1 be flat. Then ds_2 is flat if and only if h is harmonic. Moreover, $p \in S$ is a conical point with angle θ_i, $i = 1, 2$ if and only if h has in p a logarithmic singularity of residue $\delta_2 - \delta_1$.*

As a further definition, a surface S is said to have an Euclidean structure with conical singularities at $\{x_1, \ldots, x_n\} \in S$ if, given the positive numbers $\theta_1, \ldots, \theta_n$, the surface $S_0 := S - \{x_1, \ldots, x_n\}$ such that each x_i admits a neighborhood which is locally isometric to the tip of the cone V_{θ_i} [Troyanov (1986)].

6.2.2 *The 4D case*

As to the 4D case, we are interested in static metrics which allow a bifurcate event horizon. We can limit ourselves to Riemannian manifolds $\mathcal{M}_E$ that are locally isometric to $V_\theta \times \Sigma$ near the horizon. Σ represents a surface (in d-dimensional space, a d-2 surface). Riemannian metric associated with the manifold is assumed to be of the form

$$ds^2 = \exp(2\sigma) \left(\rho^2 d\phi^2 + d\rho^2 + \left(\gamma_{ij}(x) + \rho^2 h_{ij}(x)\right) dx^i dx^j + o(\rho^2)\right), \tag{6.14}$$

as in [Fursaev and Solodukhin (1995)]. It can be shown that, as far as $\theta \neq 2\pi$, there appears a conical singularity and, moreover, distributional contributions to the curvature. For example, one finds [Fursaev and Solodukhin (1995)]

$$R_\theta = R - 4\pi\delta\ \delta_\Sigma, \tag{6.15}$$

where δ_Σ is such that

$$\int_{\mathcal{M}} \varphi \delta_\Sigma = \int_\Sigma \varphi. \tag{6.16}$$

6.3 The Hawking temperature from the extendability of the metric at $r = r_+$

We consider the Euclidean version of a static spherosymmetric black hole manifold in the so-called Schwarzschild gauge (6.3), where $r \in (r_+, r_0)$, with $r_0 \leq \infty$, $\Omega \in S^2$, and near the horizon $r = r_+$,

$$g(r) = g'(r_+)(r - r_+) + O((r - r_+)^2), \tag{6.17}$$

with

$$g'(r_+) \neq 0. \tag{6.18}$$

This can be obtained by requiring e.g. $g \in C^1([r_+, r_1))$, where r_1 may delimit a local right neighborhood of r_+. Note that the requirement for g to be extendable (by continuity) at $r = r_+$, which is implicit in the above mathematical requirement, does not amount to the extendability of the metric itself (cf. (6.3)).

We shall find that near the horizon the $\tau - r$ part of the metric is conformal to a flat plane only if a special choice of an angle parameter, to be related to the inverse of the Hawking temperature, is chosen.

We refer to the diffeomorphism defined in (6.6). Taking for definiteness the positive sign in solving for $\frac{dy}{d\bar{r}}$, and also taking into account the behavior for $r \to r_+{}^+$, one finds

$$y \sim A\left(\bar{r} - \frac{2\pi}{\beta} r_+\right)^{\left(\frac{2\pi}{\beta}\right)\frac{1}{|g'(r_+)|}}, \tag{6.19}$$

where A is an integration constant. Then we get

$$H^2(y) \sim \frac{1}{A^2}|g'(r_+)|\left(\frac{\beta}{2\pi}\right)\left(\bar{r} - \frac{2\pi}{\beta} r_+\right)^{1-\left(\frac{4\pi}{\beta}\right)\frac{1}{|g'(r_+)|}}. \tag{6.20}$$

We have to investigate if the manifold is extendable to include the horizon $r = r_+ \Leftrightarrow y = 0$. The metric is conformal to the metric $y^2 d\psi^2 + dy^2$, where one easily realizes that t appearing in the Lemma 9.114 of [Besse (1987)] is replaced by y and that $\varphi(t)$ is replaced by y, and the canonical metric of S^1 is $d\psi^2$ with $\psi \in [0, 2\pi)$. By construction, as a consequence of the aforementioned Lemma, $y^2 d\psi^2 + dy^2$ is extendable. Thus, we have to analyze the conformal factor $H^2(y)$ in order to allow extendability of our original metric. Extendability is ensured if we can include in the manifold also the point $y = 0$, and then we have to require

$$\lim_{y \to 0+} H^2(y) = \lim_{y \to 0+} \frac{1}{y^2} g(r(y)) = H_0^2, \tag{6.21}$$

where H_0^2 is a finite positive constant. By inspection of (6.20) this amounts to requiring

$$1 - \left(\frac{4\pi}{\beta}\right) \frac{1}{|g'(r_+)|} = 0, \tag{6.22}$$

and, as a consequence, one has to choose

$$\beta = \frac{4\pi}{|g'(r_+)|} =: \beta_H. \tag{6.23}$$

The aforementioned trick for finding the black hole temperature relies on a regularity property of a local diffeomorphism which amounts to the extendability of the metric itself up to include the horizon $r = r_+$. The point is that, once (6.22) is ensured, the conformal factor, which depends only on y, is strictly positive and can be extended by continuity to $y = 0$.

It is easy to realize that the above regularity condition (6.22) amounts to eliminating a conical singularity, indeed if we define

$$h(y) =: \frac{1}{2} \log H^2(y) \Leftrightarrow \exp(2h(y)) = H^2(y), \tag{6.24}$$

then, by defining

$$\delta := \frac{1}{2}\left(1 - \frac{4\pi}{\beta}\frac{1}{|g'(r_+)|}\right), \tag{6.25}$$

we find near the horizon

$$h(y) \sim \frac{1}{2} \log \frac{1}{A^2}|g'(r_+)| \left(\frac{\beta}{2\pi}\right) + \delta \log\left(\bar{r} - \frac{2\pi}{\beta} r_+\right), \tag{6.26}$$

so the only way to avoid a conical singularity of weight δ is to impose $\delta = 0$, which coincides with (6.22).

The above method can be used also to confirm that the temperature does not depend on the (static) conformal factor. We refer to [Jacobson and Kang (1993)] for conformal invariance of T_H. Indeed, an overall conformal factor $\Omega^2(r)$ which is finite and non-vanishing at the horizon ($\lim_{r\to r_+} \Omega^2(r) = b_+^2 > 0$) does not modify (6.22).

Of course, we can perform also exact calculations for the conformal factor, if possible. The result does not change, as the same condition (6.25) is reached. Further contributions appear in the conformal factor, showing that the starting manifold is non-flat (as the function h in the conformal factor is not harmonic).

In concluding this section, we point out that the method can be applied as well also to the world of analogous black holes [Barcelo *et al.* (2005)], as far as the non-dispersive case is taken into account (in presence of dispersion, other techniques inherited by scattering theory are necessary). A naive application to a dielectric black hole in a analogue gravity framework is found in [Belgiorno *et al.* (2011b)].

6.4 Example: Schwarzschild black hole

The function $g(r)$ is given by

$$g(r) = 1 - \frac{r_+}{r} \tag{6.27}$$

where r_+ is the black hole radius, associated with the black hole mass M by $r_+ = 2M$.

In the framework of the so called Euclidean path integral approach, the thermodynamical manifold is postulated to be the one representing the so called Euclidean section of the maximal extension of the Schwarzschild spacetime, characterized by $\gamma = 1$ and $R = 0$

$$\tau \in [0, 4\pi r_+); \; r \in [r_+, \infty); \; \Omega \in S^2. \tag{6.28}$$

Note that the metric in (6.3) is related with a coordinate chart not covering the value $r = r_+$. This value corresponds to a set of fixed points for the Killing vector ∂_τ and can be covered e.g. by means of the Kruskal euclidean coordinates T, X, as well known [Gibbons and Hawking (1977a)].

If we pass to rescaled coordinates (6.4), (6.5), then the diffeomorphism (6.6) can be explicitly found:

$$h(y) = \frac{1}{2}\log\left(\frac{1}{A^2}\right) - \bar{r} - \frac{1}{2}\log\bar{r} + \frac{1}{2}\left(1 - \frac{4\pi}{\beta}r_+\right)\log\left(\bar{r} - \frac{2\pi}{\beta}r_+\right), \tag{6.29}$$

so that the extendability at $r = r_+$ is obtained only by imposing

$$\beta = 4\pi r_+ = 8\pi M, \tag{6.30}$$

which amounts to the disappearance of the conical singularity. Note that the Euclidean manifold, with the elimination of the conical singularity and the inclusion of $r = r_+$, is a complete Riemannian manifold, as known.

6.5 The failure of the argument for extremal black holes

Reissner-Nordström electrically charged black hole are the most direct black hole spherosymmetrical extensions of the Schwarzschild solution. They are characterized by two geometrical hair, the mass M and the charge Q. We will consider only uncharged scalar matter fields on these charged black hole backgrounds. We will here focus our study on the degenerate black hole case: when $Q^2 = M^2$ the event horizon is not a bifurcate Killing horizon and in the Schwarzschild gauge the metric in (6.3) has

$$g(r) = \frac{(r - r_+)^2}{r^2}. \tag{6.31}$$

Near the horizon there is no conical singularity. Indeed $H^2(y)$ displays a singular behavior, characterized by an essential singularity as $r \to r_+$. As to $h(y)$, a simple pole appears which cannot be removed as in previous examples. One finds

$$\begin{aligned} h(y) = &\left(1 - \frac{4\pi}{\beta} r_+\right) \log\left(\bar{r} - \frac{2\pi}{\beta} r_+\right) - \log(\bar{r}) - \bar{r} \\ &+ \left(\frac{2\pi}{\beta}\right)^2 r_+^2 \frac{1}{\bar{r} - \frac{2\pi}{\beta} r_+}, \end{aligned} \tag{6.32}$$

and the last term cannot be eliminated for any finite β. It is interesting to note that, for $\beta = \infty$, it is possible to avoid any singularity, which would suggest that extremal black holes are zero-temperature black holes. Still, the complete conformal factor includes also a term $(\frac{\beta}{2\pi})^2$, so, even if the suggestion about the periodicity is the expected one, the argument is neither rigorous nor conclusive. Furthermore, it is also evident the reason why the extendability fails: the Euclidean section of the given manifold $S^1 \times (r_+, \infty)$ is geodetically complete, and then it is non-extendable (see also Appendix B in [Belgiorno (1998)]). So there is no possibility to include $r = r_+$ for any β, and completeness holds true also for the universal covering $\mathbb{R} \times (r_+, \infty)$.

Chapter 7

Rigorous aspects of Hawking radiation

With rigorous methods, we do not obviously intend to present the Hawking evaporation effect as a result of a general theorem stated in a rigorous mathematical setting, but only for very particular situations. If one considers the Schwarzschild black hole manifold as a solution of the vacuum Einstein equations, then a semi rigorous result can be obtained. With semi rigorous we mean that, first of all, the mathematical conditions allowing for the exact statements don't take into account certain important physical conditions, like the Sachs-Bondi energy conservation. The final state of the black hole then appear as a stationary infinite flux of energy from the horizon. It is clear that such a result is unphysical, and a backreaction on the metric should be included. Moreover, the high level of symmetries required in such deductions are not realizable in a physical situation. However, this problems affects also the deductions performed in a less rigorous way. The main motivation for looking for rigorous deductions is due to the fact that the original computations by Hawking, and many other ones, make use of approximations which are uncontrollable, like, for example, the rapid variation of the refractive index during the last stage of the collapse, which is heuristically neglected by arguing that the very strong blueshift makes matter essentially transparent, but which is mathematically out of control, or how much the details of the collapse influence the final answer. The point is thus that a rigorous formulation, even though not yet including all physical conditions and regards quite specific situations, would show that the Hawking effect is essentially independent from the details of the collapse then showing the robustness of such effect.

There is a good number of works concerning rigorous deduction of the Hawking temperature. Here we will refer mainly to three papers: the work of Haag, Narnhofer and Stein [Haag *et al.* (1984)], who invoke a principle of local definiteness and of local stability to prove that there is a natural highly symmetric preferred state which is thermal at the Hawking temperature. It is really interesting to notice that if one simply works in the external Schwarzschild chart, without imposing condition on the horizon boundary, all thermal states, without any constraints on the temperature, are allowed. But if one accepts that the singularity at the horizon is only a coordinate singularity and that the solution must be extendible to the whole Kruskal spacetime, then local definiteness and local stability must also be imposed on the horizon, and this fixes the temperature to be the one determined by Hawking. Thus, if a thermal state exists, the temperature is fixed by well definiteness on the horizon.

The existence of thermal states has been proved by Kay [Kay (1985)] by studying the Klein Gordon equation in the Schwarzschild double wedge. This is constructed starting from the Boulware quantization, as a pure state on the double wedge. On each wedge, the local C^*-algebra satisfies the Ree-Schlieder property.

Finally, we will review the paper of Summers and Verch [Summers and Verch (1996)] who make use of the Tomita-Takesaki theory to prove that, under very general conditions, on a globally hyperbolic spacetime with a bifurcated Killing horizon there exists an equilibrium state which is necessarily at the Hawking temperature.

We stress that we are aware that our choice of arguments is not exhaustive (a whole book would be required only for a discussion on rigorous results in the Hawking effect as well as for the mathematics involved) and to some extent peculiar. At the end of the chapter we shall provide a list of further readings on the subject.

7.1 Local definiteness principle and local stability

In 1973 S. A. Fulling [Fulling (1973)] has shown that in general commutation brackets and equations of motion are not sufficient in order to define the theory of quantum fields. The main difficulty in curved spacetime backgrounds is due to the lack of a global Killing timelike vector field. Several inequivalent representations giving different theories can then be realized. To overcome this problem, R. Haag, H. Narnhofer and U. Stein [Haag *et al.* (1984)] in 1984 introduced the *principle of local definiteness* and of *local stability*.

7.1.1 *The local definiteness*

The physics of a quantum system is essentially encoded in the algebra of fields and their equations of motion. The states then select the initial conditions. However, even in the case of a free theory, the Lie brackets together with the equations of motion leave enough freedom for the choice of states and operators acting on them, thus defining the representation. This correspond to inequivalent choices of the vacuum state. Without a global Killing vector there is no natural choice of a vacuum state. The idea [Haag *et al.* (1984)] is that the ambiguity should disappear in a very small neighborhood of any point, since it should approximate the Minkowskian situation. This is the essence of local definiteness. The inequivalences should then appear only at global level because of topological reasons or boundary conditions. Thus, superselection rules specifying a theory should be necessary only at global level.

The problem can be stated as follows. Let us consider a family $\{\mathcal{O}_n\}_n$ of contractible open sets with compact closure, shrinking to a point x as $n \to \infty$, i.e. such that $\mathcal{O}_{n+1} \subset \mathcal{O}_n$ and $\bigcap_n \mathcal{O}_n = x$. Since there are not observables in a point, if $\{A_n\}_n$ is a family of observables in $\mathcal{O}_n$, then necessarily $\lim_{n\to\infty} A_n$ must belong in the commutant of the algebra of observables. *Local definiteness* requires that such a limit must be then a multiple of the identity.

The main idea in [Haag *et al.* (1984)] was then that one should construct a quantum field theory on the tangent space of any point x of the spacetime manifold, and then going to the point limit by taking a suitable sequence of test functions whose support tends to the origin. More concretely, for a given field ψ in a linear theory (like the ones we are interested in) one can work with the Wightman distributions in the point x defined by

$$W_x^n(f) := \lim_{\lambda\to\infty} \int_{(\mathbb{R}^4)^n} \psi(x+y_1)\ldots\psi(x+y_n) f_\lambda(y_1,\ldots,y_n) d\mu, \quad (7.1)$$

$$\lim_{\lambda\to\infty} \operatorname{supp}(f_\lambda) = 0, \quad (7.2)$$

where y^i are coordinates on the tangent space, $d\mu := g(x)^{\frac{n}{2}} d^4y_1 \ldots\ d^4y_n$, and $g = |\det(\boldsymbol{g})|$, $\boldsymbol{g}$ being the spacetime metric. From the above discussion, we see that for definiteness, being in the point limit, the expectation value $\langle W_x^n(f)\rangle_\omega$ must be independent from any allowed choice of the state ω. If this is true, the right choice of the test functions f_λ must ensure the independence from the choice of coordinates adopted on the tangent space. This can be checked by fixing a map

$$\zeta : T_xM \longrightarrow M, \quad (7.3)$$

with

$$\zeta(0) = x, \qquad \frac{d}{dt}\zeta(vt)\Big|_{t=0} = v, \ \forall v \in T_x M. \tag{7.4}$$

The independence is thus equivalent that for any positive integer n it must exist a function $N^n(s)$ such that

$$W_x^n(f) = \lim_{t\to 0} N^n(s) \int \psi(\zeta(y_1 t)) \dots \psi(\zeta(y_n t)) f_\lambda(y_1, \dots, y_n) d\mu \tag{7.5}$$

converges at a finite limit. Indeed, since only the limit is relevant, such a function must satisfy

$$\lim_{t\to 0} \frac{N^n(\lambda t)}{N^n(t)} = \lambda^{\alpha_n} \tag{7.6}$$

for some α_n. Condition (7.5) is also called the *scaling limit property*. In particular, the properties of $W_x^n(f)$ are then strictly related to Poincaré invariance. For example, the Lorentz invariance follows from invariance under parallel transport along small closed paths, necessary for well definiteness. Since this transport acts on the tangent by $SO(1,3)$ holonomy, this immediately implies the invariance of $W_x^n(f)$ under Lorentz transformations. It is also evident that invariance under Poincaré translations is equivalent to the continuity in x of the functions $W_x^n(f)$.

However, the principle of local definiteness is not sufficient to ensure the existence of a vacuum state. This requires to express the *principle of local stability*, which, for a linear free field theory, requires that the support on the cotangent space $T_x^* M$ of the Fourier defined (in the distributional sense) by

$$W_x^2(y_1, y_2) = \int \tilde{W}_x^2(p) e^{ig_{\mu\nu}(x) p^\mu (y_1^\nu - y_2^\nu)} \sqrt{g(x)} d^4 p, \tag{7.7}$$

is restricted to the future lightcone. Since for a free theory the commutator is a numerical distribution, this fix the antisymmetric part of $W_x^2(x_1, x_2)$, from which one determines the scaling factor to be $N^2(s) = s^2$. From (7.5) then one sees that

$$W_x^2(\lambda x_1, \lambda x_2) = \lambda^{-2} W_x^2(x_1, x_2), \tag{7.8}$$

and then

$$\tilde{W}_x^2(\lambda p) = \lambda^{-2} \tilde{W}_x^2(p). \tag{7.9}$$

The normalization is also completely fixed and one determines in this way the Wightman two points function, so that one gets that in the limit

$$W_x^2(z) = \frac{1}{4\pi^2} \frac{1}{g_{\mu\nu}(x) z^\mu z^\nu}, \tag{7.10}$$

where $z^\mu = x_1^\mu - x_2^\mu$ and z^0 is replaced by $z^0 + i\varepsilon$.

In this way the authors of [Haag *et al.* (1984)] show that a local quantum field theory is well defined on curved background.

7.2 Hawking temperature from local definiteness and stability

Let us consider the Schwarzschild black hole spacetime with metric

$$ds^2 = -f(x)dt^2 + g(x)dx^2 + h(x)d\Omega^2, \tag{7.11}$$

where

$$d\Omega^2 = d\theta^2 + \sin^2\theta\, d\phi^2 \tag{7.12}$$

is the angular metric, t is the coordinate along the timelike Killing vector $K = \frac{\partial}{\partial t}$, and x is a radial coordinate. The "true radial" one r is such that

$$h(x^1) = r^2, \qquad f(x) = g(x)^{-1} = 1 - \frac{2M}{r}, \tag{7.13}$$

where M is the mass of the black hole. However, for our purposes, it is convenient to leave the coordinate x arbitrary. The horizon $x = x_H$ is defined by $f(x_H) = 0$.

A Klein-Gordon field Φ with mass μ on this background must satisfy the equation

$$-f(x)^{-1}\frac{\partial^2\Phi}{\partial t^2} + g(x)^{-1}\frac{\partial^2\Phi}{\partial x^2} + g(x)^{-1}\frac{d}{dx}\log\left(h(x)\frac{f(x)^{\frac{1}{2}}}{g(x)^{\frac{1}{2}}}\right)\frac{\partial\Phi}{\partial x}$$
$$+h(x)^{-1}\Delta_\Omega\Phi - \mu^2\Phi = 0. \tag{7.14}$$

In order to define the quantum field, one has to determine the commutator

$$[\Phi^\dagger(\bar{x}), \Phi(\bar{y})] = i\Delta(\bar{x}, \bar{y}), \tag{7.15}$$

which is the unique solution of the Klein-Gordon equation (in each of the variables $\bar{x}$ and $\bar{y}$), such that

$$\Delta(\bar{x}, \bar{y})|_{x^0=y^0} = 0, \tag{7.16}$$

$$\frac{\partial}{\partial y^0}\Delta(\bar{x}, \bar{y})\Big|_{x^0=y^0} = \delta_c^{(3)}(\bar{x}, \bar{y}), \tag{7.17}$$

where $\delta_c^{(3)}(\bar{x}, \bar{y})$ is the covariant delta distribution on the hypersurface $x^0 = y^0$, with support in $\bar{x} = \bar{y}$ and

$$\int_{t_y=const.} \delta_c^{(3)}(\bar{x}, \bar{y})h(y)\sqrt{\frac{g(y)}{f(y)}}\ \sin\theta_y\, dy\, d\theta_y\, d\phi_y = 1, \tag{7.18}$$

where $(t_y, y, \theta_y, \phi_y)$ are the coordinates of $\bar{y}$.

One can solve this equation by separating variables in the form

$$\Delta(\vec{x},\vec{y}) = \frac{1}{i}\int d\omega \sum_{l,m} R_{\omega,l,m}(x,y) Y_{l,m}(\theta_x,\phi_x) Y^*_{l,m}(\theta_y,\phi_y) \times e^{-i\omega(t_x-t_y)}, \tag{7.19}$$

where Y_{lm} are the usual scalar harmonic functions on the sphere, and the radial functions $R_{\omega,l,m}$ satisfy the equations

$$\left\{ g(x)^{-1}\frac{\partial^2}{\partial x^2} + g(x)^{-1}\frac{d}{dx}\log\left(h(x)\frac{f(x)^{\frac{1}{2}}}{g(x)^{\frac{1}{2}}}\right)\frac{\partial}{\partial x} - h(x)^{-1}l(l+1) - \mu^2 + \frac{\omega^2}{f(x)} \right\} R_{\omega,l,m}(x,y) = 0, \tag{7.20}$$

and the same w.r.t. y in place of x. Note that this equation is degenerate in m, so that we can write $R_{\omega,l,m}(x,y) \equiv R_{\omega,l}(x,y)$. Since

$$\sum_{l,m} Y_{l,m}(\theta_x,\phi_x) Y^*_{l,m}(\theta_y,\phi_y) = \delta^{(2)}_{S^2}(\theta_x-\theta_y,\phi_x-\phi_y), \tag{7.21}$$

the conditions (7.16) and (7.17) are fulfilled if

$$\int d\omega R_{\omega,l}(x,y) = 0, \tag{7.22}$$

$$\int d\omega\ \omega R_{\omega,l}(x,y) = \frac{1}{h(x)}\sqrt{\frac{f(x)}{g(x)}}\delta(x-y). \tag{7.23}$$

One indeed can find such a solution in the form

$$R_{\omega,l}(x,y) = F_{\omega,l}(x)F_{\omega,l}(y), \tag{7.24}$$

where $F_{\omega,l}$ satisfy

$$\left\{ g(x)^{-1}\frac{d^2}{dx^2} + g(x)^{-1}\frac{d}{dx}\log\left(h(x)\frac{f(x)^{\frac{1}{2}}}{g(x)^{\frac{1}{2}}}\right)\frac{d}{dx} - h(x)^{-1}l(l+1) - \mu^2 + \frac{\omega^2}{f(x)} \right\} F_{\omega,l}(x) = 0. \tag{7.25}$$

At this point, after assuming the existence of an equilibrium thermal state with inverse temperature β, from the KMS condition w.r.t. the one parameter evolution α_t

$$\int \langle B\alpha_t(A)\rangle_\beta e^{-i\omega t}dt = e^{\beta\omega}\int \langle \alpha_t(A)B\rangle_\beta e^{-i\omega t}dt \tag{7.26}$$

one gets

$$\langle \Phi(t_x,\vec{x})\Phi(t_y,\vec{y})\rangle_\beta = \frac{i}{2\pi}\int_{\mathbb{R}^2} d\tau d\omega \Delta((t_x+\tau,\vec{x}),(t_y,\vec{y}))q(\beta\omega)e^{i\omega\tau}, \tag{7.27}$$

where we define

$$q(\beta\omega) := \frac{e^{\beta\omega}}{e^{\beta\omega} - 1}. \tag{7.28}$$

Using the solution (7.19) we get

$$\langle \Phi(x)\Phi(y) \rangle_\beta = \int d\omega \sum_{l,m} R_{\omega,l}(x,y) Y_{l,m}(\theta_x, \phi_x) Y^*_{l,m}(\theta_y, \phi_y) \times e^{-i\omega(t_x - t_y)} q(\beta\omega). \tag{7.29}$$

In order for the theory to be well defined, we need to impose the scaling limit condition, which, in particular, implies that the two point function in the limit of short distances (for a free field) must be the normalized one for a free massless particle. This can be accomplished by setting

$$t_x - t_y = \epsilon\tau, \qquad x - y = \epsilon z, \qquad x^\perp - y^\perp = \epsilon z^\perp, \tag{7.30}$$

where $\perp$ means the spatial directions orthogonal to the radial one, and then taking the limit $\epsilon \to 0$. To do this, we now we fix the coordinate x (y) to be the one such that $g(x) = 1$ and the horizon is at $x = 0$. This means that

$$dx = \frac{rdr}{\sqrt{r(r - 2M)}}, \tag{7.31}$$

so that x is related to the standard radial coordinate r by

$$x = \sqrt{r(r - 2M)} + M \log\left(\frac{\sqrt{r(r - 2M)} + r - M}{M}\right). \tag{7.32}$$

In particular, near the horizon we have

$$f(x) = \frac{1}{16M^2}x^2 + o(x^2), \qquad h(x) = 4M^2 + \frac{1}{2}x^2 + o(x^2). \tag{7.33}$$

In order to analyze the limit, one can first perform the summation over m in (7.29), by using

$$\sum_{m=-l}^{l} Y_{l,m}(\theta_x, \phi_x) Y^*_{l,m}(\theta_y, \phi_y) = \frac{2l+1}{4\pi} P_l(\cos\gamma_{xy}), \tag{7.34}$$

where γ_{xy} is the angle between the directions (θ_x, ϕ_x) and (θ_y, ϕ_y), so that

$$\langle \Phi(\bar{x})\Phi(\bar{y}) \rangle_\beta = \int d\omega \sum_l R_{\omega,l}(x,y) \frac{2l+1}{4\pi} \times P_l(\cos\gamma_{xy}) e^{-i\omega(t_x - t_y)} q(\beta\omega). \tag{7.35}$$

In the limit $\epsilon \to 0$, γ_{xy} becomes smaller and smaller, so that $P_l(\cos\gamma_{xy}) \to P_l(1) = 1$. We thus see that, after fixing $y > 0$ and assuming also $\epsilon z > 0$, nothing special happens for finite values of l or when integrating only over finite ranges for ω. This means that contributions to the singular part in the point limit come from the infinite values of l and ω. Notice that for large values of ω the Bose factor $e^{\beta\omega}/(e^{\beta\omega}-1)$ becomes nearly 1 and is then independent from β. This implies that the principles of local definiteness and local stability cannot fix any value for β, and any value of the temperature is allowed. Indeed, this can be proved in a rigorous way [Kay (1985)].

Things change in points reaching the horizon, that is when $y \to 0$. The point is that equation (7.25) has a Fuchsian singularity in $x=0$, t.i. at the horizon. Near $x \to 0$, in the coordinates fixed above, the equation takes the form

$$F''_{\omega,l} + \frac{1}{x}F'_{\omega,l} - \left[\frac{l(l+1)}{4M^2} + \mu^2 - \frac{16M^2\omega^2}{x^2}\right] F_{\omega,l} = 0, \tag{7.36}$$

so that the singular term in the potential is just proportional to ω^2. Using (7.23) and (7.33), we see that the solutions must be normalized so that

$$\int_{\mathbb{R}} d\omega\, \omega F_{\omega,l}(x) F_{\omega,l}(y) = \frac{1}{16M^3} x\delta(x-y). \tag{7.37}$$

A solution of the approximate equation is

$$F_{\omega,l}(x) = C_{\omega,l} K_{i4M\omega}(kx), \tag{7.38}$$

with $k = \sqrt{\frac{l(l+1)}{4M^2} + \mu^2}$. Indeed, using

$$\int_{\mathbb{R}} d\omega\omega \sinh(\pi\omega) K_{i\omega}(kx) K_{i\omega}(ky) = \pi^2 x\delta(x-y), \tag{7.39}$$

we see that

$$R_{\omega,l}(x,y) = \frac{1}{M\pi^2}\sinh(4M\pi\omega) K_{i4M\omega}(kx) K_{i4M\omega}(ky) \tag{7.40}$$

has the right normalization and is the solution we are looking for. We are interested in taking the limit $y \to 0$ and $x = \epsilon z$. To do this, notice that

$$\begin{aligned} K_{i4M\omega}(ky) &\approx_{y\to 0} \frac{i\pi}{2\sinh(4M\pi\omega)} \left(\frac{e^{i4M\omega\log\frac{kx}{2}}}{\Gamma(1+i4M\omega)} - \frac{e^{-i4M\omega\log\frac{kx}{2}}}{\Gamma(1-i4M\omega)} \right) \\ &\approx \lim_{k\to\infty} \frac{\sin(k4M\omega)}{4M\omega} = \frac{\pi}{4M}\delta(\omega). \end{aligned} \tag{7.41}$$

Using this, we get

$$\langle \Phi(\bar{x})\Phi(\bar{y}_H)\rangle_\beta \simeq \sum_l \frac{1}{M\beta}\frac{2l+1}{4\pi} P_l(\cos\gamma_{xy}) K_0(\sqrt{\frac{l(l+1)}{4M^2} + \mu^2 x}). \tag{7.42}$$

Again, we see that if we stop the summation at any finite value, no singular contribution appears in the limit $\epsilon \to 0$ ($\bar{y} \to \bar{x}_H$). Then, to evaluate the singular part, we can consider the approximate expressions for large l, end pass to the continuum limit. Indeed, from

$$k = \sqrt{\frac{l(l+1)}{4M^2} + \mu^2} \simeq \frac{l}{2M}, \tag{7.43}$$

we can replace the summation with integration in k:

$$\sum_{l=0}^{\infty} \longmapsto \int_0^{\infty} 2M dk. \tag{7.44}$$

Moreover, for large l (and small $\gamma_x \equiv \gamma_{xy_H}$) one has

$$P_l(\cos\gamma_x) \simeq_{l\to\infty} J_0(l\gamma_x) \simeq J_0(4Mk\gamma_x). \tag{7.45}$$

This gives

$$\begin{aligned} \langle\Phi(\bar{x})\Phi(\bar{y}_H)\rangle_\beta &\simeq \int_0^{\infty} dk \frac{4M}{2\pi\beta} J_0(2Mk\gamma_x) K_0(kx) \\ &= \frac{4M}{2\pi\beta} \frac{1}{x^2 + 4M^2\gamma_x^2}. \end{aligned} \tag{7.46}$$

We can write this as

$$\langle\Phi(\bar{x})\Phi(\bar{y}_H)\rangle_\beta \simeq \frac{4M}{2\pi\beta} \frac{1}{g_{\mu\nu}(\bar{y}_H) x^\mu x^\nu}. \tag{7.47}$$

Now, we can invoke the limit point formula (7.10), obtained by applying the principles of local definiteness and local stability. This shows that the theory is well defined and extendable on the points of the horizon only if

$$\beta = 8\pi M, \tag{7.48}$$

which correspond exactly to the Hawking temperature.

In conclusion, if one works in the Schwarzschild wedge, if equilibrium thermal states exist then any temperature is allowed. Only requiring that the theory should be well defined on the whole Kruskal manifold and then on the horizon fixes the value of the temperature at the Hawking one. It is worth to remark that this does not provides a complete proof of the Hawking result, since we have not proved the existence of the thermal state but only used the fact that if it exists then it must satisfy the KMS condition. A rigorous proof of the existence of such states has been provided by Kay in [Kay (1985)].

7.3 Existence of Hartle-Hawking state

The main idea in [Kay (1985)] is to look at the dynamical system outside the Schwarzschild black hole as one half of a double system in the Kruskal-Szekeres spacetime (regions I and II in Fig. 1.1).

To put this idea on a mathematically rigorous ground, it is necessary to introduce the concept of a double dynamical system.

7.3.1 *Almost free double quantum dynamical systems*

Let us consider a complex Hilbert space $\mathcal{H}$. If $(|)$ is the sesquilinear product, then $\sigma := 2\text{Im}(|)$ defines a symplectic structure on $\mathcal{H}$. Let H be the strictly positive Hamiltonian operator and $U(t) = e^{-iHt}$ the corresponding strongly continuous time evolution map. It is a symplectic group over $(\mathcal{H}, \sigma)$. The Heisenberg algebra of the field operators (and their conjugate momenta) can be described by the associated Weyl algebra, which has the advantage of being described in terms of bounded operators. In abstract, it is a map w from $\mathcal{H}$ to the C^*-algebra $B(\mathcal{H})$ of the bounded operators, satisfying the conditions $w(v_1)w(v_2) = e^{\frac{i}{2}\sigma(v_1,v_2)}w(v_1 + v_2)$ and $w(v)^* = w(-v)$ for any $v_1, v_2, v \in \mathcal{H}$. H defines a unique fundamental state ω_0 that determines a Fock representation up to unitary equivalences. Via GNS construction it is thus defined the vacuum state Ω in the Fock space $\mathcal{F}(\mathcal{H})$ and the representation ρ such that

$$\omega_0(w(v)) = (\Omega|\rho(w(v))\Omega)_{\mathcal{F}(\mathcal{H})} = e^{-\frac{1}{2}\|v\|_{\mathcal{H}}}. \tag{7.49}$$

Notice that $\rho(w(v)) =: W(v)$ generate a C^*-algebra $\mathcal{U}$ of operators acting on the Fock space. Let Γ be the usual second quantization map, such that if A is (anti-)unitary over $\mathcal{H}$ then $\Gamma(A)$ is (anti-)unitary over $\mathcal{F}(\mathcal{H})$, and satisfies $\Gamma(A)\Omega = \Omega$, $W(Av) = \Gamma(A)W(v)\Gamma(A)^{-1}$. Hence, $\alpha(t) := \Gamma(U(t))$ defines an automorphism group for $\mathcal{U}$. This define a system of free relativistic scalar particles with Hamiltonian H and symplectic structure σ.

This simple situation suggests the general definition:

Definition 7.1. Let (D, σ) be a symplectic linear space. Let $\mathcal{U}$ a C^*-algebra generated by elements of the form $W(v)$ where v belongs in the symplectic space, and such that

$$W(v_1)W(v_2) = e^{\frac{i}{2}\sigma(v_1,v_2)}W(v_1 + v_2), \tag{7.50}$$

and

$$W(v)^* = W(-v). \tag{7.51}$$

Let $U(t)$ a one parameter symplectic group over (D, σ) and define $\alpha(t)$ by

$$\alpha(t)W(v) = W(U(t)v). \tag{7.52}$$

All these ingredients define a *bosonic almost free quantum dynamical system.*

This definition does not select a free system, nor allow for determining particle states.

Definition 7.2. A *single particle structure* over a classical dynamical system $(D, \sigma, U(t))$ is a complex Hilbert space $\mathcal{H}$ with a real linear map $K : D \to \mathcal{H}$ with dense range and such that

$$\sigma(v_1, v_2) = 2\mathrm{Im}(Kv_1|Kv_2)_{\mathcal{H}} \tag{7.53}$$

for all $v_1, v_2 \in \mathbb{H}$ (symplecticity) and a strongly continuous one parameter unitary group $e^{-i\hat{H}t}$ such that

$$KU(t) = e^{-i\hat{H}t}K. \tag{7.54}$$

In this case the fundamental state ω_0 is defined by

$$\omega_0(W(v)) = e^{-\frac{1}{2}\|Kv\|^2_{\mathcal{H}}}. \tag{7.55}$$

The one particle structure is said to be *regular* if KD is in the domain of $\hat{H}^{-\frac{1}{2}}$.

The first important result is that, given a dynamical system $(D, \sigma, U(t))$, the one particle structure $(K, \mathcal{H}, e^{-i\hat{H}t})$ is univocally determined up to equivalences, t.i. up to isomorphisms $U : \mathcal{H} \to \mathcal{H}'$ such that $K' = UK$ and $e^{-i\hat{H}'t} = Ue^{-i\hat{H}t}U^{-1}$.

Now we can consider double systems. Let $\mathcal{U} = \mathcal{U}^L \otimes \mathcal{U}^R$ a C^*-algebra with two factors that commute reciprocally and let $\alpha(t) : \mathcal{U}^{L.R} \to \mathcal{U}^{L,R}$ a one parameter group of automorphisms. Let ι be an anti-involution which commute with $\alpha(t)$ and exchanges the factors, $\iota : \mathcal{U}^{L,R} \to \mathcal{U}^{R,L}$.

For a fixed $\beta \in (0, \infty)$, let us define a state ω_β over $(\mathcal{U}, \alpha(t), \iota)$ with the property to be ι-invariant and such that the associated GNS triple $(\rho, \mathcal{H}, \Omega)$ has $\mathcal{H}$ separable and has the property that Ω is cyclic only for $\rho(\mathcal{U}^R)$. Further, we assume that the unitary implementation of $\alpha(t)$ is a strongly continuous operator e^{-iHt} and that $\rho(\mathcal{U}^R)$ is in the domain of $e^{-\beta H/2}$, with

$$e^{-\beta H/2}\rho(O)\Omega = \rho(\iota O^*)\Omega, \qquad \forall O \in \mathcal{U}^R. \tag{7.56}$$

Setting

$$\alpha_R(t) := \alpha(t)|_{\mathcal{U}^R}, \tag{7.57}$$

it follows that the restriction $\omega_\beta^R = \omega_\beta|_{(\mathcal{U}^R, \alpha_R(t))}$ is a KMS state with inverse temperature β.

In general, any dynamical system is the restriction of a double system, and any *KMS* state over $(\mathcal{U}_R, \alpha_R(t))$ is the restriction of a double KMS state. The associated modular group is $\Delta^{it} = e^{-iH\beta t}$ and the anti-involution is the unique complex conjugation J such that $J\Omega = \Omega$ and $J\rho(O)J = \rho(\iota O)$.

Let us indicate with $\rho(\mathcal{U})'$ the commutant of $\rho(\mathcal{U})$ in $B(\mathcal{H})$ and with $\rho(\mathcal{U})''$ the bicommutant. Then

Theorem 7.1. *For any double KMS state ω_β on a double dynamical system $(\mathcal{U}, \alpha(t), \iota)$ having GNS triple $(\rho, \mathcal{H}, \Omega)$ (and unitary implementator $e^{-i\hat{H}t}$ for $\alpha(t)$) it holds*

$$\rho(\mathcal{U}^R)' = \rho(\mathcal{U}^L)''. \tag{7.58}$$

If we further assume that for each $v \in \mathcal{D}(\hat{H})$ such that $\hat{H}v = 0$ implies $v = \lambda\Omega$, then we get

$$\rho(\mathcal{U}^{R,L})' \cap \rho(\mathcal{U}^{R,L})'' = \{\lambda \mathbb{I} \lambda \in \mathbb{C}\}, \tag{7.59}$$

and $\rho(\mathcal{U})$ is irreducible, t.i. ω_β is a pure state.

The next step consists in defining a double linear dynamical system:

Definition 7.3. A *double linear dynamical system* $(D, \sigma, U(t), \mathcal{I})$ consists in a symplectic space (D, σ) with a one parameter symplectic group $U(t)$ and an anti-symplectic involution $\mathcal{I}$ over (D, σ) (so that $\sigma(\mathcal{I}v_1, \mathcal{I}v_2) = -\sigma(v_1, v_2)$) such that $[U(t), \mathcal{I}] = 0$, and $D = D^L \oplus D^R$, where

$$\sigma(v^L, v^R) = 0, \quad \forall v^{L,R} \in D^{L,R}, \tag{7.60}$$
$$U(t) : D^{L,R} \to D^{L,R}, \tag{7.61}$$
$$\mathcal{I}D^L = D^R. \tag{7.62}$$

We finally arrive at the definition we are interested in:

Definition 7.4. A *bosonic almost free quantum dynamical system* is one generated from a double linear dynamical system. In this case $\mathcal{U}$ is the Weyl algebra over (D, σ) and

$$\alpha(t)W(v) = W(U(t)v), \qquad \iota(W(v)) = W(\mathcal{I}v). \tag{7.63}$$

As above we can realize a double KMS one particle system: a Hilbert space $\mathcal{H}$, and a linear map $K^\beta : D \to \mathcal{H}$ such that

$$2\mathrm{Im}(K^\beta v_1 | K^\beta v_2) = \sigma(v_1, v_2)$$

and that $K^\beta D^{L,R} + \iota(K^\beta D^{L,R})$ is dense in $\mathcal{H}$, a one parameter group $e^{-i\hat{H}t}$ strongly continuous on $\mathcal{H}$, such that $\hat{H}$ does not have null eigenvalues and $K^\beta U(t) = e^{-i\hat{H}t}$, and finally a complex conjugation j over $\mathcal{H}$ such that $K^\beta D^R + jK^\beta D^R \subset \mathcal{D}(e^{-\beta\hat{H}/2})$ and $e^{-\beta\hat{H}/2}x = -jx$ for all $x \in k^\beta D^R$ (and similar with $R \to L$, $\hat{H} \to -\hat{H}$). Again, the double KMS single particle system is unique up to unitary equivalences.

It remains to see how the double KMS states can be obtained. Let $(D, \sigma, U(t), \mathcal{I})$ be a classical double linear dynamical system. Then, $(D^R, \sigma_R, U_R(t))$, where R means restriction to D^R, admits a structure $(K, \mathcal{H}, e^{-i\hat{H}t})$ with regularity condition $KD^R \subset \mathcal{D}(\hat{H}^{-\frac{1}{2}})$. If c is a conjugation on $\mathcal{H}$ that commute with $\hat{H}$, one can construct $(K^\beta, \tilde{\mathcal{H}}, e^{-i\tilde{H}t}, \iota)$ by

(1) $\tilde{\mathcal{H}} = \mathcal{H} \oplus \mathcal{H}$,

(2) for all $v \in D$ let $v = v^L + v^R \in D^L \oplus D^R$ and put

$$K^\beta v := \begin{pmatrix} \cosh z^\beta & \sinh z^\beta\, c \\ \sinh z^\beta\, c & \cosh z^\beta \end{pmatrix} \begin{pmatrix} -cK\mathcal{I}v^L \\ Kv^R \end{pmatrix}, \tag{7.64}$$

where $\tanh z^\beta := e^{-\beta\hat{H}/2}$. The regularity condition ensures that $KD^R \subset \mathcal{D}(\cosh z^\beta)$ and $KD^R \subset \mathcal{D}(\sinh z^\beta)$;

(3)

$$e^{-i\tilde{H}t} = \begin{pmatrix} e^{i\hat{H}t} & 0 \\ 0 & e^{-i\hat{H}t} \end{pmatrix};$$

(4)

$$\iota = \begin{pmatrix} 0 & -c \\ -c & 0 \end{pmatrix}.$$

Then, one finds

$$\begin{aligned} \omega^\beta(W(v)) &= \exp\left(-\frac{1}{2}\|K^\beta v\|^2_{\tilde{\mathcal{H}}}\right) \\ &= e^{-\frac{1}{2}(K\mathcal{I}v^L|\coth\frac{\beta\hat{H}}{2}K\mathcal{I}v^L) - \frac{1}{2}(Kv^L|\coth\frac{\beta\hat{H}}{2}Kv^R)} \\ &\quad \times e^{-\mathrm{Re}(K\mathcal{I}v^L|\coth\frac{\beta\hat{H}}{2}Kv^R)}. \end{aligned} \tag{7.65}$$

In the Fock representation $W(v) \mapsto W^{\mathcal{F}}(K^\beta v)$ on $\mathcal{F}(\mathcal{H}) \otimes \mathcal{F}(\mathcal{H})$, the modular automorphism is

$$\Delta^{it/\beta} = \Gamma(e^{-it\tilde{H}}) = e^{-it(-d\Gamma(\hat{H})\otimes\mathbb{I} + \mathbb{I}\otimes d\Gamma(\hat{H}))} \tag{7.66}$$

and the modular anti-involution is

$$J(v_1 \otimes v_2) = \Gamma(-c)v_1 \otimes \Gamma(-c)v_2. \tag{7.67}$$

We finally cite the theorem

Theorem 7.2. *Let $\mathcal{U}$ a factor of a von Neumann algebra acting over a Hilbert space $\mathcal{H}$, and let Ω be cyclic and separating for $\mathcal{U}$. Let Δ^{it} be the corresponding modular group. If*

$$\forall v \in \mathcal{H},\ \Delta^{it}v = v \quad \Rightarrow \quad v = \lambda\Omega, \tag{7.68}$$

and the spectrum of Δ is

$$Sp(\Delta) = [0, \infty), \tag{7.69}$$

then $\mathcal{U}$ is of type III.

Indeed, the second condition, in the formulation of Connes of the type III factors, means that it is of type III_1.

Before applying this machinery to the Schwarzschild spacetime, it is convenient to understand how it works in Minkowski spacetime.

7.3.2 *The Minkowski vacuum*

Let $M^{1,3} = (\mathbb{R}^4, \eta)$ be the Minkowski spacetime with $\eta = \mathrm{diag}(-1,1,1,1)$ in the inertial coordinates (τ, x, y, z). Let us consider a massive Klein-Gordon field ϕ and consider the $\mathcal{C}^\infty$ solutions of $(\Box + m^2)\phi = 0$ (where $\Box = -\eta^{\mu\nu}\partial_\mu\partial_\nu$), having compact support on a Cauchy surface (and, then, on any Cauchy surface). Let $q = \phi|_{\tau=0}$, $p = \dot{\phi}|_{\tau=0}$ the Cauchy data, so that if $C \simeq \mathbb{R}^3$ is the variety $\tau = 0$, the set of initial data is in $D = \mathcal{C}_0^\infty \times \mathcal{C}_0^\infty$.

Let us divide C in the left part $C^L = \{(0,x,y,z)|x<0\}$ and the right part $C^R = \{(0,x,y,z)|x>0\}$, and, correspondingly, the data sets

$$D^R := \mathcal{C}_0^\infty(C^R) \times \mathcal{C}_0^\infty(C^R), \qquad D^L := \mathcal{C}_0^\infty(C^L) \times \mathcal{C}_0^\infty(C^L), \tag{7.70}$$

$$\tilde{D} := D^L \oplus D^R \simeq \mathcal{C}_0^\infty(C^L \cup C^R) \times \mathcal{C}_0^\infty(C^L \cup C^R). \tag{7.71}$$

Thus, a given solution ϕ selects an element (q,p) in each of these spaces. On $\tilde{D}$ there is the symplectic form

$$\sigma(\phi_1, \phi_2) = \int_C (q_1 p_2 - q_2 p_1) dx dy dz. \tag{7.72}$$

The associated Weyl algebra is defined by

$$W(\phi_1)W(\phi_2) = e^{-\frac{i}{2}\sigma(\phi_1,\phi_2)} W(\phi_1 + \phi_2)$$

and the elements $W(\phi)$ generate $\mathcal{U}$, the $*$-algebra for the quantum solutions. If Δ is the fundamental solution (the advanced minus the retarded Green function) then the convolution $\phi = \Delta \star F$, with $F \in C_0^\infty(M^{1,3})$, is solution of the homogeneous equation. Then, one has $W(\phi) = \exp(i \int_{M^{1,3}} F d\tau dx dy dz)$. Let Φ be the Cauchy data corresponding to this solution. Then, $\Phi \in D^R$ when F has support in the right wedge

$$\mathcal{R} = \{(\tau, x, y, z) \in \mathbb{R}^4 | x > |\tau|\}. \tag{7.73}$$

Thus, we have the right algebra $\mathcal{U}^R$ generated by the set $\{W(\phi)|\phi \in D^R\}$. In the same way we can consider the left wedge

$$\mathcal{L} = \{(\tau, x, y, z) \in \mathbb{R}^4 | x < -|\tau|\}, \tag{7.74}$$

we can construct the algebra $\mathcal{U}^L$ generated by the set $\{W(\phi)|\phi \in D^L\}$. Finally we consider the double algebra $\tilde{\mathcal{U}}$ generated by $\{W(\phi)|\phi \in \tilde{D}\}$, associated to the double wedge $\mathcal{L} \cup \mathcal{R}$.

The symplectic group $U(t)$ is defined by the evolution map of the Cauchy data corresponding to the translations $\phi \mapsto \phi_t$, where $\phi_t(\tau, x, y, z) = \phi(\tau + t, x, y, z)$ (on the classical solutions). This way, $(\tilde{D}, \sigma, U(t))$ is a linear classical dynamical system.

Alternatively, one can introduce the one parameter symplectic group $V(t)$ of the boosts $\phi \mapsto \phi_t$ with $\phi_t(\tau, x, y, z) = \phi(\Lambda(t)(\tau, x), y, z)$,

$$\Lambda(t) = \begin{pmatrix} \cosh t & \sinh t \\ \sinh t & \cosh t \end{pmatrix}.$$

The symplectic anti-involution is $\mathcal{I}(q, p) = (\check{q}, -\check{p})$ where in general $\check{f}(x, y, z) = f(-x, y, z)$, which corresponds to the wedge reflection. We see that $(\tilde{D}, \sigma, V(t), \mathcal{I})$ is a classical double linear system. If we define the automorphism $\alpha(t)$ by

$$\alpha(t) W(\phi) = W(V(t)\alpha(\phi)), \tag{7.75}$$

and the anti-involution ι

$$\iota(W(\phi)) = W(\mathcal{I}\phi), \tag{7.76}$$

then we see that $(\tilde{\mathcal{U}}, \alpha(t), \iota)$ is a quantum double dynamical system, with dynamics $\gamma(t)$ given by $\gamma(t) W(\phi) = W(U(t)\phi)$.

The Minkowski quantization is thus defined by

$$\omega_0(W(\phi)) = \exp\left(-\frac{1}{2} \|K_0 \phi\|^2_{\mathcal{H}_0}\right), \tag{7.77}$$

where $\mathcal{H}_0 = L^2(\mathbb{R}^3, \mathbb{C})$ is the 1-particle Hilbert space, and

$$K_0\phi = \frac{1}{\sqrt{2}}(\mu^{\frac{1}{2}}q + i\mu^{-\frac{1}{2}}p)$$

with $\mu = (m^2 - \nabla^2)^{\frac{1}{2}}$. The main point is that $(K_0, \mathcal{H}_0, e^{-i\mu t})$ is a one particle structure for $(D, \sigma, U(t))$. In the GNS construction, $\rho_0(W(\phi)) = W^{\mathcal{F}}(K_0\phi)$ over $\mathcal{F}(\mathcal{H}_0)$, and $\gamma(t)$ is implemented by the unitary group with positive energy $\Gamma(e^{-i\mu t})$, whereas $\alpha(t)$ is implemented by $\Gamma(e^{-iK_0t})$ with $e^{-iK_0t} = K_0V(t)K_0^{-1}$. Finally, ι is implemented by $\Gamma(j_0)$, with $j_0\chi = c_0\check{\chi}$, c_0 being the complex conjugation on $L^2(\mathbb{R}^3, \mathbb{C})$.

Now, let us restrict ω_0 to the double wedge algebra $\tilde{\mathcal{U}}$. By using the Bisognano-Wichmann theorem ([Haag (1996)],Th4.1.1,CH.V) and the fact that $\Omega^{\mathcal{F}}$ is cyclic for $\rho_0(\mathcal{U}^R)$, as a consequence of the Ree-Schlieder theorem ([Haag (1996)], Th5.3.1,CH.II), one gets that $\omega_0|_{\mathcal{U}^R}$ is a KMS state with inverse temperature $\beta = 2\pi$ w.r.t. $\alpha(t)$.

One can also construct directly ω_0 over $\tilde{\mathcal{U}}$ in terms of elements of the double wedge (Fulling quantization). Let us first admit that there is a 1-particle structure $(K_F, \mathcal{H}_F, e^{-i\hat{H}_Ft})$ for the linear dynamical system $(D^R, \sigma, V(t))$. Then, for any $\beta > 0$ we can define the double KMS state ω_F^β on $(\tilde{\mathcal{U}}, \alpha(t), \iota)$ by setting

$$\omega_F^\beta(W(\phi)) = e^{-\frac{1}{2}\|K_F^\beta\Phi\|^2_{\mathcal{H}_F\oplus\mathcal{H}_F}} \tag{7.78}$$

for all $\Phi \in \tilde{D}$. If $\Phi \in \tilde{D}$, then $\phi = \phi^R + \phi^L$ and

$$K_F^\beta\phi = \begin{pmatrix} \cosh z^\beta & \sinh z^\beta\ c_F \\ \sinh z^\beta\ c_F & \cosh z^\beta \end{pmatrix} \begin{pmatrix} -K_F(\check{q}_L, \check{p}_L) \\ K_F(q^R, p^R) \end{pmatrix}, \qquad \tanh z^\beta := e^{-\frac{1}{2}\beta\hat{H}_F}, \tag{7.79}$$

and c_F is the temporal inversion $c_FK_F(q^R, p^R) = K_F(q^R, -p^R)$. The fundamental state is

$$\begin{aligned} \omega_F^\beta(W(\phi)) = \exp\Bigg[& -\frac{1}{2}\left(K_F(\check{q}^L, \check{p}^L)|\coth\frac{\beta\hat{H}_F}{2}K_F(\check{q}^L, \check{p}^L)\right) \\ & -\frac{1}{2}\left(K_F(q^R, p^R)|\coth\frac{\beta\hat{H}_F}{2}K_F(q^R, p^R)\right) \\ & -\mathrm{Re}\left(K_F(\check{q}^L, -\check{p}^L)|\coth\frac{\beta\hat{H}}{2}K_F(q^R, p^R)\right)\Bigg]. \end{aligned} \tag{7.80}$$

If we take $\beta \to \infty$ we get a product state over $\mathcal{U}^L \otimes \mathcal{U}^R$ which restriction, say, to $\mathcal{U}^R$ is the Fulling state

$$\omega_F(W(\phi)) = e^{-\frac{1}{2}\|K_F\phi\|^2_{\mathcal{H}_F}}, \tag{7.81}$$

which is the ground state for $(\mathcal{U}^R, \alpha(t))$.

Since in a free theory there are not phase transitions, one expects that the states obtained in different ways are the same. Indeed:

Theorem 7.3. *On $\tilde{\mathcal{U}}$ it holds $\omega_0 = \omega_F^{2\pi}$.*

7.3.3 *The Hartle-Hawking state*

We now consider a Klein-Gordon field of mass m on a mass M Kruskal-Szekeres manifold $\mathcal{M}_M \simeq (\mathbb{R}^2 \times S^2, ds^2 \equiv g)$, with

$$ds^2 = 32M^3 \frac{e^{-\frac{r}{2M}}}{r}(-dT^2 + dX^2) + r^2 d\Omega^2, \tag{7.82}$$

where $d\Omega^2$ is the metric on the unit sphere, r is determined by

$$T^2 - X^2 = \left(1 - \frac{r}{2M}\right) e^{\frac{r}{2M}},$$

and the region of interest is $T^2 - X^2 < 1$. In particular, we are interested in the external regions $\mathcal{R}$ $(X > |T|)$, $\mathcal{L}$ $(X < -|T|)$, and $\mathcal{L} \cup \mathcal{R}$. The Cauchy surface $T = 0$ is $C \simeq \mathbb{R} \times S^2$ its parts C^R and C^L corresponding to $x > 0$ and $x < 0$, are Cauchy surfaces for $\mathcal{R}$ and $\mathcal{L}$ respectively. As above, from this we can construct the Cauchy data sets D, D^L, D^R and $\tilde{D}$. We then consider the Klein-Gordon equation

$$(\Box_g + m^2)\phi = 0 \tag{7.83}$$

with a choice of Cauchy data. Thus [Hawking and Ellis (2011)]:

Proposition 7.1. *The Klein-Gordon equation with a choice of the Cauchy data is a well-posed problem: given $\Phi = (q,p) \in D$, there is a unique solution $\phi \in \mathcal{C}_0^\infty(\mathcal{M}_M)$ such that*

$$\phi|_C = q, \qquad \partial_T \phi|_C = p.$$

Moreover, the support of ϕ is the dependence domain of Φ.

On D we can construct a symplectic form in the usual way, as the integral of the zeroth component of the conserved current $j^\mu(\phi_1, \phi_2) = (\det g)^{\frac{1}{2}}(\phi_1 \partial_\nu \phi_2 - \phi_2 \partial_\nu \phi_1)$ on a spacelike surface. Then, $\mathcal{U}$ is constructed on (D, σ) and becomes the $*$-algebra for the quantum solutions, defining

$$W(\Phi) = \exp\left(i \int_{\mathcal{M}_M} F\sqrt{g} dT dX \sin\theta d\theta d\phi\right), \tag{7.84}$$

with $F \in \mathcal{C}_0^\infty(\mathcal{M}_M)$, and Φ being the Cauchy data of the solution $\Delta \star F$, Δ being the fundamental solution seen as an operator $\mathcal{C}_0^\infty \to \mathcal{C}^\infty$, [Kay

(1985); Dimock (1980)]. From the causal support of Δ, one sees that Φ belongs in D^R when F is in $\mathcal{R}$, and then these data define $\mathcal{U}^R$ generated by $\{W(\Phi)|\Phi \in D^R\}$. In the same way one defines $\mathcal{U}^L$ and then $\tilde{\mathcal{U}}$.

Now, in $\mathcal{M}_M$ does not exist the temporal translation. However, one can define $V(t)$ via $\Phi \mapsto \Phi_t$, associated to $\phi_t(T, X, \theta, \phi) = \phi(\Lambda(t)(T, X), \theta, \phi)$ with

$$\Lambda(t) = \begin{pmatrix} \cosh \frac{t}{4M} & \sinh \frac{t}{4M} \\ \sinh \frac{t}{4M} & \cosh \frac{t}{4M} \end{pmatrix} \tag{7.85}$$

which on $\mathcal{R}$ is the translation of the Schwarzschild time. The above proposition ensures that $V(t) : D \to D$ preserves σ (and the conservation of j^μ). Finally, $\mathcal{I}(q, p) = (\check{q}, -\check{p})$, where $\check{h}(X, \theta, \phi) = h(-X, \theta, \phi)$ comes from the reflection of the wedge.

Thus, $(\tilde{D}, \sigma, V(t), \mathcal{I})$ is a double classical linear system. Setting $\alpha(t)W(\Phi) = W(V(t)\Phi)$ and $\iota W(\Phi) = W(\mathcal{I}\Phi)$ we finally get a double quantum dynamical system $(\tilde{\mathcal{U}}, \alpha(t), \iota)$.

Now, in order to construct the analogue of ω_0 there is not the translation $\gamma(t)$. However, we can construct the double KMS state ω_H over $(\tilde{\mathcal{U}}, \alpha(t), \iota)$. Provisorily, let us assume the existence of a one particle structure $(K_B, \mathcal{H}_B, e^{-i\hat{H}_B t})$ for the dynamical linear system $(D^R, \sigma, V(t))$. Then, for any $\beta > 0$ we can define a state ω_B^β, analogous to the Fulling state.

Definition 7.5. We define the Hartle-Hawking state by $\omega_H := \omega_B^{8\pi M}$.

Then, the main result of Kay is

Theorem 7.4. *Define the von Neumann algebras $\mathcal{A}_I = \rho_H(\mathcal{U}^I)''$, $I = L, R$, where ρ_H is the GNS representation associated to ω_H. Then*

(1) *ω_H on $\tilde{\mathcal{U}}$ is a pure state;*
(2) *the GNS vacuum Ω is cyclic for $\rho_H(\mathcal{U}^I)$ (Ree-Schlieder property);*
(3) *$\mathcal{A}'_L = \mathcal{A}_R$ (duality property);*
(4) *$\mathcal{A}_I$ are factors: $\mathcal{A}'_I \cap \mathcal{A}_I = \{\mathbb{C}\mathbb{I}\}$;*
(5) *the factors $\mathcal{A}_I$ are of type III_1 in the Connes classification.*

It is worth to remark that the condition $\beta = 8\pi M$ is just a definition and is nowhere required in the proof of the theorem. Thus, this way one can construct a thermal state on the Schwarzschild wedge for any possible value of β. The temperature is fixed by requiring prolongability through the horizon, t.i. the arguments of Haag, Narnhofer and Stein.

However, the argument proving the existence of such thermal states is based on the assumption on the existence of a one particle structure

$(K_B, \mathcal{H}_B, e^{-i\hat{H}_B t})$. The last step is thus to prove the existence of such a structure.

7.3.4 *The Boulware one particle structure*

In order to construct the one particle state $(K_B, \mathcal{H}_B, e^{-i\hat{H}_B t})$, it is convenient to change coordinates, passing to the new ones (t, r, θ, ϕ), related to the old ones by

$$T = e^{\frac{x}{4M}} \sinh \tfrac{\tau}{4M}, \tag{7.86}$$

$$X = e^{\frac{x}{4M}} \cosh \tfrac{\tau}{4M}, \tag{7.87}$$

so that $x \equiv r^*$ and, as said before, the boost transformation reduces to τ-translation: $\Lambda(t) : (\tau, x, \theta, \phi) \mapsto (\tau + t, x, \theta, \phi)$. Thus, the Klein-Gordon equation takes the form

$$\begin{pmatrix} \dot{\phi} \\ \ddot{\phi} \end{pmatrix} = \begin{pmatrix} 0 & 1 \\ -\hat{h} & 0 \end{pmatrix} \begin{pmatrix} \phi \\ \dot{\phi} \end{pmatrix}, \tag{7.88}$$

where

$$\hat{h} = -\partial_x^2 + \left(1 - \frac{2M}{r}\right)\left(\frac{2M}{r^3} - \frac{\Delta_\Omega}{r^2} + m^2\right), \tag{7.89}$$

the dot indicates derivation w.r.t. t, and Δ_Ω is the Laplacian on the unit sphere. In particular, one has

$$\int_{C^R} f^* \hat{h} f dx \sin\theta d\theta d\phi \geq \int_{C^R} \left(1 - \frac{2M}{r}\right) \frac{2M}{r^3} |f|^2 dx \sin\theta d\theta d\phi, \tag{7.90}$$

$\forall f \in \mathcal{C}_0^\infty(C^R)$.

In these coordinates, the system $(D^R, \sigma, V(t))$ is conveniently replaced by the system $(\bar{D}, \bar{\sigma}, \bar{V}(t))$ with $\bar{D} \simeq \mathcal{C}_0^\infty(\mathbb{R} \times S^2) \times \mathcal{C}_0^\infty(\mathbb{R} \times S^2)$, and $\Phi = (q, p) \in D^R$ becomes $\bar{\Phi} = (\bar{q}, \bar{p}) \in \bar{D}$, with $\bar{q} = q$ and $\bar{p} = e^{\frac{x}{4M}} p$. Moreover

$$\bar{\sigma}(\bar{\Phi}_1, \bar{\Phi}_2) = \sigma(\Phi_1, \Phi_2) = \int_{C^R} (q_1 \bar{p}_2 - q_2 \bar{p}_1) dx \sin\theta d\theta d\phi \tag{7.91}$$

and $\bar{V}(t)\bar{\Phi} = \overline{V(t)\Phi}$.

From the previous analysis we then know that the initial data $(\phi(0), \dot{\phi}(0)) = (q, \bar{p})$ determine a unique solution $\bar{\Phi}(\tau) = \bar{V}(\tau)(q, \bar{p})$, which is of class $\mathcal{C}^\infty$ and has compact support for any region with finite τ. Moreover, $\bar{V}(t)$ preserves $\bar{\sigma}$ and the **energy norm**

$$\|\bar{\Phi}\|_{\hat{h}}^2 := \frac{1}{2} \int_{C^R} (q \hat{h} q + \bar{p}^2) dx \sin\theta d\theta d\phi. \tag{7.92}$$

We can then complete $\bar{D}$ to a Hilbert space $\mathcal{H}_D$ and extend $\bar{V}(t)$ by closure with the energy norm. Thus $\bar{V}(t)$ is strongly continuous on $\mathcal{H}_D$ and has strong derivative $-\hat{H}$,

$$\hat{H} = \begin{pmatrix} 0 & 1 \\ -\hat{h} & 0 \end{pmatrix}. \tag{7.93}$$

So $\hat{H}$ is essentially self adjoint on $\bar{D} \subset \mathcal{H}_D$ and if $\hat{H}_c$ is its closure, then $\bar{V}(t) = e^{-\hat{H}_c t}$. Moreover, we have that

$$\hat{H}^2 = -\begin{pmatrix} \hat{h} & 0 \\ 0 & \hat{h} \end{pmatrix} \tag{7.94}$$

is essentially self adjoint on $\bar{D}$ so that, after restriction over the Cauchy data $(q, \bar{p})$, we get that $\hat{h}$ is essentially selfadjoint over $C_0^\infty(C^R) \subset L^2(C^R, dx \sin\theta d\theta d\phi)$. In particular, $\hat{h}$ is positive on $C_0^\infty(C^R)$ and so is its closure. Indeed,

Proposition 7.2. *$\hat{h}$ is strictly positive.*

Let us consider the restricted domain

$$\mathcal{D}_r := \mathcal{D}(\hat{h}^{\frac{1}{2}}) \oplus L^2(C^R, dx \sin\theta d\theta d\phi) \subset \mathcal{H}_D. \tag{7.95}$$

On $\mathcal{D}_r$ it is possible to extend $\bar{\sigma}$, and

$$e^{-\hat{H}t}|_{\mathcal{D}_r} = \begin{pmatrix} \cos(\hat{h}^{\frac{1}{2}}t) & \hat{h}^{-\frac{1}{2}} \sin(\hat{h}^{\frac{1}{2}}t) \\ -\hat{h}^{\frac{1}{2}} \sin(\hat{h}^{\frac{1}{2}}t) & \cos(\hat{h}^{\frac{1}{2}}t) \end{pmatrix}. \tag{7.96}$$

Further restriction to $\mathcal{D}_{red} = \mathcal{D}(\hat{h}^{\frac{1}{2}}) \oplus \mathcal{D}(\hat{h}^{-\frac{1}{2}})$ allows to introduce the operator

$$\bar{K} : (q, \bar{p}) \mapsto \frac{1}{\sqrt{2}}(\hat{h}^{\frac{1}{4}} q + i\hat{h}^{-\frac{1}{4}} \bar{p}) \tag{7.97}$$

in $L^2(C^R, dx \sin\theta d\theta d\phi, \mathbb{C})$. Thus $\bar{K}$ exchanges $e^{-\hat{H}t}$ with $e^{-i\hat{h}^{\frac{1}{2}}t}$ and is symplectic.

One has that

Proposition 7.3. *$(\bar{K}, L^2(C^R, dx \sin\theta d\theta d\phi, \mathbb{C}), e^{-i\hat{h}^{\frac{1}{2}}t})$ is a regular one particle structure in $(\mathcal{D}(\hat{a}^{\frac{1}{2}}) \oplus \mathcal{D}(\hat{a}^{-\frac{1}{2}}), \bar{\sigma}, e^{-\hat{H}t})$.*

The final result is

Theorem 7.5. *$\bar{D} \subset \mathcal{D}(\hat{a}^{\frac{1}{2}}) \oplus \mathcal{D}(\hat{a}^{-\frac{1}{2}})$.*
Moreover, the restriction of $(\bar{K}, L^2(C^R, dx \sin\theta d\theta d\phi, \mathbb{C}), e^{-i\hat{h}^{\frac{1}{2}}t})$ to $(\bar{D}, \bar{\sigma}, \bar{V}(t))$ is a regular one particle structure.

This completes the proof of the existence of thermal states on the Schwarzschild double wedge.

7.4 Hawking temperature for a spacetime with bifurcated Killing horizon

By applying the Tomita-Takesaki theory to a quantum field theory on a curved spacetime with a bifurcated Killing horizon, it is possible to prove establish sufficient model independent conditions granting that the equilibrium fundamental state is at the corresponding Hawking temperature [Summers and Verch (1996)]. From section 1.3, we recall that a spacetime with bifurcated horizon is a quintuple $(\mathcal{M}, g, \tau_t, \Sigma, h)$, where $(\mathcal{M}, g)$ is a time oriented Lorentzian globally hyperbolic manifold, τ_t is a one parameter group of nontrivial isometries for $(\mathcal{M}, g)$, Σ is a bidimensional connected spacelike submanifold, contained in a spacelike Cauchy surface, punctually invariant under the action of τ_t, and h is the bifurcated Killing horizon, the union of two $\mathcal{C}^\infty$ three-dimensional submanifolds, generated by null geodesic emanated orthogonally from Σ. We also recall the points on the null geodesics can be parameterized as (U,p) (or (V,p)) where p is the starting point of the geodesic on h, so that $\tau(t)(U,p) = (e^{\kappa t}U, p)$ and $\tau(t)(V,p) = (e^{-\kappa t}V, p)$, where κ is the constant surface gravity on the Killing horizon. Indeed, since p is invariant, $\tau_t(U,p) =: (\tilde{\tau}(U), p)$ defines an action of the group on $U \in \mathbb{R}^+$. If ξ is the Killing vector field generating τ_t and χ is the vector field generating the geodesics parameterized by U, then we know that it exists a function $\alpha(U,p)$ such that $\xi(U,p) = \alpha(U,p)\chi(U,p)$, from which one finds

$$\frac{d}{dt}\tilde{\tau}_t(U) = \alpha(\tilde{\tau}_t(U), p), \tag{7.98}$$

so that $\kappa = \xi^\mu \nabla_\mu \ln \alpha$ implies

$$\kappa = \frac{d}{dt}\tilde{\tau}_t(U) \frac{1}{\alpha(\tilde{\tau}_t(U), p)} \frac{\partial}{\partial x}\alpha(x,p)\bigg|_{x=\tilde{\tau}_t(U)}. \tag{7.99}$$

Employing the group properties, from this it follows that $\tilde{\tau}_t : u \mapsto u + t$, from which the assert follows.

Beyond this one can define the affine translations

$$\ell_b(U,p) = (U - b, p), \qquad \ell_b(V,p) = (V - b, p). \tag{7.100}$$

7.4.1 *Modular inclusion and Hawking temperature*

Let $\mathcal{B}$ a basis for the topology of $\mathcal{M}$. Let h_A the component of h generated by (U,p) (and h_B generated by (V,p)). Let $\mathcal{K}$ be the set of open sets of h_A. It is invariant under the Killing flux and the affine translations. We can further separate h_A in h_A^L and h_A^R if $U < 0$ or $U > 0$ respectively. Thus $\mathcal{K}_X = \{h_A^X \cap \mathcal{O} | \mathcal{O} \in \mathcal{K}\}$, $X = L, R$ (and similar for B). Let $\mathcal{U}$ a C^*-algebra and $\{\mathcal{U}(\mathcal{O})\}_{\mathcal{O}\in\mathcal{K}}$ be a net of C^*-algebras over h^A. Then, we can define the restrictions (C^*-subalgebras)

$$\mathcal{U}_A^X = \{\mathcal{U}(\mathcal{O} \cap h_A^X) | \mathcal{O} \in \mathcal{K}\}, \qquad X = L, R. \tag{7.101}$$

Let $\mathcal{B}$ a basis for the topology of $\mathcal{M}$, $\mathcal{A}$ a C^*-algebra and $\{\mathcal{A}(\mathcal{O})\}_{\mathcal{O}\in\mathcal{B}}$ the corresponding net. Then:

Definition 7.6. $\{\mathcal{U}(\mathcal{V})\}_{\mathcal{V}\in\mathcal{K}}$ is said a *symmetry improving restriction* to h_A of the net $\{\mathcal{A}(\mathcal{O})\}_{\mathcal{O}\in\mathcal{B}}$ if

(1) it is covariant w.r.t. the Killing flux and the affine translation, t.i. there exist continuous representations α_t, λ_b, with $\mathbb{R} \ni t \mapsto \alpha_t \in \mathrm{Aut}\mathcal{U}$, $\mathbb{R} \ni b \mapsto \lambda_b \in \mathrm{Aut}\mathcal{U}$ such that if $\mathcal{V} \in \mathcal{K}$then

$$\alpha_t(\mathcal{U}(\mathcal{V})) = \mathcal{U}(\tau_t\mathcal{V}), \qquad \lambda_b(\mathcal{U}(\mathcal{V})) = \mathcal{U}(\ell_b(\mathcal{V})); \tag{7.102}$$

(2) one has the inclusion $\mathcal{L} \subset \{\mathcal{O} \cap h_A | \mathcal{O} \in \mathcal{B}\}$ and for all $\mathcal{V} \in \mathcal{K}$, $\mathcal{V} = \mathcal{O} \cap h_A$ for some $\mathcal{O} \in \mathcal{B}$, then $\mathcal{U}(\mathcal{K})$ is a subalgebra $\mathcal{A}(\mathcal{O})$.

Obviously the same can be done with h_B. Notice that the affine translations act on $\mathcal{U}$ only and not on $\mathcal{A}$. Indeed, we did never required for $\mathcal{A}(\mathcal{O})$ to be covariant. We now need a theorem of Borchers [Borchers (1996)] (Th.II.9 therein):

Theorem 7.6. *Let $\mathcal{N}$ a von Neumann algebra acting on a Hilbert space $\mathcal{H}$ and let $\Omega \in \mathcal{H}$ be cyclic and separating for $\mathcal{N}$. Let Δ and J be respectively the modular operator and the modular conjugation correspondent to $(\mathcal{N}, \Omega)$. Let $U(t)$, $t \in \mathbb{R}$ be a strongly continuous one parameter group with a positive generator, which leaves Ω invariant. If $U(t)\mathcal{N}U(t)^\dagger \subset \mathcal{N}$ for $t \geq 0$, then*

$$\Delta^{iT} U(t) \Delta^{-iT} = U(e^{-2\pi T} t) \quad \textit{and} \quad JU(t)J = U(-t) \qquad \forall T, t \in \mathbb{R}. \tag{7.103}$$

An inverse theorem is also true ([16],[5] in [Summers and Verch (1996)]). The main result of Summers and Verch is

Theorem 7.7. *Let ω be a state on a net $\{\mathcal{A}(\mathcal{O})\}_{\mathcal{O}\in\mathcal{B}}$, which admits a symmetry improving restriction $\{\mathcal{U}(\mathcal{V})\}_{\mathcal{V}\in\mathcal{K}}$ over h_A (or h_B) with the properties*

(1) $\mathcal{V}_1 \neq \mathcal{V}_2$ implies $\mathcal{U}(\mathcal{V}_1) \neq \mathcal{U}(\mathcal{V}_2)$;
(2) $\omega|_{\mathcal{U}}$ is a ground state w.r.t. λ_b (so the GNS representation of λ_b has positive spectrum);

(3) $\omega|_{\mathcal{U}_R}$ (or $\mathcal{U}_L$) is a KMS state with nonvanishing inverse temperature β w.r.t. the Killing flux α_t.

Then $\beta = 2\pi/\kappa$ over h_A ($\beta = -2\pi/\kappa$ over h_B).

The conditions (2) and (3) come from the Hadamard condition and the invariance under the Killing flux of the almost free states of the Klein-Gordon field, on the bifurcated horizon. This has been studied by Kay and Wald [Kay and Wald (1991)]. Thus, for a Killing horizon, if the hypothesis of the theorem can be satisfied, the Hawking temperature arises in a natural way. However, if there are more disconnected bifurcated horizons (as for a Schwarzschild de Sitter spacetime) such a global state may not exist.

For a single bifurcated horizon it always exists, as one can see by working out the case of a massive complex Klein-Gordon field [Summers and Verch (1996)]:

Proposition 7.4. *On any globally hyperbolic Lorentzian manifold with a single bifurcated Killing horizon it always exists a state ω and a net of algebras $\{\mathcal{A}(\mathcal{O})\}_{\mathcal{O}\in\mathcal{B}}$, which satisfy the hypothesis of the above theorem.*

7.5 Further readings for black holes of the Kerr-Newman family

We limit ourselves to indicate the reader a list of references concerning rigorous results for the Hawking effect, and we limit ourselves to black holes of the Kerr-Newman family. We first quote two seminal contributions [Gibbons (1975); Wald (1975)]. See also [Bisognano and Wichmann (1975, 1976); Sewell (1982)]. As the Hadamard condition plays a fundamental role, we refer to [Radzikowski (1996b,a); Fewster and Verch (2013); Dappiaggi *et al.* (2011, 2017)]. As to the Hawking effect, we refer also to [Bachelot (1999); Melnyk (2004); Hafner (2009)].

PART 2
Second Part

Chapter 8

The roots of analogue gravity

We have seen that the Hawking phenomenon is quite robust from a theoretical point of view, in the sense that if does not exist a complete incontrovertible exact proof of the Hawking effect, able to take into account the back reaction and eventual quantum gravity effects, there are several ways for analyzing it, all leading to the same result (for a quasi stationary black hole): the black hole must radiate away field modes (particles) at a temperature determined by the surface gravity κ. If this strongly motivates for thinking to Hawking radiation as a physical phenomenon, it is obviously not enough, since this only one half of the medal. To be really considered a physical phenomenon it has to be verified experimentally. However, the task of measuring the Hawking phenomenon via astrophysical observations is essentially hopeless: a black hole with mass M is expected to emit like a black body with a temperature

$$T_H = \frac{\hbar c^3}{8\pi G k_B M}, \tag{8.1}$$

where G is the Newton gravitational constant and k_B the Boltzmann constant, and a corresponding maximal power

$$P = A\sigma T_H^4, \tag{8.2}$$

where

$$\sigma = \frac{\pi^2 k_B^4}{60 c^2 \hbar^3} \tag{8.3}$$

and A is the area of the horizon. For example, for a Schwarzschild black hole, if $M_o = 1.989 \cdot 10^{30} Kg$ is the mass of the Sun, we have

$$T_H = \frac{\hbar c^3}{8\pi G k_B M_o} \frac{M}{M_o} = 6.168 \cdot 10^{-8} K \frac{M_o}{M}, \tag{8.4}$$

$$P = \frac{\hbar c^6}{15 \cdot 2^{10} \pi G^2 M_0^2} \frac{M_o^2}{M^2} = 9.003 \cdot 10^{-29} J/s \frac{M_o^2}{M^2}. \tag{8.5}$$

Black holes coming from the collapse of stars (and further accretion) will have masses larger than the solar one so that the above values are higher bounds for the temperature and the power emitted. Any possible signal would be completely washed out, for example, by the cosmological microwave background. The alternative would be to look for primordial micro black holes roaming in the universe, if exist. If these were the only possibilities, than Hawking radiation would remain essentially a theoretical speculation.

Luckily, there is another possibility: trying to reproduce the peculiar characteristics of an event horizon in a physical system simple enough to be realized in a laboratory. The prototype of such a configuration has been discovered by Unruh in the seminal paper [Unruh (1981)], which gave birth to a new line of research: analogue gravity. We don't delve in the very extensive literature about analogue gravity, which represent the topic treated in many papers referring mostly to the case of non-dispersive systems, as our focus is simply associated with black holes in dielectric systems. For a more general view and a quite updated list of references, see [Barcelo *et al.* (2005)]. We also refer to [Visser (1998b)] for a discussion about the limits of the so-called analogue gravity framework in reproducing aspects of black hole physics. We limit ourselves in the following section to basic ideas inherited by [Unruh (1981)]

8.1 Experimental black hole evaporation in water

Let us consider a non relativistic irrotational non viscous fluid characterized by a velocity field $\vec{v} = \vec{\nabla}\psi$, a pressure p and a density ρ, described by the equations

$$\rho[\partial_t \vec{v} + (\vec{v} \cdot \vec{\nabla})\vec{v}] = - \vec{\nabla} p - \rho \vec{\nabla} \phi, \tag{8.6}$$

$$\partial_t \rho + \vec{\nabla} \cdot (\rho \vec{v}) = 0, \tag{8.7}$$

where ϕ is the gravitational potential We will assume the fluid to be barotropic: $p = p(\rho)$. Employing irrotationality, the Euler equation can be rewritten as

$$\vec{0} = \vec{\nabla}(\partial_t \psi + \frac{v^2}{2} + \phi) + \frac{1}{\rho} \partial_\rho p \vec{\nabla} \rho, \tag{8.8}$$

so that if $G(\rho)$ is a primitive of the function $\frac{1}{\rho}\partial_\rho p(\rho)$, we can write it as

$$\partial_t \psi + \frac{v^2}{2} + \phi + G = 0. \tag{8.9}$$

Let us now fix a background solution $\psi_0(t, \vec{x})$, $\rho_0(t, \vec{x})$ of the above system of equations, and consider perturbations

$$\psi = \psi_0 + \delta\psi, \tag{8.10}$$

$$\phi = \phi_0 + \delta\phi. \tag{8.11}$$

Then, at the first order in the perturbation, the continuity equation and equation (8.9) take respectively the form

$$\partial_t \delta\rho + \vec{\nabla}\left[\delta\rho \vec{v}_0 + \rho_0 \vec{\nabla}\delta\psi\right] = 0, \tag{8.12}$$

$$\partial_t \delta\psi + \vec{v}_0 \cdot \vec{\nabla}\delta\psi + \frac{1}{\rho_0^2}\partial_\rho p_0 \delta\rho = 0, \tag{8.13}$$

where $\vec{v}_0 = \vec{\nabla}\psi_0$, and $\partial_\rho p_0 = \partial_\rho p|_{\rho=\rho_0}$.

From this system we get a unique equation for $\delta\psi$, which we write in the form

$$\frac{1}{\rho_0}\left[\partial_t\left(\frac{\rho_0}{\partial_\rho p_0}\partial_t\delta\psi\right) + \partial_t\left(\frac{\rho_0}{\partial_\rho p_0}\vec{v}_0 \cdot \vec{\nabla}\delta\psi\right) + \vec{\nabla}\cdot\left(\frac{\rho_0}{\partial_\rho p_0}\vec{v}_0\partial_t\delta\psi\right)\right.$$
$$\left. - \vec{\nabla}\cdot(\rho_0\vec{\nabla}\delta\psi) + \vec{\nabla}\cdot\left(\vec{v}_0\frac{\rho_0}{\partial_\rho p_0}\vec{v}_0\cdot\vec{\nabla}\delta\psi\right)\right] = 0. \tag{8.14}$$

The main observation is that this equation is exactly the Klein-Gordon equation for a massless scalar field on a nontrivial spacetime metric

$$g_{\mu\nu}\nabla^\mu\nabla^\nu\Phi = 0, \tag{8.15}$$

where $g_{\mu\nu}$ are the components of the effective metric

$$ds^2 = \frac{\rho_0}{c_0}\left[(c_0^2 - v_0^2)dt^2 + 2dt\, \vec{v}_0 \cdot d\vec{x} - d\vec{x}\cdot d\vec{x}\right], \tag{8.16}$$

in which $c_0 = \sqrt{\partial_\rho p_0}$ is the local sound velocity.

Now, since $\vec{v}_0$ is irrotational, it is locally well defined the new time coordinate

$$\tau = t + \int_\gamma \frac{\vec{v}_0 \cdot d\vec{x}}{c_0^2 - v_0^2}, \tag{8.17}$$

because the integral depends only on the initial and final point of the path γ. Moreover, always locally, we can consider coordinates $\vec{x}_\perp$ on spatial surfaces orthogonal to γ.[1] With this new coordinate the metric takes the form

$$ds^2 = \frac{\rho_0}{c_0}\left[(c_0^2 - v_0^2)d\tau^2 - \frac{(\vec{v}_0 \cdot d\vec{x})^2}{c_0^2 - v_0^2} - d\vec{x}\cdot d\vec{x}\right], \tag{8.18}$$

and, introducing the coordinate q such that

$$dq = \frac{\vec{v}_0 \cdot d\vec{x}}{v_0}, \tag{8.19}$$

we finally have

$$ds^2 = \frac{\rho_0}{c_0}\left[(c_0^2 - v_0^2)d\tau^2 - \frac{c_0^2 dq^2}{c_0^2 - v_0^2} - d\vec{x}\cdot d\vec{x}\right]. \tag{8.20}$$

From this expression, we clearly see that if the flux $\vec{v}_0$ is suitably chosen, there may exist a surface described by the equation $q = q_0$, where $v_0 = -c_0$. This surface is then an horizon. If we linearize near q_0, say $v_0 \approx -c_0 + \alpha(q - q_0) + \ldots$ (where c_0 is valued at q_0), reducing down to the two dimensional τ-q space we get

$$ds^2 \approx \frac{\rho_0}{c_0}\left(2c_0\alpha(q - q_0)d\tau^2 - \frac{dq^2}{2\alpha(q - q_0)}\right), \tag{8.21}$$

to be compared with the t-r reduction of the Schwarzschild metric near the horizon

$$ds^2_{Schw} \approx \frac{r - 2M}{2M}dt^2 - \frac{2M dr^2}{r - 2M}. \tag{8.22}$$

At this point, with the usual methods, we can deduce that if we quantize the above perturbations we must expect that from the horizon will be emitted quanta with a thermal spectrum of temperature

$$T = \frac{\hbar}{2\pi k_B}\left.\frac{\partial v_0}{\partial q}\right|_{q=q_0}. \tag{8.23}$$

The novelty of this discovery was not only the possibility to reproduce the semiclassical result of Hawking in a laboratory, but also to look at possible effect simulating quantum gravity since the limit of continuum is broken by the microscopic structure. Moreover, the "evaporating modes" are not of an external field, but of the same constituting the "spacetime".

As to the methods for studying analogue black holes, in absence of dispersive effect, as in the above deduction, all the methods displayed in

[1] Typical configurations are the ones with a flux in a fixed direction, like for example the direction z, in which case $d\gamma = dz$ and $\vec{x}_\perp \equiv (x, y)$, or in the radial direction, so that $d\gamma = dr$ and the orthogonal surfaces are concentric spheres with, say, polar coordinates.

the first part of the book can be applied, and analogue Hawking radiation can be deduced without particular efforts. There is a huge literature on the subject, see [Barcelo *et al.* (2005)] and references therein. Problems arise when dispersion is taken into account. This topic is discussed in the following.

8.2 Analogue systems and dispersion: The Gospel according to Unruh

Dispersive terms in the Lagrangian for the physical model to be implemented implies that the dispersion relation is no more such that, e.g. in 2D, the wave vector k and the frequency ω are proportional. Linearity is lost, and even in the limit of small dispersion effects one face with a drastic change in the physics at hand. In absence of dispersion, indeed one has exactly the same features which occur in the standard case, with the presence of a logarithmic divergence in the configuration space in the Hawking mode as the horizon coordinate is approached. This ultraviolet problem is one of the 'hard problems' in the physics of Hawking radiation, and is associated with the fact that Hawking radiation requires the presence of modes displaying arbitrarily high, up to transplanckian frequencies (see e.g. [Jacobson (1991)], and see also [Jacobson (1993); Brout *et al.* (1995b)]). In the traditional view, the phenomenon seems to be unavoidably involved with very high energy particles. A naive argument allowing to realize this point is that e.g. massless particles arising from a Schwarzschild black hole and arriving at null infinity, very near the horizon are very blueshifted because of the standard redshift relation between frequency at infinity and frequency at the horizon (infinite in the limit) [Wald (1984)]. In other terms, propagating back a wave packet perceived at infinity, as it is reach the horizon it is affected by unboundedly growing frequencies.

In the view proposed by Unruh (see e.g. [Unruh (1995); Schutzhold and Unruh (2008)]), at least at the level of analogue gravity systems, Hawking radiation in a dispersive medium should be a low energy phenomenon, unaffected by ultrashort wavelengths, but still preserving the standard relation between temperature and surface gravity for the nondispersive analogue gravity model one is considering (about the presence of high energy modes, see also [Jacobson (1996)]). Thermality is associated with tearing apart of quantum fluctuations near the black hole horizon (compaction in the case of white hole horizon, which plays the main role in experiments, cf. the last chapter), in a exponential way, with part of the wave packet with negative energy remaining in the black hole, and part, with positive energy, emerging as Hawking quantum [Schutzhold and Unruh (2011)]. The Hawking mode perceived by the distant observer emerges from a peculiar process of 'mode conversion' (according to the terminology coined by T.Jacobson [Jacobson (1996)] in analogy also with plasma physics, see e.g. [Swanson (1998)]) occurring at the horizon, and involving substantially three modes: A final state positive norm mode H (the Hawking mode), characterized by a small value of the wave vector, and two initial states P, N with high values of the wave vectors (but finite in any event), the first with positive norm and the second with negative one. In the simplest situation, a fourth positive norm mode B appears in the initial state, but it is often negligible. (The time-reversed process, occurring in the white hole case, is much more perspicuous from a physical point of view, and easier to set up in a scattering framework.) Thermality emerges as the fact that the ratio between the negative norm and the positive norm contributions to the initial state is thermal (see also [Jacobson (1999)]):

$$\frac{|J_x(N)|}{|J_x(P)|} = \exp(-\beta\omega), \tag{8.24}$$

where J_x stands for the current in the spatial direction (say x) for the mode at hand, and β is the inverse Hawking radiation of the corresponding nondispersive model. There are several conceptual changes with respect to the nondispersive case:

- The horizon is replaced by the so-called group horizon, i.e. in the scattering process, at least of the level of the eikonal approximation, one requires that from the dispersion relation a turning point (TP) emerges, corresponding to a zero of the group velocity. One refers also to it as blocking horizon. This is considered a necessary feature for thermality [Schutzhold and Unruh (2011)]. This horizon a priori is frequency dependent, and this marks one of the most striking differences with respect to the standard horizon of black hole physics.
- Thermality itself could be jeopardized by a dependence of the temperature on the frequency. In the case of low dispersion, there are several examples where this does not happen (conversely, no system displaying such a dependence in actual measurements has be yet found).
- Thermality does not occur for all frequencies: there exists a finite interval $(\omega_{min}, \omega_{max})$ where the phenomenon can occur (in standard situations, $\omega_{min} = 0$). In the complementary set of frequencies, scattering processes of different nature take place.

- One distinguishes between subluminal and superluminal dispersion (see e.g. [Corley (1998); Schutzhold and Unruh (2002); Unruh and Schutzhold (2005)]), with reference to the fact that high frequency behavior of the modes involves a velocity which, with respect to the limit velocity in the medium (e.g. sound velocity in fluids or light velocity in dielectrics), is lower (subluminal case) or higher (superluminal case). Some differences occur, e.g. the Hawking mode originates from two short wavelength modes, far outside the black hole in the subluminal case, and far inside the black hole in the superluminal one (see e.g. [Corley (1998)]).
- One also has to consider the so-called sub-critical case: i.e. when there is no horizon but still there is a possibility of amplification (i.e. pair production). The opposite case is called trans-critical (also called super-critical), and the physical parameter which distinguishes between them in the case of fluid flows is the so-called Froude number [Michel and Parentani (2014)] i.e. the ratio $F := v/c$ between the fluid velocity and the velocity c of shallow waves. The best correspondence with black hole physics occurs when the trans-critical case is verified, i.e. when F crosses 1 (see e.g. [Euvé *et al.* (2015)] and references therein). This is of main concern for the case of white hole physics, which is the one occurring in actual experiments. The point is that analytical calculations display the existence in the sub-critical case of a critical frequency ω_{min} such that frequencies with $\omega < \omega_{min}$ are partially transmitted, and transmission becomes non-negligible. A sharp transition from transmission ($\omega < \omega_{min}$) to blocking ($\omega > \omega_{min}$) takes place in a restricted range of frequencies. Actually, numerical simulations show that transmission occurs irrespective of ω [Euvé *et al.* (2015)], with transmission which is more efficient for low frequencies. Only for suitable frequencies above ω_{min} blocking is effective. See below for the associated references, and see the last chapter for a discussion in relation with experiments.

This picture a priori could discourage the expert of the traditional Hawking effect with respect to the actual feasibility of the analogous Hawking radiation, because form one hand none of the above features is shared with standard Hawking radiation (with the exception of the dependence of the temperature on the frequency in the picture by Parikh and Wilczek discussed in chapter 4); on the other hand, in presence of dispersion, none of the standard strategies we discussed in the previous chapters applies unaltered to the physics at hand, but a much more involved scattering approach (at least from the point of view of analytical calculations) has to be set up. Still, despite the very limited number of participants in studies on dispersive analogue gravity, a series of robust results (also analytical) have been produced showing that the approach not only is viable, but also experimentally falsifiable, and some important experimental results promise to be on the right direction for revealing analogue Hawking radiation. See in particular the last chapter.

8.3 A sample model for dispersive fluids: Essentials

We take into account in this section the model which has been considered in analytical approaches to analogous Hawking radiation developed by S. Corley [Corley (1998)] and then improved in [Coutant *et al.* (2012b)]. The original paper by Corley represents the first analytical achievement after the numerical achievements in [Unruh (1995)] and [Corley and Jacobson (1996)]. What emerges, is that the surface gravity κ deduced at the level of the non-dispersive model, still plays a role also in presence of dispersion. Indeed, it enters in the correct way the temperature expression and, roughly speaking, it discriminates between particles in the spectrum participating to the Hawking process and particles which instead are involved in further scattering processes which can play a role in actual experiments. Indeed, particle frequencies $\omega \leq \kappa$ are mostly involved in the Hawking process, whereas particles with $\omega \gg \kappa$ participate to a distinguished scattering process which is non-Hawking.

The starting point is represented by the action which corresponds to the fluid model in absence of dispersion:

$$S_0 = \frac{1}{2} \int dtdx \left[((\partial_t + v\partial_x)\psi)^2 + \psi \partial_x^2 \psi \right], \tag{8.25}$$

where $v(x)$ is the fluid flow velocity, and which has still a full correspondence with the geometrical model. Dispersive effects are introduced by means of terms like

$$S_d = \frac{1}{2} \int dtdx \left[\pm \psi \frac{1}{k_0^2} \partial_x^4 \psi \right]^2, \tag{8.26}$$

whose analytical study is contained in [Corley (1998)]. The total action $S = S_0 + S_d$ can be summarized by

$$S = \frac{1}{2} \int dtdx \left[((\partial_t + v\partial_x)\psi)^2 + \psi G(\partial_x)\psi \right], \tag{8.27}$$

where $G(\partial_x)$ is a suitable function, e.g.

$$G_{\pm}(\partial_x) = \partial_x^2 \pm \frac{1}{k_0^2}\partial_x^4, \tag{8.28}$$

which correspond to subluminal (G_+) or superluminal (G_-) dispersion relations, in the sense that in the latter case high frequency modes are involved with superluminal propagation.[2] The further class of models

$$G^2(\partial_x) = \left(\partial_x + \frac{1}{2\Lambda^2}\partial_x^3\right)^2 \tag{8.29}$$

is mainly studied in [Coutant *et al.* (2012b)].

Equations of motion are

$$[(\partial_t + \partial_x v)(\partial_t + v\partial_x) - G(\partial_x)]\,\psi = 0, \tag{8.30}$$

By assuming a velocity profile $v(x)$ asymptotically constant and $\psi = e^{i(\omega t - kx)}$, one finds in the eikonal approximation

$$(\omega - vk)^2 = G^2(k), \tag{8.31}$$

which corresponds to the asymptotic dispersion relation (DR henceforth) for the class of models at hand. The last equation can be also be rewritten in terms of the frequency $\Omega = \omega - vk$ in the reference frame of the fluid, as $\Omega^2 = G^2(k)$.

There is a conserved inner product, which can be defined as

$$(H, K) = i\int dx(H^*(\partial_t + v\partial_x)K - K(\partial_t + v\partial_x)H^*), \tag{8.32}$$

and plays a fundamental role in the individuation of particles and antiparticles in the quantum theory, as it can be associated with a the notion of norm for the states.

The nondispersive limit is obtained as $k_0 \to \infty$ or as $\Lambda \to \infty$. The metric one finds is

$$ds^2 = dt^2 - (dx - v(x)dt)^2, \tag{8.33}$$

which displays an horizon for $v(x) = -1$. The profile for the velocity is chosen to be smooth, and such that the horizon occurs at $x = 0$. For example,

$$v(x) = -1 + \tanh(x). \tag{8.34}$$

[2]We use G in place of F appearing in [Corley (1998)] in order to deserve F only for the Froude number.

The profile is assumed to be such that, near the horizon, a linear approximation holds:

$$v(x) = -1 + \kappa x + O(x^2), \tag{8.35}$$

where κ is defined by

$$\kappa = \frac{dv}{dx}(0), \tag{8.36}$$

and is positive in the black hole case and negative in the white hole one. The region where the linear approximation holds true is also called linear region, and plays a relevant role in the dispersive case, where matching of WKB modes and near horizon ones is to be performed in the linear region (as far as both the approximations are legitimate in the region itself).

A further important role is deserved to the asymptotic value of $v(x)$: e.g. in the superluminal case [Coutant *et al.* (2012b)]

$$v(-\infty) := -1 - D_{as} \tag{8.37}$$

as D_{as} determines the maximal value ω_{max} of ω such that no flux of particles occurs for $\omega > \omega_{max}$. Indeed, the dispersion relation is such that no turning point occurs for $\omega > \omega_{max}$.

From the DR one can recover the presence of a turning point, to be identified with a group horizon, as follows. From the DR one can infer in the geometrical optics approximation

$$\omega - vk = G(k), \tag{8.38}$$

which amounts to the cubic equation

$$\frac{1}{2\Lambda^2}k^3 + (1 + v(x))k - \omega = 0 \Leftrightarrow H(\omega, k) = 0. \tag{8.39}$$

A turning point is found by imposing

$$H = 0, \qquad \partial_k H = 0, \tag{8.40}$$

i.e., together with (8.39), we have

$$\frac{3}{2\Lambda^2}k^2 + (1 + v(x)) = 0. \tag{8.41}$$

One finds

$$k = -(\omega\Lambda^2)^{1/3}, \tag{8.42}$$

and also

$$\kappa x_{gh} = -\frac{3}{2}\left(\frac{\omega}{\Lambda}\right)^{2/3}. \tag{8.43}$$

We sketch how analytical calculations are developed. There are fundamentally two steps:

- A WKB analysis is produced, which holds in the asymptotic region and also in part of the linear region, except for a neighborhood of the horizon $x = 0$. Exact solutions for a quartic equation (but also different dispersion relations can be allowed) are not in general known, and then WKB wave functions are obtained in the limit as ω is a perturbation with respect to k [Corley (1998); Coutant *et al.* (2012b)]. Then analytical approximation both for the phases and for the amplitudes can be found, also in the linear region.
- A near horizon approximation is developed in the following way. One considers the Fourier transform of the complete equation of motion, by deserving to the dispersive term the substitution $x \mapsto i\partial_k$. As we are in the linear region, only the first term in the Maclaurin expansion of $v(x)$ is considered [Corley (1998); Coutant *et al.* (2012b)]. Saddle point techniques and contour integrals around branch cut(s) are common tools in this calculation A single branch cut appears in the aforementioned calculations.
- Matching of the states obtained in the near horizon approximation and asymptotic states in the asymptotic DR is then considered. This allows to obtain the S-matrix for the complete scattering process. Again thermality can be identified by studying the number of negative norm states identified in the process, as discussed in the previous section [Corley (1998); Coutant *et al.* (2012b)].

We prefer to refer the reader to the original references, and will give an explicit example of such kind of calculations in the case of dispersive dielectric case, in Chapter 10.

8.4 Analogue gravity in BEC

The original proposal of analogue Hawking radiation was in stationary inviscid fluid. Nevertheless, it is not obvious to expect to be able discovering pure quantum effects by looking at a classical fluid. Moreover, the temperatures predicted in fluid analogue models remain so small that one would need to work at very low temperature to avoid the overwhelming presence of noise. A way out is to adopt quantum fluids in a regime where the fluidodynamical approximation works correctly.

A possibility is thus to consider Bose-Einstein condensates (BEC) [Pitaevskii and Stringari (2003)]. In this section, we limit ourselves to quote a seminal paper [Garay *et al.* (2000)]. Identical bosonic particles have the property that at very low temperatures are allowed to reach the fundamental state all together, and, in practice, a condensate of boson is get when the fraction of particles reaching the ground state among N remains positive in the thermodynamic limit $N \to \infty$. Thus, the condensate behaves essentially as a one particle in the sense that can be described as a whole in terms of a one particle wave function. If the particles system is sufficiently dilute, the interaction among particles does not depend on details but just on the scattering length a_s. The low density approximation is obtained when the condition $na_s^3 \ll 1$ is satisfied by the particle density function n. In this limit, atoms with mass m subject to an external potential $V(\vec{x},t)$ must satisfy the Gross-Pitaevskii equation

$$i\hbar\partial_t\psi(\vec{x},t) = \left[-\frac{\hbar^2}{2m}\Delta + V(\vec{x},t) + g|\psi(\vec{x},t)|^2\right]\psi(\vec{x},t), \tag{8.44}$$

where

$$g_s = \frac{4\pi\hbar^2 a_s}{m} \tag{8.45}$$

is the scattering coupling. This is a nonlinear Schrödinger equation, which, despite the diluite limit approximation, yet evidences a quantum behaviour and further considerations are necessary in order to distinguish a classical fluid background. In order to see this it is convenient to employ the Madelung representation of the wave function:

$$\psi(\vec{x},t) = n(\vec{x},t)^{\frac{1}{2}} e^{-i\frac{\phi(\vec{x},t)}{\hbar}}. \tag{8.46}$$

If we introduce the classical velocity field

$$\vec{v}(\vec{x},t) = -\frac{\vec{\nabla}\phi}{m} \tag{8.47}$$

then the Gross-Pitaevskii equation is equivalent to

$$\partial_t n + \mathrm{div}(n\vec{v}) = 0, \tag{8.48}$$

$$m\partial_t\vec{v} + \vec{\nabla}\left(\frac{m}{2}\vec{v}^2 + V(\vec{x},t) + g_s n - \frac{\hbar^2\Delta n^{\frac{1}{2}}}{2mn^{\frac{1}{2}}}\right) = \vec{0}. \tag{8.49}$$

The term

$$V_q(\vec{x},t) := -\frac{\hbar^2\Delta n^{\frac{1}{2}}}{2mn^{\frac{1}{2}}} \tag{8.50}$$

is a quantum correction to the potential acting on the fluid described by the velocity field $\vec{v}$. In a second quantization formalism, let us separate the real fields n and ϕ in a classical part and a quantum fluctuation

$$n = n_c + n_q, \qquad \phi = \phi_c + \phi_q, \tag{8.51}$$

where c stays for classic and q for quantum. The problem of the presence of the quantum correction to the potential is not much that it modifies the external potential, but the fact that it contribute to the linearized equation for the flux potential ϕ_q/m by an operatorial term. This operatorial term however becomes irrelevant for modes whose wavelengths are larger than the coherence length

$$\lambda \gg \lambda_c(\vec{x},t) = \frac{\hbar}{mc_s} \tag{8.52}$$

where

$$c_s(\vec{x},t) = \sqrt{\frac{g_s n(\vec{x},t)}{m}} \tag{8.53}$$

is the speed of sound. For wavelengths satisfying this condition the linearized quantum modes satisfy the equation

$$\partial_\mu \left(\sqrt{|g|} g^{\mu\nu} \partial_\nu \phi_q \right) = 0 \tag{8.54}$$

where the metric $g_{\mu\nu}$ is determined by the classical solution n_c, ϕ_c of (8.49), including the quantum potential correction, as

$$g_{00} = -\frac{n_c(\vec{x},t)}{mc_s(\vec{x},t)}(c_s^2(\vec{x},t) - \vec{v}_c^2(\vec{x},t)), \quad g_{0i} = g_{i0} = -\frac{n_c(\vec{x},t)}{mc_s(\vec{x},t)}\vec{v}_{ci}(\vec{x},t),$$
$$g_{ij} = \frac{n_c(\vec{x},t)}{mc_s(\vec{x},t)}\delta_{ij}. \tag{8.55}$$

This is the classical form of a dumbhole metric with horizon, thus allowing for analogue Hawking radiation. The main difference with respect to a classical fluid is that here the quantum modes are perfectly well defined from a physical point of view, and, moreover, a BEC fluid can work at very low temperatures.

It is important to notice that the external potential $V(\vec{x},t)$ has not only the aim of fixing the background black hole but also to ensure the stability of the system versus perturbations. Robustness in this direction can be obtained ensuring the absence of vortexes. For this reason the best experimental set up is obtained after confining the BEC fluid to an essentially one dimensional dynamics. Therefore, the external potential is usually the sum of two contributions, a part devoted to confine the system down to an almost one dimensional one, and a shaping part determining the black hole configuration.

8.5 Further readings concerning analogue gravity in presence of dispersion

In the following chapters, we shall focus exclusively on black hole analogues in dielectric media. Our choice is admittedly of limited extent, as it does not represent the mainstream of analogue gravity studies, which concern water, BEC, and so on. Nevertheless, the scope of the book is to provide the reader some aspects of analogue gravity systems, and our experience has been devoted to the case of dielectric systems, where our studies have been developed in the recent years. The following list of papers is far to be complete, still we think that the main ones are present. Works devoted to dispersive dielectrics will be referred to mainly in Chapter 10.

Apart from the papers quoted above, Unruh has given a series of relevant contributions to the field, often in collaboration with R. Schutzhold. Of primary interest is the paper [Weinfurtner *et al.* (2011)], concerning the measurement of Hawking radiation in an actual experiment involving water flow. We indicate [Schutzhold and Unruh (2013)], concerning the feasibility of WKB approximation in the near horizon region in dispersive models of sonic black holes. About Hawking radiation with the electromagnetic field, see [Schutzhold and Unruh (2005); Unruh and Schutzhold (2003); Linder *et al.* (2016)].

We also quote [Corley and Jacobson (1996)], concerning numerical simulations in dispersive models, and the seminal paper about black hole lasers [Corley and Jacobson (1999)], which is involved in one of the most promising measurements of the Hawking effect. We don't delve into this important topic, referring the reader to the existing literature.

A series of further important contributions is associated with Renaud Parentani and his coworkers, including some works to which he coworked in the following list. We refer the reader to a series of works concerning dispersive models of acoustic black holes and white holes for fluid models and for the case of Bose-Einstein condensates (BEC). We begin our list with [Balbinot *et al.* (2005)], and quote [Macher and Parentani (2009a,b); Coutant and Parentani (2010); Mayoral *et al.* (2011); Finazzi and Parentani (2011); Coutant *et al.* (2012b); Zapata *et al.* (2011); Finazzi and Parentani (2012); Coutant *et al.* (2012a); Coutant and Parentani (2014b); Michel and Parentani (2013, 2015b); Aurégan *et al.* (2015); Robertson and Parentani (2015)], where various aspects of black hole and white hole radiation in various analogous systems (fluids, BEC, gas) and black hole lasers are discussed. Even the interesting topic represented by undulation is taken into account. A further series of studies is devoted to the problem of sub-critical and trans-critical flows, also in relation with the Weinfurtner et al. experiment quoted above [Michel and Parentani (2014); Euvé *et al.* (2015); Michel and Parentani (2015a); Euvé *et al.* (2016); Robertson *et al.* (2016)].

Another relevant series of paper is to be indicated. With reference to subcritical flows, see [Coutant and Weinfurtner (2016)]. We also address to [Coutant and Weinfurtner (2018b,a)]. Superradiant scattering is studied in [Richartz *et al.* (2015)]. Measurements of supperradiance are discussed in [Cardoso *et al.* (2016)] and in [Torres *et al.* (2017)]. Other interesting aspects of analogue gravity systems are discussed in [Bruschi *et al.* (2013)]. See also [Faccio and Wright (2017a,b)].

Early analytical calculations appears in [Corley (1998); Saida and Sakagami (2000); Himemoto and Tanaka (2000)], and have been developed in [Coutant *et al.* (2012b)]. See also [Leonhardt and Robertson (2012); Robertson (2012); Philbin (2016)]. A first experiment with water waves was reported in [Rousseaux *et al.* (2008)], and further discussion on horizon effects is found in [Rousseaux *et al.* (2010)]. A series of important studies involving black holes and white hole in BEC involves [Balbinot *et al.* (2008); Carusotto *et al.* (2008); Recati *et al.* (2009); Balbinot *et al.* (2010); Mayoral *et al.* (2011); Larre *et al.* (2012)]. Polariton superfluids are discussed in [Gerace and Carusotto (2012)]. As a further reference we also indicate the book [Faccio *et al.* (2013)].

Chapter 9

Hawking radiation in a non-dispersive nonlinear Kerr dielectric

The aim of this chapter is to show how event horizons can appear in certain nonlinear optical systems and how this can be employed for measuring analogue Hawking radiation. The simplest way in order to understand such a phenomenon is to suppose that the light propagates in a very weakly dispersive medium in such a way that dispersion can be ignored at least at a first stage. However, in optics, dispersion is always relevant and almost non dispersive regimes can be eventually observed only at determinate frequency ranges for a given material, so that it is practically impossible to ignore dispersion effects. For this reason, despite the title, we will not ignore dispersion in the present chapter, but we will include it generically in a naive way, t.i. by putting it at hand in the equations previously deduced for an hypothetical dispersionless material, demanding a full treatment of the Hawking radiation in dispersive dielectrics to the next chapter. Surely, this procedure can be expected to be too much ingenuous, since one could expect that the presence of nontrivial dispersion relations may affect the physical processes at the basis of the investigated phenomenon, a problem that is common to all analogue gravity systems. For example, as we will see, the same concept of horizon will be not so clear in a dispersive dielectric.

However, there are at least two reasons for considering such a strategy: first, it would be very hard to understand the analogy directly by including the dispersion *ab initio*. As we will see, there are important differences w.r.t. the non dispersive case, so that without a "transition" from non dispersive to dispersive it would be impossible to appreciate where the roots of Hawking radiations manifest themselves in the analogue phenomenon. For these reasons one needs to understand what really characterizes the process in a way that it can be recognized also in presence of dispersion.

The second reason is that, as we will see, the naive approach keep all physical reasoning at a quite elementary level but provides predictions that seem not so far from what one observes experimentally.

More precisely, here will be considered the so called non linear Kerr media, in which a soliton with intensity I generates through the nonlinear Kerr effect a refractive index perturbation (RIP), $\delta n = n_2 I$, where n_2 is the Kerr index. The original idea to obtain a black hole in a dielectric medium tanks to a traveling dielectric perturbation generated by means of the Kerr effect is due to [Philbin *et al.* (2008)]. Similarly to the acoustic analogy, it will shown that the RIP modifies the spacetime geometry as seen by co-propagating light rays so that, if the RIP is locally superluminal, i.e. if it travels faster than the phase velocity of light in the medium, a trapping horizon is formed and Hawking radiation is to be expected. In general, the RIP generates a stationary (but in general non static) metric, so that, under certain limits, an ergoregion occurs.

We will then concentrate on the case of a RIP propagating with constant velocity v in the x direction, which will provide a static black hole configuration with an horizon infinitely extended in the transverse y and z directions. In this case, the Hawking radiation can be inferred even without recourse to the analogous model characterized by a curved spacetime geometry. Indeed, since Hawking radiation is not a current phenomenon in nonlinear optics, if it occurs, it must have a description also in terms of the standard tools of quantum electrodynamics, without any reference to the geometrical picture. Next, we will discuss the effects of dispersion in the spirit described above. This will be done in a two dimensional reduction of the model. This will allow to discuss the concepts of phase velocity horizons and group velocity horizons, which can both play a relevant role in the physical process. Moreover, it will be shown that, whereas in the dispersionless case the Hawking spectrum is expect to be a blackbody spectrum, in a dispersive medium it is expected the excitation of a limited spectral region, dependent on v, so that one cannot more think to be able to recognize the shape of a blackbody spectrum.

9.1 Classical analysis of the effective spacetime geometry

Consider a dielectric at rest, say, in an inertial lab. We want to provide a covariant description of it in order to be able to describe all our results in any inertial frame. Let v^μ the four-velocity of the dielectric and $E^\mu :=$

$v_\mu F^{\mu\nu}$ the covariant electric field. We will consider a dielectric medium for which the permittivity and the permeability have the form

$$\epsilon^{\alpha\beta} = \varepsilon(E)(\eta^{\alpha\beta} - v^\alpha v^\beta), \tag{9.1}$$

$$\mu^{\alpha\beta} = \mu_0\mu_r(\eta^{\alpha\beta} - v^\alpha v^\beta), \tag{9.2}$$

respectively. Here $E := \sqrt{-E^\mu E_\mu}$ is the "invariant electric field amplitude", that is the modulus of the electric field measured in the laboratory frame. These expressions are equivalent to say that the permittivity is a function of the invariant electric field alone. In other words, in the frame where the dielectric is at rest only a dielectric field induces a polarization and any nonlinear effects depend exclusively by the intensity of the latter. The magnetic permeability is assumed to be constant and, as usual, is described by the product μ of the vacuum permeability μ_0 and the relative permeability μ_r of the medium. Notice that (9.1) and (9.2) hold true for any nonlinear, homogeneous dielectric material, no matter how the strong, background pulse is generated. These approximations are justified by the fact that these conditions are fulfilled by a good number nonlinear media. In particular we will consider a Kerr medium, which is an isotropic medium whose permittivity has the form

$$\varepsilon = \varepsilon_0\varepsilon_r = \varepsilon_0(1 + \chi^{(1)} + \chi^{(3)}E^2), \tag{9.3}$$

where ε_0 is the vacuum permittivity. For these reasons, these materials are also called $\chi^{(3)}$ dielectrics, to be compared with the $\chi^{(2)}$ materials whose permittivity is linear in the dielectric field. Notice, that a $\chi^{(2)}$ material cannot be isotropic if the $\chi^{(2)}$ coefficient is non null. Usually, the nonlinear contribution in a Kerr medium is very small, so that very strong fields are necessary in order to make the non linear effects exploitable. Thus, we can think to separate the electric field into two components: a very strong electric background field, which fixes the electromagnetic properties of the dielectric, and a weak electromagnetic field which is modulated independently and that takes the role of a probe for the background geometry. We will be interested in the situation when the strong background field as a spatially localized pulse, which is traveling through the medium with constant velocity. We can consider the weak field as small rapidly oscillating fluctuations of the electromagnetic field on top of the strong background. In such situation it has been proved in [De Lorenci and Klippert (2002); Novello *et al.* (2003); Novello and Bergliaffa (2003)] that the small fluctuations feel an effective curved spacetime with a "bimetric", in the sense that in the limit of geometrical optics there appear two classes of rays, the ordinary and the extra-ordinary ones, according to the polarization of the background field, similarly to what happens in birefringent media (uniaxial crystals). These rays appear to move in a curved background with different metrics depending on the polarization of the ray w.r.t. the one of the background. The analogue metrics deduced in [De Lorenci and Klippert (2002); Novello *et al.* (2003); Novello and Bergliaffa (2003)], are of Gordon type[1] [Gordon (1923)]:

$$g_{\mu\nu}^{(o)} = \eta_{\mu\nu} - \frac{1}{c^2} v_\mu v_\nu \left[1 - \frac{1}{n^2}\right] \tag{9.4}$$

for the ordinary rays, and

$$g_{\mu\nu}^{(eo)} = \eta_{\mu\nu} - \frac{1}{c^2} v_\mu v_\nu \left[1 - \frac{1}{n^2(1+\xi)}\right] + \frac{\xi}{\xi+1} p_\mu p_\nu \tag{9.5}$$

for the extra-ordinary ones, where $\xi := \frac{E}{\varepsilon}\frac{d\varepsilon}{dE}$ is the it nonlinear parameter, $p_\mu := \frac{E_\mu}{E}$ is the *background field polarization* four-vector, and $n := \sqrt{\varepsilon_r \mu_r}$ is the refractive index. $g^{(o)}$ is the standard Gordon form [Gordon (1923)]. The essential difference between the two metrics is in the fact that for the extraordinary metric the refracting index is $n_{eo} = n\sqrt{(1+\xi)}$. If

$$n_\ell = \sqrt{\mu_r \varepsilon_r} = \sqrt{\mu_r \left(1 + \chi^{(1)}\right)} \tag{9.6}$$

is the refractive index at the linear regime, we see that

$$n = \sqrt{\mu_r \varepsilon_r} \simeq n_0 + \frac{1}{2} \frac{\sqrt{\mu_r} \chi^{(3)} E^2}{(1+\chi^{(1)})^{\frac{1}{2}}} =: n_0 + \delta n, \tag{9.7}$$

whereas

$$n_{eo} = \sqrt{\mu_r \varepsilon_r (1+\xi)} = n_0 + \frac{3}{2} \frac{\sqrt{\mu_r} \chi^{(3)} E^2}{(1+\chi^{(1)})^{\frac{1}{2}}} = n_0 + 3\delta n, \tag{9.8}$$

this meaning that the extraordinary rays experience a disturbance of the refractive index three times stronger than the ordinary ones.

Now, we will investigate some classical properties of the Gordon metric, with particular interest in null geodesics describing the trajectories of monochromatic waves, which we will call *wave rays*. It is important to mention that these have not to be identified with the light rays and this is the reason we have deserved here a different name to them. Indeed, the light rays describe the trajectories of the photons and not just of monochromatic waves. Congruences of null geodesics describe constant phase surfaces.

[1] We restore c henceforth.

9.1.1 *Wave rays geometry in the RIP frame*

Consider a localized electromagnetic pulse that, in the lab frame, moves along the positive x axis with uniform velocity v. Assumed to be intense enough, it will generate a localized refractive index perturbation (RIP) with the same shape as the intensity profile, moving accordingly to the pulse. We will assume cylindrical symmetry around the propagation axis. As a consequence, the RIP can be described as a function of $x - vt$ and of the transversal coordinate $\rho = \sqrt{y^2 + z^2}$. In the laboratory frame we can describe the geometry experienced by weak electromagnetic propagating fields in terms of the Gordon metric with $n = n_0 + \delta n(x - vt, \rho)$. It is however clear that in a frame comoving with the RIP the metric must appear explicitly stationary. Indeed, in this frame, with coordinates (t', x', ρ, ϕ), the Gordon metric takes the form

$$
\begin{aligned}
ds^2 =& \gamma^2 \frac{c^2}{n^2} \left(1 + n\frac{v}{c}\right) \left(1 - n\frac{v}{c}\right) dt'^2 + 2\gamma^2 \frac{v}{n^2} \left(1 - n^2\right) dt' dx' \\
&- \gamma^2 \left(1 + \frac{v}{nc}\right) \left(1 - \frac{v}{nc}\right) dx'^2 - d\rho^2 - \rho^2 d\phi^2,
\end{aligned}
\tag{9.9}
$$

with

$$
n := n_0 + \delta n(x', \rho). \tag{9.10}
$$

We assume that the function δn, describing the shape of the ring, is a smooth function over $\mathbb{R} \times \mathbb{R}_{\geq}$, rapidly decaying at infinity and with a single maximum of height η, say, in $(0, 0)$. The symmetries of this metric are evident, since, obviously, the vector fields $\partial_{t'}$ and ∂_ϕ are both Killing vector fields for the entire class of metrics. A specific metric in this class is selected by a specific choice for the function δn. It is worth to notice that the isotropy of the refractive index is broken by the velocity v: this is made evident in the comoving frame where the contraction of lengths in the x is generated by the boost w.r.t. the lab frame. We also point out that if there is an explicit dependence of the RIP shape on ρ, then the metric is stationary but not static. Indeed, let $t_\mu \equiv (g_{00}, g_{01}, 0, 0)$ be the covariant components of the timelike Killing vector $\boldsymbol{t} = \partial_{t'}$. The Frobenius's theorem (see e.g. [Wald (1984)], App. B) states that the vector field $\boldsymbol{t}$ is surface orthogonal (t.i. defines a foliation of hypersurfaces orthogonal to it) if and only if it satisfies the equation

$$
t_{[\mu} \nabla_\nu t_{\sigma]} = 0. \tag{9.11}
$$

But in our case, for $\mu = 0$, $\nu = 1$, we have $t_0(\nabla_1 t_\sigma - \nabla_\sigma t_1) + t_1(\nabla_\sigma t_0 - \nabla_0 t_\sigma) = 0$, i.e., for $\sigma = \rho$, $g_{00}(-\partial_\rho g_{01}) + g_{01}\partial_\rho g_{00} = 0$, which is not satisfied unless the RIP is independent from ρ. If interested, the reader could verify that this failing is equivalent to say that it is not possible to change coordinates in such the way that the new metric coefficients continue to be independent on the (new) time coordinate and that moreover $g_{0i} = 0$ for $i \neq 0$. Indeed, this is an informal way to defining stationary and static metrics: a stationary metric is such that it exists a set of coordinates x^μ such that the coefficients of the metric do not depend on x^0, whereas the metric is stationary if, moreover, g_{0i}, or, equivalently, it is invariant under time inversion.

For the moment we concentrate on the general case of stationary, non-static metrics. We will consider the particular static case more extensively later. In general, stationary metrics allows not only for horizons but also also *ergoregions*, which are regions delimited by ergosurfaces, which is the set of points where the timelike Killing vectors become null (it becomes spacelike in the ergoregion). In stationary coordinates, the ergosurfaces are thus described by the equation

$$g_{00} = 0. \tag{9.12}$$

In our case it is

$$1 - n\frac{v}{c} = 0. \tag{9.13}$$

Not all the metrics in the above class admit an ergoregion, since this equation admits a solution only for a RIP whose velocity v satisfies the condition

$$\frac{1}{n_0 + \eta} \leq \frac{v}{c} < \frac{1}{n_0}. \tag{9.14}$$

Notice that the Killing vector becomes null also at an event horizon, however, in the case of an ergosurface the surface itself is not a null one as can be seen by looking at its normal vector. Then, in general and ergosurface does not coincide with an horizon. Notice that for $\rho = 0$ the ergoregion for a RIP becomes lightlike so it meets an event horizon, if there, in the antipodal points which solve the equation

$$\delta n(x', 0) = \frac{c - n_0 v}{v}. \tag{9.15}$$

In general, it is not equivalently simple to identify a nontrivial event horizon in effective geometries.

9.1.2 *Horizons in effective geometries*

From the general theory of black holes we know that there is a rigorous and quite abstract definition of event horizon in a Lorentzian spacetime. In an analogue system, thanks to the existence of a global time associated to the real flat Minkowski metric η, there is an equivalent but more practical and intuitive definition, which we will employ here. Obviously we continue to assume the medium to be dispersionless. Analogue spacetime in general contain both white and black horizons. For definiteness, following [Cacciatori *et al.* (2010)] we will work with the white horizon, all the constructions for the black case being completely similar. To simplify the discussion we will work in the laboratory frame, where we recall that the RIP is moving in the positive x direction with velocity v, and will refer to Fig. 9.1.

We start by looking at the ergosurface, which from (9.13) consists in all the points where the local velocity of the light coincides with the velocity of the RIP. For simplicity we will assume that that at any fixed time t_0

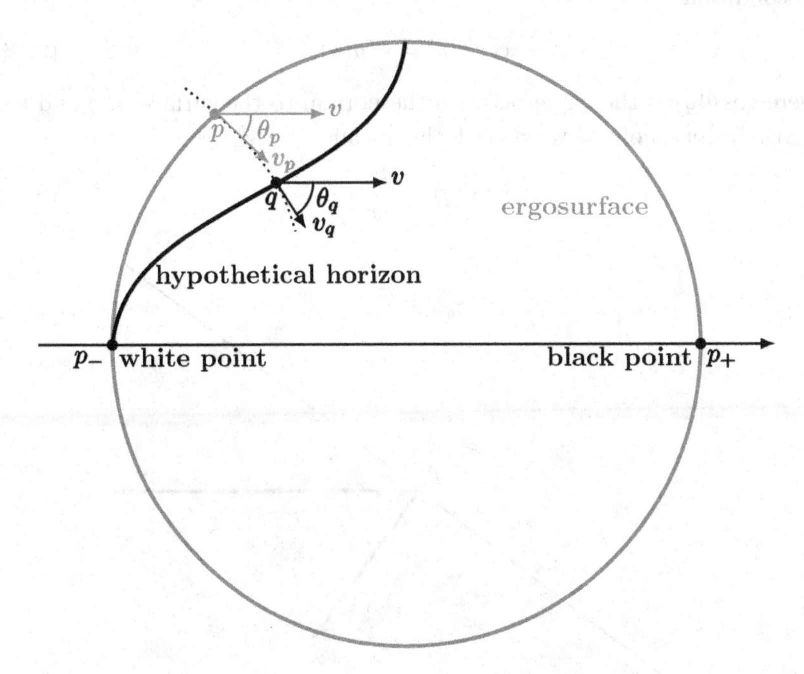

Fig. 9.1 Picture of the ergosphere and of the hypothetical horizon at a fixed point of time in the laboratory frame. The small segments at p and q represent infinitesimal portions of null geodesics (light rays) passing, respectively, through P and Q at that given point of time.

the ergosurface (intersected with the hyperplane $t =_0$) has a convex shape. Consider a light ray which is running after the RIP in the x direction and that tends to reach the point p on the ergosurface. If the plane tangent to p (in a given instant) forms an angle θ with respect to the plane $y - z$, then, an observer in p at rest in the laboratory will see the tangent plane move away with velocity $v_\theta = v\cos\theta$, see Fig. 9.2. It is thus clear that if the wave is moving in the right direction it can cross the ergosurface in p and enter the ergoregion, being $v_\theta < c/n(p)$.

Thus, with reference to Fig. 9.1, only the point p_- of the ergosurface belongs on the white hole horizon, whereas p_+ is the only point of the black hole horizon which intersects the ergosphere. This means that an eventual white hole horizon must lie inside the ergosurface. Assuming axial symmetry as above, let us suppose that the eventual white hole horizon is described (at least locally) by an equation of the type $\rho = h(x - vt)$. Then, at any point q (where $n(q) > n(p) = c/v$) of this surface it must be satisfied the condition

$$v\cos\theta(q) = c/n(q), \tag{9.16}$$

where $\cos\theta(q)$ is the angle between the normal to the surface in q and the x axis. In terms of the function h this means

$$\cot\theta(q) = \frac{dh}{dx}(q), \tag{9.17}$$

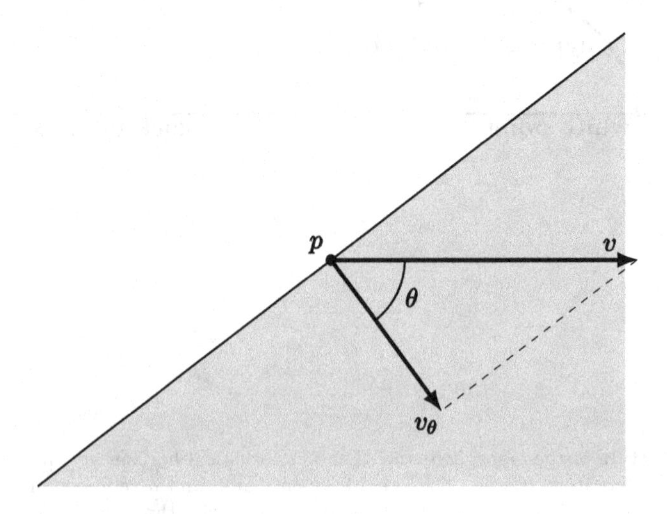

Fig. 9.2 Normal velocity v_θ compared to the pulse velocity, v.

and then

$$\frac{dh}{dx} = \pm \frac{1}{\sqrt{\frac{v^2}{c^2}n^2 - 1}}. \tag{9.18}$$

If we now move to the comoving frame, using coordinates (x', ρ), for the horizon we obtain the equation

$$\frac{dh}{dx'} = \pm \frac{\sqrt{1 - \frac{v^2}{c^2}}}{\sqrt{\frac{v^2}{c^2}n^2 - 1}}. \tag{9.19}$$

We can obtain this equation in a global form starting from (9.13). Setting $f = 1 - n\frac{v}{c}$, we see that the ergosurface is the 0 level-set of the function f. In the same way we can expect the horizon to be described as the 0 level-set of a function G, so that $G = 0$ is an implicit equation that determine the function h locally and the horizon globally. Since the horizon has to be a lightlike surface, it must satisfy the equation $g^{\mu\nu}(\partial_\mu G)(\partial_\nu G) = 0$. We work directly in the comoving frame where staticity is manifest: $\partial_{t'} G{=}0$. Moreover, we impose the cylindrical symmetry so that the imposition for the normal to the event horizon to be lightlike is

$$\frac{1}{1 - \frac{v^2}{c^2}} \left(1 - n^2(x', \rho)\frac{v^2}{c^2}\right)(\partial_{x'} G)^2 + (\partial_\rho G)^2 = 0. \tag{9.20}$$

This equation has a simple solution in the static case. Indeed, in this case $(\partial_\rho G)^2 = 0$, and the horizon consists of two planes parallel to the $y - z$-plane, located in the points $p_\pm$, where $1 - n^2(x')\frac{v^2}{c^2} = 0$, $\rho = 0$. In the general case it is not easy to determine a solution of this equation. Since a necessary condition is $1 - n^2(x, \rho)\frac{v^2}{c^2} \leq 0$, we see that the horizon must be contained inside the ergosurface. Locally the implicit equation must be equivalent to $\rho - h(x') = 0$, and we then we can assume $F(x', \rho) = \rho - h(x')$ so that equation (9.20) becomes $g^{11}(\frac{dh}{dx'})^2 + g^{22} = 0$. This is equivalent to equation (9.19). This form does not simplify too much the problem of finding explicit solutions. The fact that the antipodal points belong to the horizons suggest that they could be taken as starting points for the Cauchy problems. The solution should thus be expected to extend inside the ergoregion. However, it can be argued that generically we do not expect for such solution to exist and that in general horizons for such model are expected to be generated by points of the ergosurface whose tangent plane is orthogonal to the x direction. We will not discuss further this technical point here.

Thus, for spheroidal ergosurfaces the "horizons" reduce just to the antipodal point, let us call them the white point p_- and the black point p_+. It is clear that they cannot be strictly considered as event horizons since they have not a normal vector unless they belong on a flattened spheroid, see Fig. 9.3.

We will generically interested in "true" horizons, but it is natural to wonder whether a "trapping point" still emits analogue Hawking radiation. We will see in the next section that the isolated horizon points are attractors, a phenomenon which support the thesis for which they behave as standard horizons and emit particles with a thermal spectrum. However, since we will interested in working with extended horizons, we will not elaborate this question further.

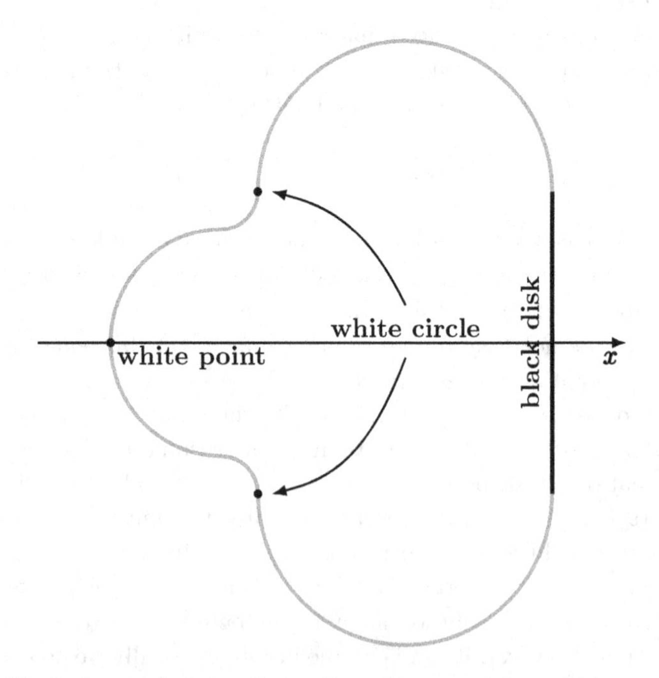

Fig. 9.3 The horizon surfaces are the portions of an ergosurface, which are orthogonal to the direction of motion. Assuming circular symmetry around the x axis, we recognize a disk representing a black horizon, a circle representing a white horizon, and a white point.

9.1.3 *Null geodesics in dispersionless and in dispersive dielectrics*

In order to improve our confidence with analogue horizons, before specializing to static configurations, it is convenient to proceed studying the behavior of null geodesics by means of some numerical analysis. To this end we prefer to work in the laboratory frame where the Gordon metric has the simplest form

$$ds^2 = \frac{c^2}{n^2(x - vt, y, z)} dt^2 - dx^2 - dy^2 - dz^2. \tag{9.21}$$

For the numerical simulations we will consider a gaussian RIP propagating with velocity v along the x axis, say

$$n = n_0 + \delta n, \tag{9.22}$$

$$\delta n = \eta e^{\frac{-y^2 - (x-vt)^2}{\sigma^2}}. \tag{9.23}$$

The amplitude η and the largeness σ are assumed to be constant so that the pulse do not change it shape during the motion (even though this can happen in realistic experiments). Employing the cylindrical symmetry, we can reduce to consider only one transversal direction. Forgetting, for example, the z direction, the geodesic equations for null rays take the form

$$\frac{d^2x}{d\lambda^2} = -\frac{2}{\sigma^2}\frac{\delta n}{n^3}(x - vt)c^2\left(\frac{dt}{d\lambda}\right)^2, \tag{9.24}$$

$$\frac{d^2y}{d\lambda^2} = -\frac{2}{\sigma^2}\frac{\delta n}{n^3}yc^2\left(\frac{dt}{d\lambda}\right)^2, \tag{9.25}$$

$$\frac{d^2t}{d\lambda^2} = \frac{2}{\sigma^2}\left[\frac{\delta n}{n^2}v(x - vt)\left(\frac{dt}{d\lambda}\right)^2 - y\frac{dy}{d\lambda}\frac{dt}{d\lambda} - (x - vt)\frac{dx}{d\lambda}\frac{dt}{d\lambda}\right], \tag{9.26}$$

where λ is an affine parameter. There is an obvious solution of this system of equations, which describes a ray along the x axis, described by

$$\frac{dx}{dt} = 1/n, \qquad y(t) = 0. \tag{9.27}$$

For less trivial solutions it is necessary to resort to numerical analysis. This can be done, for example, by using Matlab's ode45 function. See also Fig. 9.4.

9.1.3.1 *Trapping of light*

We first observe that we can normalize the affine parameter so that the four-velocity of the geodesic coincide with the wave vector:

$$(c\dot{t}, \dot{x}, \dot{y}, \dot{z}) = (k^0, k^1, k^2, k^3) = (n^2\omega, k_x, k_y, k_z). \tag{9.28}$$

It is worth to notice that this way ω is identified as the zero-th component of the covariant four-momentum k^{μ}, which corresponds to measuring frequencies as the inverse of the lapse of coordinate time between two successive crests. In other words, the frequency is measured in terms of the Minkowskian time as measured in the rest frame of the laboratory. This is relevant since experimental measurements are done in the laboratory referring to the Minkowski time and not with respect to the proper curved time $d\tau = ds$.

We can determine the variation of the frequency ω along the geodetic simply by deriving the equality $\frac{dt}{d\lambda} = n^2\omega$, w.r.t. λ. We get

$$\frac{d\omega}{d\lambda} = \frac{1}{n^2}\left(\frac{d^2t}{d\lambda^2} - 2\omega n\frac{d\delta n}{d\lambda}\right). \tag{9.29}$$

We assume to work with fused silica, a material which has been adopted for the experiment we will discuss later, in which a laser beam is injected with initial wavelength $\lambda_{in} = 527$ nm. In order to simplify the qualitative

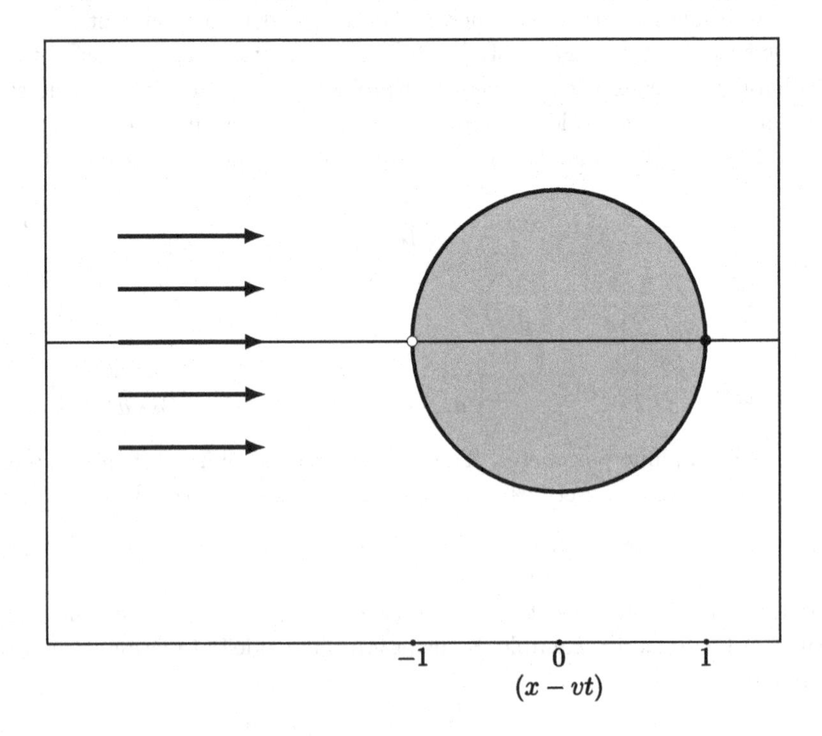

Fig. 9.4 Sketch showing the initial conditions of the simulation.

analysis we consider a RIP with amplitude $\eta = 0.1$ and largeness $\sigma = 1m$. This is quite unrealistic, since in a laboratory one can produce pulses of largeness in the range $\sigma \sim 1 - 100\ \mu m$. However, the qualitative behavior of the geodesics is left invariant by the choice of σ and the simulations are simpler with large values.

In order to check the predicted presence of stopping points, the velocity v of the RIP, which is right moving, varies in a range depicted in Fig. 9.5.

The rays solving the geodesic equations start away from the RIP with initial conditions

$$\left(t_{in}, x_{in}, y_{in}, \frac{dt_{in}}{d\lambda}, \frac{dx_{in}}{d\lambda}, \frac{dy_{in}}{d\lambda}\right) = (t_{in}, x_{in}, y_{in}, \dot{t}_{in}, \dot{x}_{in}, 0),$$

where the constraints $\dot{x}_{in}/\dot{t}_{in} = 1/n_0$ and $\dot{t}_{in} = n_0^2 \omega_{in}$.

With these data, the numerical simulation shows that there exists a range of values for the velocities for which the rays are trapped near the

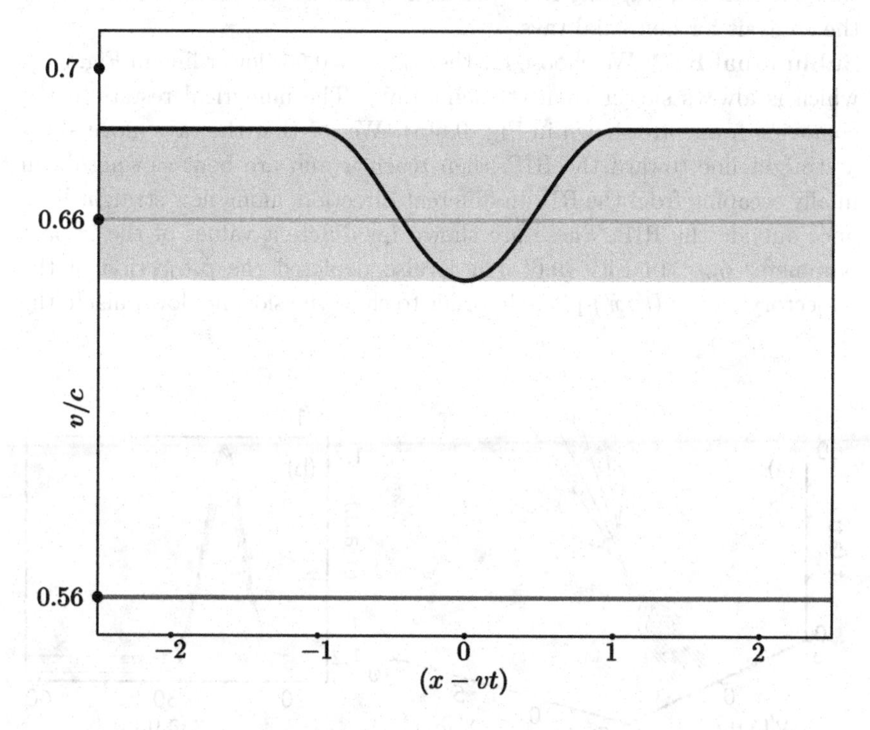

Fig. 9.5 Light ray phase velocity (c/n). The shaded region highlights the range of velocities of the RIP for which the rays are trapped. The two straight lines indicate the values of v/c used in the simulations.

pulse, exactly as predicted by equation (9.14). For higher values of v the ray never reaches the pulse, whereas for values of v below the range the rays which enter the pulse are bent and then escape. This is clearly illustrated in Fig. 9.5: the black curve is the velocity of a ray which moves along the x axis, versus $x - vt$. The horizontal straight lines represent the constant velocity of the RIP. In the case of the lower line v is too small for generating a trapping region: the velocity of a ray is slowed down when enters the RIP, but not enough for being stopped (relatively to the RIP itself). So, the ray passes beyond the RIP and goes over. In this case we say that the RIP is a *subluminal RIP*. In the case of the upper line, initially the ray velocity is larger than v, but when the ray reaches the RIP it slow down until, asymptotically stops at the first intersection point between the upper straight line and the black curve, where the velocity of the ray equals v. This looks as a trapping point at least for axial rays. In this case we say that the RIP is a *trapping RIP*. Let us now discuss the results of numerical the analysis for non axial rays.

Subluminal RIP: We choose for the RIP $v = 0.56$ (lower line in Fig. 9.5), which is always slower than the light rays. The numerical results in the comoving frame are shown in Fig. 9.6(a). We see that the rays move along a straight line toward the RIP, then reach it and are bent slowing down finally escaping from the RIP in different direction, along new straight lines once outside the RIP. These are shown for different values of the impact parameter y_{in}. In Fig. 9.6(a) it is also depicted the projection of the trajectory on the (t', x')-plane in order to show the slowing down inside the

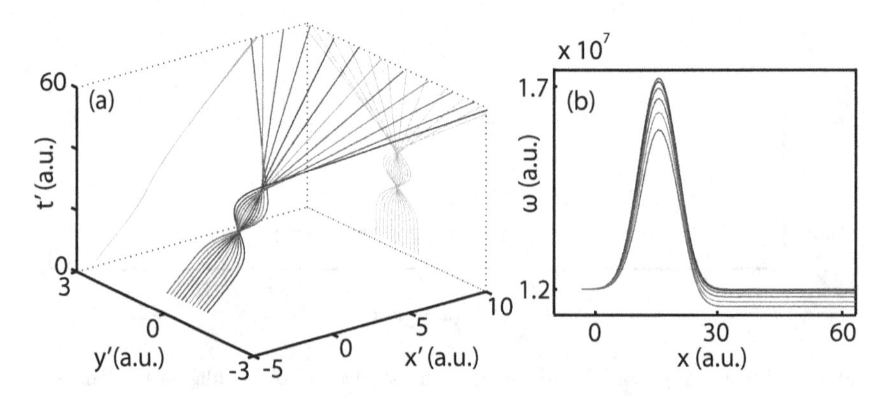

Fig. 9.6 Numerical results in the comoving reference frame (a) and frequency evolution in the laboratory reference frame (b) for the subluminal case.

RIP. In Fig. 9.6(b) we can read the evolution of the frequency ω along the x direction in the laboratory reference frame. This shows a blueshift when the rays move inside the RIP, and a Doppler redshift when the rays emerge from the RIP after being bent in different directions.

Trapping RIP: The upper line in Fig. 9.5 corresponds to a velocity $v = 0.66c$ for the RIP. The corresponding results for the numerical simulation in the laboratory frame is shown in Fig. 9.7(a). The rays move along straight lines until they reach the RIP. Then, they start slowing down and are bent until at a certain point all they acquire the same velocity as the RIP and can no longer escape from it. In the comoving frame the rays appear to slow down continuously, while oscillating in the transverse direction, until stop asymptotically at the point p_-, the white point. Indeed, we see that the slope of the projection $t' = t'(x')$ correspondingly diverges. In the laboratory an observer would see the rays reach asymptotically the point p_- which is moving with constant velocity v. In Fig. 9.7(b) it is shown the change of frequency of the ray, which experiences an infinite blue shift, as expected.

Despite a numerical analysis has been necessary, all these results are not far from intuition and can be inferred, at least qualitatively, by looking carefully at the equations if we neglect dispersion. In the dispersive case the numerical analysis becomes unavoidable. We include here the results of modifying the geodesic equations after including corrections due to non-trivial dispersion. As we will see, the trapping phenomenon is not canceled by the presence of dispersion.

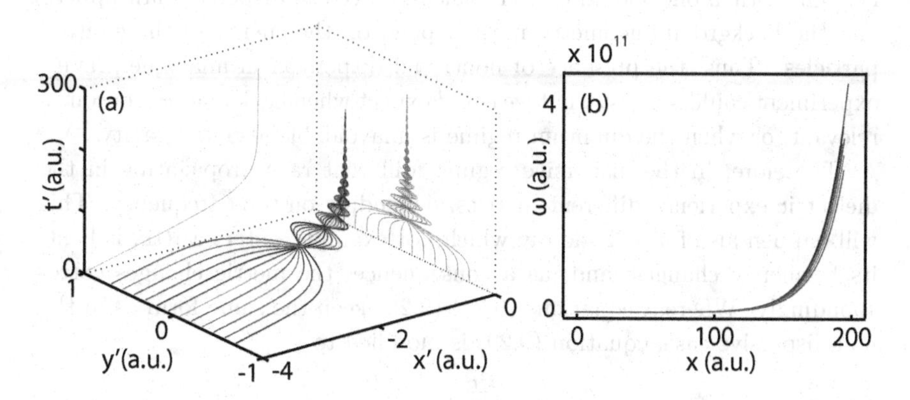

Fig. 9.7 Numerical results in the comoving reference frame (a) and frequency evolution in the laboratory reference frame (b) for the trapping case.

9.1.4 *Trapping in dispersive regime*

Up to now we have supposed to work with a dispersionless dielectric material, which results in assuming that refractive index has the form $n(\mathbf{r}, t)$ independently on the frequency. This is not much realistic since, in general, the refractive index of a medium is a quite complicated function of the frequency ω of the incident wave, so that we should write $n = n(\mathbf{r}, t, \omega)$. When the refractive index grows with frequency, one speaks about *normal dispersion*, which is the case we will consider. For such a material dependence of the n from ω is phenomenologically described by the Sellmeier equation. By using that $ck = \omega n$, in general it is convenient (and, indeed, standard) to expand n as a function of ω around a fixed value ω_0 as

$$n(\omega) = \frac{c}{\omega}\left(k(\omega_0) + \sum_{n=1}^{\infty} \frac{1}{n!}\frac{d^n k}{d\omega^n}\bigg|_{\omega_0} (\omega - \omega_0)^n\right). \tag{9.30}$$

A very interesting point for us is that also the metric is affected by the properties of the material, so that, through n, it will depend on the frequency of the probing ray:

$$ds^2 = \frac{c^2}{n^2(\mathbf{r}, t, \omega)} dt^2 - d\mathbf{r}^2. \tag{9.31}$$

At a first sight this can look strange and not favorable for analogue gravity. But, at the opposite, this is exactly one of the advantages of analog gravity: for evaporating black holes at the very late stage one expects that the back reaction due to the emission of particle becomes relevant and, in particular, when a regime when the quantum gravity effects (if present) are no more negligible, then one should expect that a non trivial dispersion law appear and the background geometry must depend on the energy of the emitted particles. Thus, the presence of nontrivial dispersion in analogue gravity experiment could suggest what we could expect when back reaction becomes relevant, or when the quantum regime is unavoidable also for gravity.

Therefore, in the dispersive regime, different rays propagating in the dielectric experience different metrics, depending on their frequency. This will happen also for a single ray which, for example, enters a RIP, is bent, its frequency changes, and, as a consequence, the metric changes correspondingly. Whereas equations (9.24)-(9.26) keep the same form as in the non-dispersive case, equation (9.29) is modified to

$$\frac{d\omega}{d\lambda} = \frac{\frac{d^2 t}{d\lambda^2} - 2\omega n \frac{d\delta n}{d\lambda}}{n(n + 2\omega dn_0/d\omega)}. \tag{9.32}$$

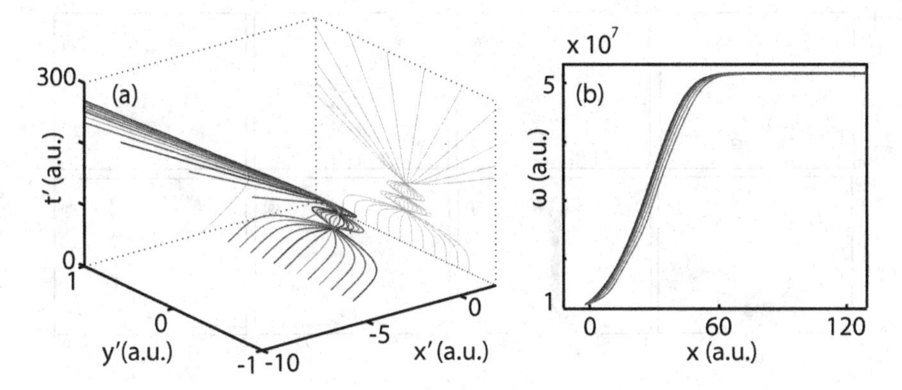

Fig. 9.8 Numerical results in the comoving reference frame (a) and frequency evolution in the laboratory reference frame (b) for the trapping case when we account for dispersion.

If the RIP is subluminal there is not a qualitative change with respect to the non-dispersive case so that the behavior of the rays is essentially the same as depicted in Fig. 9.6. On the opposite, for a trapping RIP the qualitative behavior of the null geodesics changes in an important way when dispersion is present, as shown in Fig. 9.8. Again, there are rays which are trapped, but this time this does not happen indefinitely, as clearly appears in Fig. 9.8(a) where is shown the viewpoint of the comoving reference frame. In particular, we see that at a certain point the velocity becomes even negative so that the rays have become slower than the RIP. This is a consequence of normal dispersion, for which $\frac{d^2k}{d\omega^2} > 0$. This implies that, differently from the dispersionless case, k grows faster than linearly with ω. So, when a ray reaches the RIP, its frequency grows, and, as a consequence the phase velocity $v_r = \omega/k = 1/n$ decreases more rapidly. So the ray begin to lag behind with respect to the RIP, and once it is out of the RIP it will keep its final frequency. In this process the frequency grows only by a finite amount, Fig. 9.8(b), and the ray will leave the RIP with a finite blueshift.

The trapping time: If the velocity of the RIP is downside the range of trapping velocities, then the rays will enter the RIP for a finite lapse of time in both the dispersive and the dispersionless cases. Let us call this the trapping time. If we increase the velocity of the RIP, for example, from $v = 0.5c$ up to $v = 0.65c$ (i.e. approaching the trapping case), the trapping time increases more and more. We show this behavior for the dispersionless case Fig. 9.9(a-g). We compare the trapping case versus the RIP velocity in

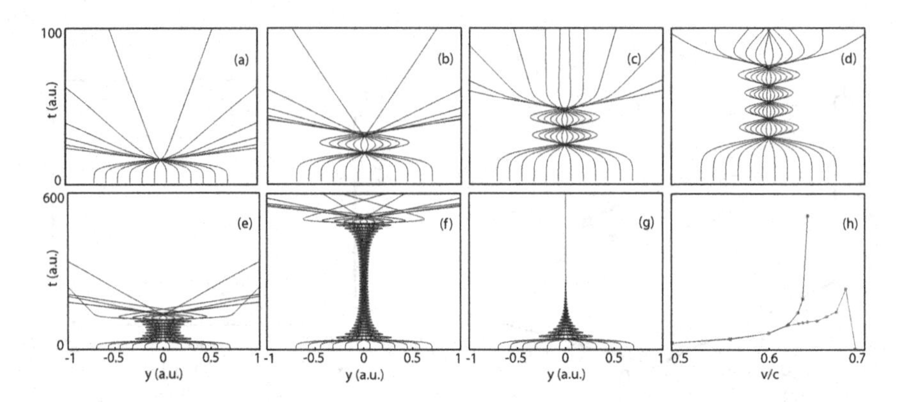

Fig. 9.9 Pictures (a) to (g) represent the trapping dynamics in the absence of dispersion for different RIP velocities: (a) $v = 0.5$; (b) $v = 0.56$; (c) $v = 0.6$; (d) $v = 0.62$; (e) $v = 0.63$; (f) $v = 0.64$; (g) $v = 0.65$. Picture (h) represents the interaction time versus the velocity of the RIP in case of no dispersion (upper curve) and in case of dispersion (lower curve).

Fig. 9.9(h), in the dispersion (lower curve) and in the dispersionless (upper curve) case. In the dispersionless case the trapping time tends to infinity as the velocity of the RIP approaches the trapping case. This divergence can be interpreted as a metastable state for the trapping of the photon inside the pulse. For axial geodesics the trapping time can be estimated to be

$$t_{trap} = \frac{2}{c}\int_0^r \frac{n d\xi}{1 - n(\xi)\frac{v}{c}},$$

where r is the longitudinal extension of the RIP. From this, we see that in absence of dispersion t_{trap} diverges whereas, in the dispersive case, it is always finite.

9.2 Hawking radiation in a static dielectric black hole

In order to investigate the effects of field quantization, we will now concentrate on the static case, and will refer mainly to [Belgiorno *et al.* (2011b)]. In order to avoid unnecessary technical complications and keep all the analysis as simple as possible, it is convenient to replace the electric field with a scalar field Φ, which, then, must satisfy the wave equation

$$\frac{1}{c^2}n^2(x - vt)\partial_t^2\Phi - \partial_x^2\Phi - \partial_y^2\Phi - \partial_z^2\Phi = 0, \qquad (9.33)$$

where (t, x, y, z) are inertial coordinates in the laboratory frame. We are thinking the refractive index $n^2(x - vt)$ to be obtained by means of an

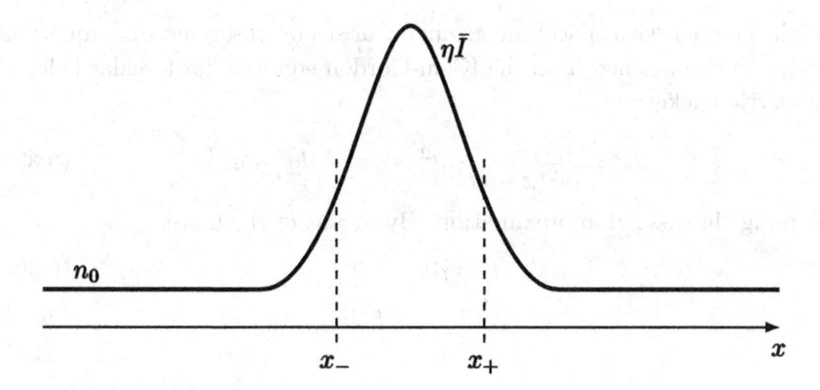

Fig. 9.10 Scheme of the RIP geometry. x_+ and x_- indicate the black hole and white hole horizon positions, respectively.

intense laser pulse in a nonlinear dielectric medium via Kerr effect, but clearly it is not relevant how it is really generated. As discussed previously, staticity corresponds to the fact that the RIP does not depend on the transverse coordinates, so that it is infinitely extended in the transverse dimensions. Clearly, such an approximation is not realistic and cannot be obtained in an actual experiment. One could restrict to a finite geometry, for example by considering a dielectric in fibre optics, but this would not change the substance of the Hawking effect, thus the approximation is well justified and facilitates the comparison with the geometrical description of the effect, beyond simplifying all calculations, including the quantum field theoretical treatment. Moreover, it is possible to produce, e.g., super-Gaussian-like[2] pulses that, at least locally, fall within the approximations adopted here. We assume

$$n(x) = n_0 + \delta n = n_0 + \eta \bar{I}(x), \tag{9.34}$$

where η is positive, $\eta \ll 1$ representing the smallness of the Kerr index, and $\bar{I}$ denotes the normalized intensity of the pulse, a smooth rapidly decaying function a single maximum at $x = 0$, of height 1. A scheme of the RIP is shown in Fig. 9.10.

[2] A super-Gaussian function is a function of the form

$$f(x) = Ae^{-(x/\sigma)^{2m}}$$

where m is an integer with $m > 1$.

The contact with spacetime geometry arises by observing that equation (9.33) can be obtained from the Klein-Gordon equation for a scalar field in the metric background

$$ds^2 = \frac{c^2}{n^2(x-vt)}dt^2 - dx^2 - dy^2 - dz^2, \tag{9.35}$$

assuming the eikonal approximation. By means of the boost

$$t' = \gamma(t - vx), \tag{9.36}$$

$$x' = \gamma(x - vt), \tag{9.37}$$

$$y' = y, \tag{9.38}$$

$$z' = z, \tag{9.39}$$

where

$$\gamma = \frac{1}{\sqrt{1-\frac{v^2}{c^2}}} \tag{9.40}$$

is the usual relativistic gamma factor, we get the metric in the comoving frame

$$\begin{aligned} ds^2 =& \gamma^2 \frac{c^2}{n^2}(1+n\frac{v}{c})(1-n\frac{v}{c})dt'^2 + 2\gamma^2\frac{v}{n^2}(1-n^2)dt'dx' \\ &- \gamma^2(1+\frac{v}{nc})(1-\frac{v}{nc})dx'^2 - dy'^2 - dz'^2. \end{aligned} \tag{9.41}$$

This form of the metric is manifestly stationary. We however know that in this case the hypothesis of the Frobenius theorem are satisfied, hence it exists a coordinates transformation that brings the metric in an explicit static form. Indeed, consider the change of time coordinate

$$dt' = d\tau - \alpha(x')dx', \tag{9.42}$$

where

$$\alpha(x') = \frac{g_{01}(x')}{g_{00}(x')}. \tag{9.43}$$

The metric becomes

$$ds^2 = \frac{c^2}{n^2(x')}g_{\tau\tau}(x')d\tau^2 - \frac{1}{g_{\tau\tau}(x')}dx'^2 - dy'^2 - dz'^2, \tag{9.44}$$

which is manifestly static (w.r.t. the time τ), where

$$g_{\tau\tau}(x') := \gamma^2\left(1+n(x')\frac{v}{c}\right)\left(1-n(x)\frac{v}{c}\right). \tag{9.45}$$

This form of the metric is immediately comparable with the Schwarzschild one. If we look at the $\tau - x'$-part, we see that the more relevant different represented by the factor $\frac{1}{n^2}$ in front of the temporal component of the metric. The horizons are determined by the condition $g_{\tau\tau} = 0$:

$$1 - n(x)\frac{v}{c} = 0, \tag{9.46}$$

or, equivalently,

$$n_0 + \eta \bar{I}(x) = c/v. \tag{9.47}$$

When

$$\frac{1}{n_0 + \eta} \le \frac{v}{c} < \frac{1}{n_0}. \tag{9.48}$$

there will be two horizons: a black hole horizon in p_+, described by the (hyper)plane of equation x'_+, located on the falling edge of the RIP (i.e. $dn/dx|_{x'_+} < 0$) and a white hole horizon in p_-, described by the (hyper)plane of equation $X' = x'_-$, on the rising edge (i.e. $dn/dx|_{x'_-} > 0$).

We can also compare the metric of the dielectric black hole in the co-moving frame with the standard form of an acoustic black hole metric, by suitable identifications. Let us look at the $x' - t'$-part of the metric, to be compared with the two dimensional acoustic metrics studied in [Barcelo *et al.* (2004)]. We need to look for a transformation carrying the metric

$$ds^2_{(2)} = -\frac{\gamma^2}{n^2}\left[-(c^2 - n^2v^2)dt'^2 - 2v(1-n^2)dt'dx' + \left(n^2 - \frac{v^2}{c^2}\right)dx'^2\right] \tag{9.49}$$

into the form

$$ds^2_{acoustic} = -\Omega^2\left[-(\tilde{c}^2 - \tilde{v}^2)d\tilde{t}^2 - 2\tilde{v}d\tilde{t}d\tilde{x} + d\tilde{x}^2\right], \tag{9.50}$$

where $\tilde{v}$ is the velocity of the fluid, in general depending on $\tilde{t}, \tilde{x}$, whereas $\tilde{c}$ is the local speed of sound, assumed to be constant. We can get the correct mapping by means of the identifications

$$\tilde{c} = c, \tag{9.51}$$

$$\tilde{v} = \gamma^2 v\frac{n^2 - 1}{n}, \tag{9.52}$$

$$\Omega^2 = \frac{1}{\gamma^2}\frac{1}{n^2 - \frac{v^2}{c^2}}, \tag{9.53}$$

$$\frac{d\tilde{x}}{dx'} = -\gamma^2\frac{n^2 - \frac{v^2}{c^2}}{n}, \tag{9.54}$$

$$\tilde{t} = t'. \tag{9.55}$$

This mapping is more than just formal, in the sense that in the pulse frame the velocity vector $\tilde{v}\mathbf{u}_x$ is such that $\nabla \wedge \tilde{v}\mathbf{u}_x = 0$, i.e. the corresponding fluid is irrotational as it should be. It is straightforward to check that the horizon condition $\tilde{v} = c$ is equivalent to the condition $n = \frac{c}{v}$.

We can also look at the causal structure of this spacetime. Take

$$d\tilde{u} = dt' - \frac{d\tilde{x}}{\tilde{c} + \tilde{v}}, \tag{9.56}$$

$$d\tilde{w} = dt' + \frac{d\tilde{x}}{\tilde{c} - \tilde{v}}, \tag{9.57}$$

as in [Barcelo *et al.* (2004)]. We get the Penrose diagram for the dielectric black hole spacetime, depicted in Fig. 9.11.

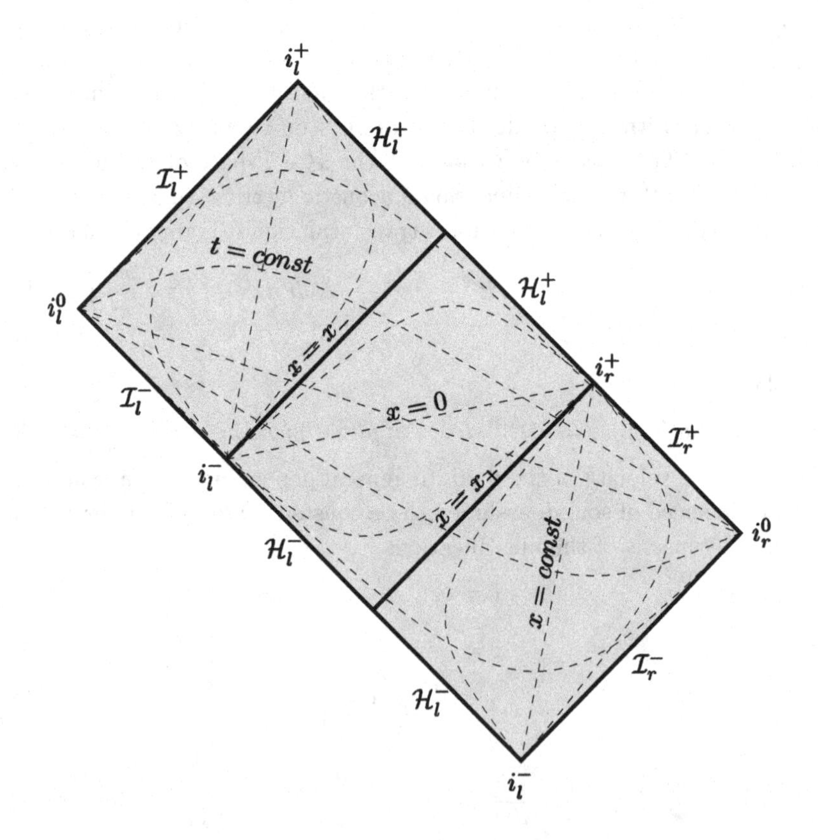

Fig. 9.11 Penrose diagram for the analogue metric (9.35).

9.2.1 *The "standard" Hawking prediction*

At this point, one can apply the Hawking tool in order to predict that the dielectric black hole horizon will emit particles with a temperature proportional to the "surface gravity" κ_+ at the black hole horizon:

$$\kappa_+^2 := -c^4 \frac{1}{2} g^{ab} g^{cd} (\nabla_a \xi_c)(\nabla_b \xi_d)|_{x=x_+} = -\frac{c^4}{2} \left[\frac{-2\gamma^4}{n^4} \left(\frac{dn}{dx} \right)^2 \right]_{x=x_+}, \quad (9.58)$$

which gives

$$\kappa_+ = \gamma^2 v^2 \left| \frac{dn}{dx}(x_+) \right| . \quad (9.59)$$

Via the above mapping, this indeed coincides with the value

$$\kappa_+ = \tilde{c} \left| \frac{\partial}{\partial \tilde{x}} (\tilde{c} - \tilde{v}) \right|_{x=x_+} \quad (9.60)$$

valid for acoustic black holes (cf. [Visser (1998a)]). One can deduce formula (9.59) in several ways. Then, the temperature is given by the now familiar formula

$$T_+ = \frac{\kappa_+ \hbar}{2\pi k_b c} = \gamma^2 v^2 \frac{\hbar}{2\pi k_b c} \left| \frac{dn}{dx}(x_+) \right| , \quad (9.61)$$

where we restore $\hbar$, c and k_b. Finally, using

$$\frac{c}{v} = n(x_+) = n_0 + k\eta, \quad (9.62)$$

with $k = \bar{I}(x_+) \in (0, 1)$, equations (9.59) and (9.61) can be conveniently rewritten in the form:

$$\kappa_+ = \frac{c^2}{(n_0 + k\eta)^2 - 1} \left| \frac{dn}{dx}(x_+) \right| , \quad (9.63)$$

$$T_+ = \frac{\hbar}{2\pi k_b c} \frac{1}{(n_0 + k\eta)^2 - 1} \left| \frac{dn}{dx}(x_+) \right| . \quad (9.64)$$

We can now be a little bit more specific.

9.2.1.1 *Gaussian pulse*

Let us now specialize for instance to the case of a Gaussian pulse

$$\bar{I}(x) = \exp\left(-\frac{x^2}{2\sigma^2} \right) \quad (9.65)$$

with η a small parameter. Thus, the horizons given by equation (9.46)) occur for

$$x_{\pm} = \pm\sigma\sqrt{-2\log\left[\left(\frac{c}{v} - n_0\right)\frac{1}{\eta}\right]}. \tag{9.66}$$

The corresponding surface gravity and the temperature are given by

$$\kappa_+ = \gamma^2 v^2 \frac{k\eta}{\sigma}\sqrt{2\log\frac{1}{k}}, \tag{9.67}$$

$$T_+ = \frac{\kappa_+}{2\pi} \sim \frac{\hbar}{2\pi k_b c}\frac{1}{n_0^2 - 1}\frac{k\eta}{\sigma}\sqrt{2\log\frac{1}{k}}. \tag{9.68}$$

Typical values of the parameters are $\sigma \sim 10^{-5}m, \eta \sim 10^{-3}, n_0 \sim 1.45$. Then, for $k = \frac{1}{2}$, say, we get

$$T_+ \sim 2 \cdot 10^{-2} K, \tag{9.69}$$

which is greater than in standard cases of sonic black holes. Moreover, this value for the temperature is not to be considered as characteristic, since somewhat larger values may be obtained. Notice that this is the temperature in the frame of the pulse and a further transformation is needed in order to recover the temperature in the laboratory frame.

9.2.1.2 *Shockwave model*

A way for producing pulses is by means of spontaneous laser pulse dynamics in transparent media with third order nonlinearity. In this case the rear part (trailing edge) of the filament is characterized by a shockwave profile, so, the same behavior is expected for the refractive index. Hence, in our model, beyond the scale 2σ, which, roughly speaking, represents the overall spatial extension of the RIP, we must introduce at least a further scale representing the 'thickness' of the region where the refractive index undergoes its most rapid variation. In practice, we can adopt the following profile: let us define

$$H(x) := 1 + \tanh\left(\frac{\sigma + x}{\delta_{wh}}\right)\tanh\left(\frac{\sigma - x}{\delta_{bh}}\right), \tag{9.70}$$

where $\delta_{wh}, \delta_{bh} > 0$ are length scales describing the 'thickness' of the region over which a rapid variation of the refractive index occurs, and $2\sigma > 0$ measures the overall spatial extension of the RIP. We then choose the refractive index profile

$$n(x) = n_0 + \eta\frac{H(x)}{\max_x H(x)}, \tag{9.71}$$

so that n converges asymptotically to n_0 and the RIP is correctly normalized. From a physical point of view, we point out that, in the presence of a shock front typical values of δ_{wh}, η may be order of $\delta_{wh} \sim 10^{-6}$m and $\eta \sim 5 \cdot 10^{-3}$ respectively, leading to $T_+ \sim 1$ K.

9.2.2 *The electrodynamics point of view*

We want now provide a dynamical picture for particle production, which does not relying only on the analogue gravity picture, in order to obtain the phenomenon in a non geometrical setting by means of standard tools of quantum field theory [Belgiorno *et al.* (2011b)].

Consider a massless scalar field propagating in a dielectric medium and satisfying Eq. (9.33). This equation arises in the eikonal approximation for the components of the electric field perturbation in a nonlinear Kerr medium [De Lorenci and Klippert (2002); Leonhardt and Piwnicki (2000)].

In order to develop the model, we can proceed in a way analogous to the original situation envisaged by Hawking in the seminal paper that we have reported in Chapter 2. There one considers an initial spherically symmetric astrophysical object, with no particles present at infinity, followed by a collapsing phase which ends up eventually to an evaporating Schwarzschild black hole. In the same way, we can evaluate the mean value of the number of quanta calculated by means of final creation and annihilation operators on the initial vacuum state, which correspond to the absence of the RIP. We can do this in two steps:

- we first look first for a complete set of solutions of equation (9.33);
- then we perform a comparison between an initial situation, consisting of an unperturbed dielectric without a laser pulse and with a uniform and constant refractive index n_0, and a final situation where a laser pulse induces a superluminal RIP in the medium.

9.2.2.1 In *states*

We have first to consider the solution of the wave equation before the creation of the strong pulse. The equation is

$$\frac{n_0^2}{c^2}\partial_t^2\Phi - \partial_x^2\Phi - \partial_y^2\Phi - \partial_z^2\Phi = 0, \tag{9.72}$$

where, assuming the material is non dispersive, the refractive index is constant. It is thus easy to determine a complete set of independent positive frequency plane wave solutions given by (with $\omega_\ell > 0$),

$$F^{\text{in}}_{\mathbf{k}_\ell} = \exp\left(i\mathbf{k}_\ell \cdot \mathbf{x} - i\omega_\ell t\right), \tag{9.73}$$

where the obvious dispersion relation holds:

$$\frac{n_0^2}{c^2}\omega_\ell^2 - k_{x\ell}^2 - k_\perp^2 = 0. \tag{9.74}$$

and where $k_\perp^2 = k_y^2 + k_z^2$. Here we have used the symbol ℓ to evidence that we are working in the laboratory frame. The transverse modes are not affected by a boost in the x direction and thus do not require a specification.

9.2.2.2 Out *states*

For the final states it is convenient to recast equation (9.33) in terms of the retarded and advanced variables, respectively $u = x - vt$ and $w = x + vt$:

$$\frac{n^2(u)v^2}{c^2}\left(\partial_u^2\Phi + \partial_w^2\Phi - 2\partial_u\partial_w\Phi\right) - \left(\partial_u^2\Phi + \partial_w^2\Phi + 2\partial_u\partial_w\Phi\right) \\ -\partial_y^2\Phi - \partial_z^2\Phi = 0. \tag{9.75}$$

Next, we make the monochromatic ansatz

$$\Phi(u, w, y, z) = A(u)e^{ik_w w + ik_y y + ik_z z}, \tag{9.76}$$

which leads to the equation

$$A''(u) + 2ik_w\frac{c^2 + n^2(u)v^2}{c^2 - n^2(u)v^2}A'(u) - \left(k_w^2 + \frac{k_y^2 + k_z^2}{1 - n^2(u)\frac{v^2}{c^2}}\right)A(u) = 0. \tag{9.77}$$

If $n(u)$ is an analytic function, we see that the coefficients of $A'(u)$ and of $A(u)$ have a first order pole where

$$1 - n(u)\frac{v}{c} = 0. \tag{9.78}$$

For example, we assume n of the form (9.34), and satisfying the condition (9.48) for the occurrence of the event horizons. Since $x = \gamma u$, the black hole and white hole horizons are located respectively at

$$u_\pm = \frac{1}{\gamma}x_\pm, \tag{9.79}$$

and we are left with a second order linear differential equation with Fuchsian singular points at $u = u_\pm$ (with $u_- < u_+$). The general solution of Eq.(9.75) can be determined, in the neighborhood of the singular points $u_\pm$, by the standard methods of integration by series: for instance, let us consider the root u_+, and, correspondingly, rewrite equation (9.77) in the form

$$A''(u) + \frac{P(u)}{u-u_+}A'(u) + \frac{Q(u)}{u-u_+}A(u) = 0, \tag{9.80}$$

where we have introduced the functions $P(u)$ and $Q(u)$ (holomorphic in the disk $|u-u_+| < |u_+ - u_-|$), and the prime indicates derivation w.r.t. u:

$$P(u) = 2ik_w \frac{c^2 + n^2(u)v^2}{c^2 - n^2(u)v^2}, \tag{9.81}$$

$$Q(u) = -\left(k_w^2 + \frac{k_y^2 + k_z^2}{1 - n^2(u)\frac{v^2}{c^2}}\right). \tag{9.82}$$

Near u_+ we can expand (9.77) to leading order in $u - u_+$, as

$$A''(u) - \frac{2ik_w c}{vn'(u_+)}\frac{1}{u-u_+}A'(u) + \frac{(k_y^2+k_z^2)c}{2vn'(u_+)(u-u_+)}A(u) = 0, \tag{9.83}$$

which provides the indicial equation $\alpha(\alpha-1) - \alpha\frac{2ik_w c}{vn'(u_+)} = 0$, whose roots are $\alpha_1 = 0$ and $\alpha_2 = 1 + \frac{2ik_w c}{vn'(u_+)}$. This means that in the neighborhood of u_+, Eq. (9.77) has two linearly independent solutions of the form

$$A^{(i)}(u) = (u-u_+)^{\alpha_i}\sum_{n=0}^{\infty} c_n^{(i)}(u-u_+)^n, \tag{9.84}$$

$i = 1, 2$, where coefficients can be obtained recursively from the equation

$$c_n^{(i)}(\alpha_i + n)(\alpha_i + n - 1 + p_0) + \sum_{r=0}^{n-1}((\alpha_i + r)p_{n-r} + q_{n-r-1})c_r^{(i)} = 0. \tag{9.85}$$

Here p_k and u_k $(k = 0, 1, 2, \ldots)$ are the coefficients of the expansion of $P(u)$ and $Q(u)$ in u_+. The above series define two holomorphic functions in the disc $|u-u_+| < |u_+ - u_-|$. In particular, for $\alpha = \alpha_2$ we get a solution that for $u > u_+$ has the form

$$F_{\alpha_2}(u, w, y, z) = \xi(u)e^{i\frac{2k_w c}{vn'(u_+)}\log(u-u_+) + ik_w w + ik_y y + ik_z z}, \tag{9.86}$$

with $\xi(u)$ holomorphic in a neighborhood of $u = u_+$ and vanishing when $u \to u_+$: indeed, we have chosen it in order to absorb the factor $(u - u_+)$ associated with the real term in α_2, so that

$$\xi(u) = (u-u_+)\eta(u), \tag{9.87}$$

such that $\eta(u)$ is holomorphic in a neighborhood of $u = u_+$ and $\eta(u) = c_0 + c_1(u - u_+) + \dots$ as $u \to u_+$. Notice the presence of the logarithmic divergence of the phase as u approaches u_+.

Since we will see that exactly this logarithmic phase in the asymptotic expansion is the crucial point, it is worth to discuss here a simpler deduction of it. By appealing to the WKB approximation, we write Φ as in Eq.(9.76) with

$$A(u) = \exp\left(i\int^u k_u(u')du'\right). \tag{9.88}$$

This gives the dispersion relation

$$n^2\frac{v^2}{c^2}(k_u - k_w)^2 - (k_u + k_w)^2 - k_\perp^2 = 0. \tag{9.89}$$

By solving for k_u as a function of $k_w, \mathbf{k}_\perp$ we get at a second degree equation with solution

$$k_u^\pm = -\frac{k_w}{1 - n^2\frac{v^2}{c^2}}\left[1 + n^2\frac{v^2}{c^2} \pm 2n\frac{v}{c}\sqrt{1 - \frac{k_\perp^2}{k_w^2}\frac{1 - n^2\frac{v^2}{c^2}}{4n^2\frac{v^2}{c^2}}}\right]. \tag{9.90}$$

The validity of these solutions as propagating waves requires the constraint

$$\frac{k_\perp^2}{k_w^2} \leq \frac{4\frac{n^2v^2}{c^2}}{1 - \frac{n^2v^2}{c^2}}. \tag{9.91}$$

Notice that, despite its structure, this constraint does not limit in general the angle for the emission of photons, since in approaching the horizon, the right hand side of the latest equation becomes infinite.

We focus our attention on k_u^+, which is singular at the horizons, whereas k_u^- is regular. Near u_+ the dependence on $\mathbf{k}_\perp$ is washed out, and we have

$$k_u^+ \sim -\frac{2k_w}{1 - n\frac{v}{c}}. \tag{9.92}$$

Then,

$$A(u) \sim \exp\left[i\frac{2ck_w}{vn'(u_+)}\log(u - u_+)\right], \tag{9.93}$$

and we recover the logarithmic divergence in the phase as $u \downarrow u_+$, deduced in the exact way. This behavior is entirely analogous to what happens for the covariant wave equation at the horizon of a Schwarzschild black hole and which we know to be the typical behavior of outgoing modes approaching the horizon. Notice however that in our case the logarithmic divergence appears without any reference to geometry. Its exact deduction shows that indeed it emerges naturally as an exact asymptotic consequence of equation (9.75).

Notice also that the dependence on $\vec{k}_\perp$, which is evident away from the horizon, is not relevant for the Hawking phenomenon [Visser (2003)]. Finally, note that k_u has the opposite sign of k_w, and $|k_u| > |k_w|$.

Let F^+ be the solutions associated to the root k_u^+. For $u \gg u_+$ they behave as

$$F^+ \sim e^{ik_u u + ik_w w + ik_y y + ik_z z}, \tag{9.94}$$

so that it can be written as

$$F^+ = F^+_{\mathbf{k}_\ell} \sim \exp\left(i\mathbf{k}_\ell \cdot \mathbf{x} - i\omega_\ell t\right), \tag{9.95}$$

with $\mathbf{k}_\ell = (k_{x\ell}, k_y, k_z)$, and

$$\omega_\ell = v(k_u - k_w), \tag{9.96}$$

$$k_{x\ell} = k_u + k_w, \tag{9.97}$$

so that

$$k_u = \frac{1}{2}\left(k_{x\ell} + \frac{\omega_\ell}{v}\right), \tag{9.98}$$

$$k_w = \frac{1}{2}\left(k_{x\ell} - \frac{\omega_\ell}{v}\right). \tag{9.99}$$

To get the solutions that are asymptotically of positive frequency and outgoing with respect to the dielectric perturbation, we choose $\omega_\ell > 0$ and $k_{x\ell} > 0$. Moreover, for definiteness we fix $k_w < 0$ so that (9.96) and (9.97) are satisfied for $k_u > -k_w$. Finally, ω_ℓ satisfies the asymptotic dispersion relation $n_0^2\omega_\ell^2 = {\mathbf{k}_\ell}^2 c^2$. A similar analysis can be performed for the solution which is regular at $u = u_+$, which we denote by $F^+_{\mathbf{k}_\ell, reg}$. Similarly, we denote by $F^-_{\mathbf{k}_\ell}$ and by $F^-_{\mathbf{k}_\ell, reg}$ the singular and, respectively, regular solutions at u_-.

If θ, defined by $k_x = |\mathbf{k}_l| \cos(\theta)$, is the angle of emission, then equation (9.99) can be written as

$$k_w = -\frac{\omega_l}{2v}\left(1 - \frac{v}{c} n_0 \cos(\theta)\right). \tag{9.100}$$

9.2.2.3 *Quantization*

For quantize the scalar field one separates the monochromatic solutions $\{F^+, F^+_{reg}\}$ corresponding to all possible values of $\mathbf{k}_\ell$, into positive and negative frequency components $\{f_{\mathbf{k}_\ell}, f^*_{\mathbf{k}_\ell}\}$. These are mutually orthogonal w.r.t. the index $\mathbf{k}_\ell$ relative to the Klein-Gordon product

$$(\Psi, \Phi) = i \int_{\mathbb{R}^3} dudydz \left(\Psi^* \frac{\partial \Phi}{\partial w} - \Phi \frac{\partial \Psi^*}{\partial w}\right), \tag{9.101}$$

and form a complete set of solutions. The scalar product is a function of w since it is indeed impossible to determine in a standard way a conserved inner product for equation (9.33)). The quantum field $\hat{\Phi}$ associated to Φ is thus defined by the expansion

$$\hat{\Phi} = \int d^3\mathbf{k}_\ell \left(a_{\mathbf{k}_\ell} f_{\mathbf{k}_\ell} + a^\star_{\mathbf{k}_\ell} f^\star_{\mathbf{k}_\ell} \right), \tag{9.102}$$

where the creation and annihilation operators $a_{\mathbf{k}_\ell}$, $a^\star_{\mathbf{k}_\ell}$ satisfy the usual algebra

$$[a_{\mathbf{k}_\ell}, a^\star_{\mathbf{k}_\ell'}] = \delta_{\mathbf{k}_\ell \mathbf{k}_\ell'}. \tag{9.103}$$

The evaluation of the average number of emitted quanta is obtained by computing the Bogoliubov coefficients that relate the creation and annihilation operators of the *in* states to the ones of the *out* states. This is the next step.

9.2.2.4 *The spectrum*

In order to determine the spectrum of emitted particles one can proceed by following the lines of the original paper of Hawking. In doing this, it is worth to keep in mind a difference: in the analogue case we have not a real collapse, but we just emulate it by starting from a dielectric material with no signal present in it, and then building a RIP inside the matter. Evaluating the expectation value of number operator, associated to the so obtained final algebra of annihilator and creator operators, on the initial vacuum state, we will get an evaluation of the generated quanta.

We start noticing that up to a shift, we can assume that the black hole horizon is at $u = 0$. Thus, a photon starting at $u < 0$ cannot reach the observer outside the horizon, at $u > 0$, so that states with $u < 0$ cannot be available to the front observer. This means that we can consider only the $u > 0$ part of the F^+ solution, so we multiplying it by a Heaviside function $\theta(u)$. We need to compute the Bogoliubov coefficients $\alpha_{\mathbf{k}\mathbf{k}'}$, which relate the positive frequencies between the initial *in* states

$$F^{\rm in}_{\mathbf{k}} = e^{ik_x x + ik_y y + ik_z z - i\omega t}, \tag{9.104}$$

and the final *out* states $F^+_{\mathbf{k}}$:

$$F^+_{\mathbf{k}} = \theta(u - u_+)\xi_{\mathbf{k}}(u) e^{i\left(\frac{2c}{n'(u_+)v} k_w \log(u - u_+) + k_w w + k_y y + k_z z\right)}, \tag{9.105}$$

where, we have introduced also the analytic part $\xi(u)$ defined in Section 9.2.2.2. It will be evident that, at least in the large frequency limit, this

$\xi(u)$ cannot affect the thermal character of particle emission. For our aim it is not necessary to know the exact expression of the coefficient $\alpha_{\mathbf{k}\mathbf{k}'}$, we only need to write it in the form

$$\alpha_{\mathbf{k}\mathbf{k}'} \propto \delta^2(\mathbf{k}_\perp - \mathbf{k}'_\perp) \int_0^\infty \xi(u) u^{i\frac{2c}{n'(u_+)v}k_w} e^{-ik'_u u} du, \tag{9.106}$$

where we have shifted the variable u so that u_+ is mapped on 0. Since $2k'_u = (k'_x + \omega'/v) > 0$, the second exponential factor is rapidly decreasing at infinity when $\mathrm{Im}(u) < 0$. Thus, we can consider the path Γ starting from 0 to $R > 0$ along the real line, then following the arc of radius R clockwise until $-iR$, and finally coming back from $-iR$ to 0 along the imaginary axis. The integrand is analytic inside the region bounded by the path, if we integrate over Γ in place of $\mathbb{R}_>$, the new integral vanishes for any positive value of R. Then, taking the limit $R \to +\infty$ we get

$$\alpha_{\mathbf{k}\mathbf{k}'} \propto e^{\frac{\pi}{2}\frac{2c}{n'(u_+)v}k_w} \frac{\delta^2(\mathbf{k}_\perp - \mathbf{k}'_\perp)}{i(k'_u)^{1+i\frac{2c}{n'(u_+)v}k_w}} \int_0^\infty dt\, \xi\left(\frac{-it}{k'_u}\right) t^{i\frac{2c}{n'(u_+)v}k_w} e^{-t}. \tag{9.107}$$

We can consider the limit as $k'_u \gg 1$ and approximate ξ for small values of its argument:

$$\alpha_{\mathbf{k}\mathbf{k}'} \propto e^{\frac{\pi}{2}\frac{2c}{n'(u_+)v}k_w} \frac{\delta^2(\mathbf{k}_\perp - \mathbf{k}'_\perp)}{(k'_u)^{2+i\frac{2c}{n'(u_+)v}k_w}} \Gamma(2 + i\frac{2c}{n'(u_+)v}k_w). \tag{9.108}$$

For computing $\beta_{\mathbf{k}\mathbf{k}'}$, it is sufficient to revert the sign of k'_u and $\mathbf{k}'$ in Eq. (9.106). Since now the term relative to the integration on the arc vanishes for $\mathrm{Im}(u) > 0$, this time we can integrate along the positive imaginary axis. Rotating the path counter-clockwise, we get

$$\beta_{\mathbf{k}\mathbf{k}'} \propto e^{-\frac{\pi}{2}\frac{2c}{n'(u_+)v}k_w} \frac{\delta^2(\mathbf{k}_\perp + \mathbf{k}'_\perp)}{(k'_u)^{2+i\frac{2c}{n'(u_+)v}k_w}} \Gamma(2 + i\frac{2c}{n'(u_+)v}k_w), \tag{9.109}$$

By comparing equations (9.108) and (9.109), we finally obtain that the Bogoliubov coefficients $\alpha_{\mathbf{k}_\ell\mathbf{k}_\ell'}$ and $\beta_{\mathbf{k}_\ell\mathbf{k}_\ell'}$ relating, respectively, the positive and negative frequency components between the initial *in* and final *out* states satisfy the fundamental relation

$$\sum_{\mathbf{k}_l'} |\alpha_{\mathbf{k}_l\mathbf{k}_l'}|^2 = e^{\frac{4\pi c k_w}{n'(u_+)v}} \sum_{\mathbf{k}_l'} |\beta_{\mathbf{k}_l\mathbf{k}_l'}|^2. \tag{9.110}$$

On the other hand, the expectation value of the number operator of the outgoing photons in the initial vacuum state is

$$\langle 0\ in | N^{out}_{\mathbf{k}_\ell} | 0\ in \rangle = \sum_{\mathbf{k}_\ell'} |\beta_{\mathbf{k}_\ell\mathbf{k}_\ell'}|^2, \tag{9.111}$$

from which, by using the completeness relation for the Bogoliubov coefficients and the backscattering coefficient

$$G(\omega_\ell, \mathbf{k}_\perp) = \sum_{\mathbf{k}_\ell'} \left(|\alpha_{\mathbf{k}_\ell \mathbf{k}_\ell'}|^2 - |\beta_{\mathbf{k}_\ell \mathbf{k}_\ell'}|^2 \right), \tag{9.112}$$

we obtain

$$\langle N_{\mathbf{k}_l} \rangle = \frac{G(\omega_\ell, \mathbf{k}_\perp)}{\exp\left(\frac{\hbar \omega_l}{k_B T}\right) - 1} \tag{9.113}$$

where

$$T = v^2 \frac{\hbar}{2\pi k_B c} \frac{1}{1 - \frac{v}{c} n_0 \cos\theta} \left| \frac{dn}{du}(u_+) \right|. \tag{9.114}$$

Here we have restored the fundamental constant c, k_B and $\hbar$. This is a thermal-like distribution written in terms of asymptotic physical frequencies.

If we compare this temperature with the value T_+ obtained in the analogue Hawking model in the comoving frame, given in equation (9.61), we see that they ar related by

$$T = \frac{1}{\gamma} \frac{1}{1 - \frac{v}{c} n_0 \cos(\theta)} T_+. \tag{9.115}$$

This is the correct transformation law for the temperature in passing from the comoving frame to the laboratory frame. This can be get from Wien's displacement law which gives the wavelength of maximum emission of a black body as a function of the temperature

$$\lambda_{\max} T = b = 2.9 \times 10^{-3} m \times K. \tag{9.116}$$

In terms of frequency of maximum emission it is

$$\omega_{l\max} = \frac{2\pi c}{b} T. \tag{9.117}$$

Under a boost connecting the lab frame to the pulse frame, the frequency transforms according to the relativistic Doppler formula in a medium with refractive index n_0

$$\omega = \omega_l \gamma \left(1 - \frac{v}{c} n_0 \cos\theta\right), \tag{9.118}$$

where ω is measured in the comoving frame, and θ is the emission angle w.r.t. the x axis in the laboratory frame. Equation (9.118) together with (9.117) gives (9.115).

Let us finally compute some explicit numbers: keeping the same values of the parameters we used to get Eq. (9.69), for $\theta = 0$ and for a typical value $v \sim \frac{2}{3}c$, we would obtain

$$T \sim 78K, \tag{9.119}$$

which is much greater than the values of T for a typical acoustic black hole. For a shock front, this temperature may increase significantly to $T \sim 2000K$ (cf. subsection 9.2.1.2).

As to the greybody factor $G(\omega_\ell, \mathbf{k}_\perp)$ determined by the backscattering effect, in our four dimensional problem cannot be neglected *a priori*. Indeed, once emitted near the horizon, the photons can be reflected back in to the horizon, with a certain probability, because of the presence of a non-vanishing potential they encounter in their propagation. We can compute the greybody factor, as the square modulus of the transmission coefficient through the potential barrier. It is not difficult to show that [Belgiorno *et al.* (2011b)]

$$G(\omega_\ell, \mathbf{k}_\perp) \simeq 1. \tag{9.120}$$

9.3 Effects of optical dispersion: Preliminary heuristics

We have studied the Hawking effect in a dielectric black hole by completely ignoring the material dispersion. A full discussion will be given in the next chapter. Herein, we limit ourselves to phenomenological and heuristic considerations [Belgiorno *et al.* (2011b)]. In optical systems the dispersion law is always relevant and cannot be ignored. In order to get a complete account for it, one should include dispersion from the beginning and reconsider the problem in the new setting. However, it is interesting to infer some qualitative corrections by trying to include dispersive effect *a posteriori*. This will allow to point out the main physical consequences of optical dispersion. In this respect, there is agreement with dispersive models in sonic black holes.

In a macroscopic description of a linear homogeneous dielectric media, the optical dispersion is described by the relation

$$n^2(\omega_l)\, \omega_l^2 - {\mathbf{k}_l}^2 c^2 = 0. \tag{9.121}$$

This dispersion relation is obeyed by the monochromatic components of the field. For example, in [Huttner and Barnett (1992)] it is argued that a phenomenological quantization of the electromagnetic field in a dielectric medium can be justified on the grounds of a more rigorous approach. In this case one decomposes the quantum field

$$\Phi(\mathbf{x}, t) = \int_0^\infty d\omega_\ell \phi(\omega_\ell, \mathbf{x}, t), \tag{9.122}$$

in terms of the monochromatic components $\phi(\omega_l, \mathbf{x}, t) = \varphi_+(\omega_\ell, \mathbf{x})e^{-i\omega_\ell t} + \varphi_-(\omega_\ell, \mathbf{x})e^{i\omega_\ell t}$. Even by allowing a spatial dependence for the refractive index, one has

$$\nabla\varphi(\omega_\ell, \mathbf{x}) + \omega_\ell^2 \frac{n^2(\omega_\ell, \mathbf{x})}{c^2}\varphi(\omega_\ell, \mathbf{x}) = 0. \tag{9.123}$$

Assuming $\varphi(\omega_\ell, \mathbf{x}) \propto \exp(i\mathbf{k}_\ell \cdot \mathbf{x})$, for an homogeneous medium one finds that (9.123) is satisfied only if

$$|\mathbf{k}_\ell| = \omega_\ell \frac{n(\omega_\ell)}{c}, \tag{9.124}$$

which states that each monochromatic component travels at its phase velocity (as should be obvious).

We now specify the dispersion law to the so-called 1-resonance model, which is described by the Sellmeier equation

$$n^2(\omega_l) = 1 + \frac{\omega_c^2}{\omega_0^2 - \omega_l^2}, \tag{9.125}$$

where ω_0 is the resonance frequency and ω_c is the coupling constant, also called plasma frequency. In particular, the function $n^2(\omega_l)$ is a Lorentz scalar, invariant under boosts. Equation (9.125) is a quartic equation in ω_ℓ:

$$\omega_\ell^4 - (\omega_L^2 + c^2\mathbf{k}_\ell{}^2)\omega_\ell^2 + \omega_0^2 c^2\mathbf{k}_\ell{}^2 = 0, \tag{9.126}$$

where we have introduced the longitudinal frequency $\omega_L^2 \equiv \omega_0^2 + \omega_c^2$. It solutions

$$\omega_{\ell\pm}^2 = \frac{1}{2}\left[\omega_L^2 + c^2\mathbf{k}_\ell{}^2 \pm \sqrt{(\omega_L^2 + c^2\mathbf{k}_\ell{}^2)^2 - 4\omega_0^2 c^2\mathbf{k}_\ell{}^2}\right], \tag{9.127}$$

determine two branches, which occur for $0 \le \omega_{\ell-} < \omega_0$ and $\omega_L \le \omega_{\ell+} < \infty$ respectively, with a forbidden region in the interval (ω_0, ω_L).

For practical purposes it is often very useful an approximate expression, provided by the so called Cauchy formula

$$n(\omega_\ell) = n_0 + B_0\omega_\ell^2, \tag{9.128}$$

where n_0 is the would-be refractive index in absence of optical dispersion and B_0 is a suitable constant; this is obtained from (9.125) for $\omega_\ell \ll \omega_0$ and works well for fused silica in the visible frequency interval. For our aims, *a posteriori*, its limited bandwidth validity will not be a real problem, since it will turn out that also Hawking effect takes place in a limited frequency window.

We have to take into account that in presence of the RIP we also deal with nonlinear effects. It is natural to wonder if the above modeling of the dispersion effects continue to be valid in such case. More precisely, we are involved with linearized quantum fields around an effective geometry, so it appear natural to proceed by considering linear dispersion effects. We make the following ansatz:

$$n(x-vt,\omega_\ell) = n_0(\omega_\ell) + \eta\, f(x-vt). \tag{9.129}$$

This way, we are neglecting optical dispersion effects in the RIP, keeping trace of it only in the background value n_0, which does not depend on spacetime variables, but only on ω_ℓ.

We can *a posteriori* justify this ansatz by taking into account what happens in the usual modelization of dielectric media when inhomogeneities are considered. A possibility is to account for inhomogeneities by including space-dependent density, polarizability, resonant frequencies, and so on. By neglecting dissipation in a polariton model, one can get a dielectric susceptibility which depends both on space and frequency, in such a way that the Sellmeier formula changes only because of an explicit dependence of ω_c, ω_0 on space variables. In the case of fused silica, in the visible region, where the Cauchy approximation works well, one can even neglect the spatial dependence of ω_0, and, recalling $u = x - vt$, we can assume [Suttorp and van Wonderen (2004)]

$$\omega_c^2(u) = \omega_{c0}^2 + \delta\, \omega_{c1}^2(u), \tag{9.130}$$

with $\delta \ll 1$ a small parameter. For

$$n_0(\omega_\ell) := \sqrt{1 + \frac{\omega_{c0}^2}{\omega_0^2 - \omega_\ell^2}}, \tag{9.131}$$

using the Cauchy approximation and neglecting terms of order $\delta\omega_\ell^2$, the Sellmeier equation leads to

$$\begin{aligned} n(\omega_l, u) &= n_0(\omega_\ell)\sqrt{1 + \delta\, \frac{1}{n_0^2(\omega_\ell)} \frac{\omega_{c1}^2(u)}{\omega_0^2 - \omega_\ell^2}} \\ &\sim n_0(\omega_\ell) + \frac{\delta}{2n_0\omega_0^2}\omega_{c1}^2(u). \end{aligned} \tag{9.132}$$

This validates our ansatz (9.129), with straightforward identifications.

In the visible frequency region, it is also often convenient to adopt the perturbative ansatz

$$n(\omega_\ell, u) = n_0 + \eta A_1(u) + (B_0 + \eta B_1(u))\omega_\ell^2 \sim n_0 + B_0\omega_\ell^2 + \eta A_1(u), \tag{9.133}$$

where $n_0 + B_0\omega_l^2 \equiv n_0(\omega_\ell)$ and where we neglect terms of order $\eta\omega_\ell^2$. The latter approximation is very useful, because allows us to write

$$n(\omega_\ell, u) = n_0(\omega_\ell) + \eta f(u) = n(u) + B_0\omega_\ell^2, \tag{9.134}$$

so isolating the contribution of the optical dispersion.

As for the dispersion relation, in laboratory we get

$$n(\omega_\ell, u)\omega_\ell = \pm|\mathbf{k}_\ell|c, \tag{9.135}$$

i.e.

$$n(u)\omega_\ell + B_0\omega_\ell^3 \pm |\mathbf{k}_\ell|c = 0. \tag{9.136}$$

In the frame of the pulse:

$$n(x)\gamma(\omega + vk_x) \pm \sqrt{\gamma^2(k_x + v\omega)^2 + k_\perp^2}\, c + B_0\gamma^3(\omega + vk_x)^3 = 0, \tag{9.137}$$

which is the same dispersion relation as in absence of optical dispersion except for the cubic term $\propto B_0$.

In what follows we will consider the two dimensional restriction, where $k_\perp = 0$, in which case, graphical solutions of the above equation are displayed in Fig. 9.12.

Notice that

$$\omega_\ell = \gamma(\omega + vk_x), \tag{9.138}$$

so that positive frequencies in the laboratory correspond to the region above the dashed line

$$\omega = -vk_x \tag{9.139}$$

in the figure.

9.3.1 *Horizons in dispersive media*

In presence of a nontrivial dispersion law, where in general are present more than quadratic terms, the eikonal approximation does no more provide simply a background metric and a definition of an horizon cannot be simply

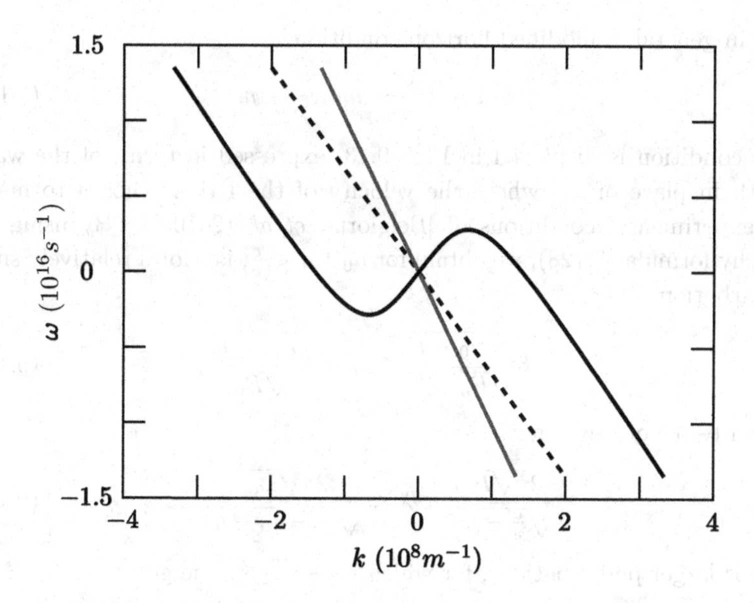

Fig. 9.12 Branches relative to the dispersion relation (9.137) in the comoving frame. The dashed line corresponds to $\omega = -vk_x$.

a geometric one. It can *a priori* for example involve phase velocity or group velocity or both, see for example [Robertson and Leonhardt (2010)].

Let us start by some consideration on what happens when one involves the phase velocity, which will be denoted as $v_{\ell,\varphi}$ in the laboratory reference frame and as v_φ in the pulse frame:

$$v_{\ell,\varphi} = v \Leftrightarrow v_\varphi = 0. \tag{9.140}$$

In the laboratory frame one has

$$v_{\ell,\varphi} = \pm \frac{c}{n(\omega_\ell, u)}, \tag{9.141}$$

which, due to (9.140), implies the horizon condition

$$n(\omega_\ell, u) = \frac{c}{v}. \tag{9.142}$$

This generalize the condition obtained without dispersion to the dispersive case. For the Gaussian model it provides

$$u_\pm = \pm \frac{\sigma}{\gamma} \sqrt{-2 \log \left(\left(\frac{c}{v} - n_0(\omega_\ell) \right) \frac{1}{\eta} \right)}, \tag{9.143}$$

and, in general, a modified horizon condition

$$n_0(\omega_\ell) < \frac{c}{v} \le n_0(\omega_\ell) + \eta. \tag{9.144}$$

This condition is displayed in Fig. 9.13, expressed in terms of the wavelength in place of ω_ℓ, where the velocity of the RIP is chosen to match the experimental conditions of [Belgiorno *et al.* (2010a)]. By using the Cauchy formula (9.128), we obtain for $n_0 + \eta < \frac{c}{v}$, i.e. for a relatively small perturbation

$$\frac{\sqrt{\frac{c}{v} - n_0 - \eta}}{\sqrt{B_0}} \le \omega_\ell < \frac{\sqrt{\frac{c}{v} - n_0}}{\sqrt{B_0}}, \tag{9.145}$$

and in terms of wavelength,

$$\frac{2\pi\sqrt{B_0}}{\sqrt{\frac{c}{v} - n_0}} < \lambda \le \frac{2\pi\sqrt{B_0}}{\sqrt{\frac{c}{v} - n_0 - \eta}}. \tag{9.146}$$

For larger perturbations, for which $n_0 + \eta \ge \frac{c}{v}$, one gets

$$0 \le \omega_\ell < \sqrt{\frac{\frac{c}{v} - n_0}{B_0}}, \tag{9.147}$$

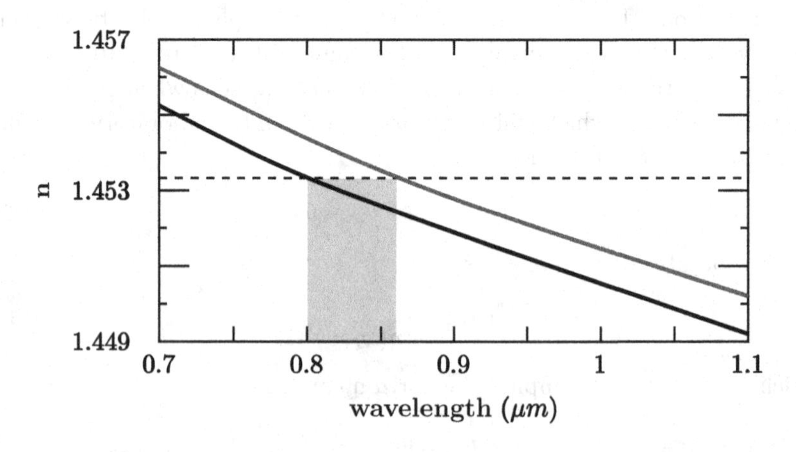

Fig. 9.13 Prediction of the Hawking emission spectral range accounting for the RIP velocity and the material refractive index n. The two curves show, for the case of fused silica, n_0 (lower curve) and $n_0 + \delta n$ (upper curve) with $\delta n = 1 \times 10^{-3}$. The shaded area indicates the spectral emission region predicted for a Bessel filament that has an effective refractive index $n = c/v_B$ where the Bessel peak velocity is $v_B = v_G/\cos\theta$ with cone angle $\theta = 7$ deg and $v_G = d\omega/dk$ is the usual group velocity.

and in terms of wavelength,

$$2\pi c\sqrt{\frac{B_0}{\frac{c}{v}-n_0}} < \lambda < \infty. \tag{9.148}$$

For the parameters of Fig. 9.13, equation (9.147) is holds when $\eta > 0.013$. This is a rather large value but yet not completely unrealistic for the refractive index variation. We now se an important difference between the dispersionless case and the dispersive one: in absence of dispersion a horizon is created only if v is tuned with extreme care in such the way that equation (9.144) with dispersionless n_0 is satisfied. For typical values of $\delta n \sim 10^{-3} - 10^{-4}$ (e.g. in fused silica, $n_2 \sim 3 \cdot 10^{-16}$ cm^2/W and $I \sim 10^{13}$ W/cm^2), this would be no minor feat.

Conversely, in the presence of dispersion, $n(\omega)$ and equation (9.48) define a horizon and a spectral emission region for *any* value of v. The velocity of the RIP v is determined by the group velocity of the intense laser pulse from which it originates, i.e. $v = 1/n_g$ where the group index $n_g = n - \lambda dn/d\lambda$ is a function of the wavelength λ. Then, equation (9.48) may be re-expressed in terms of the group index as

$$n_g - \eta \leq n_H(\omega) < n_g, \tag{9.149}$$

where n_H indicates the refractive index of the Hawking photons. Thus, the Hawking photons will in general be emitted only in a bounded spectral window, in contrast to the dispersion-less case in which, once v is properly tuned so as to achieve the horizon condition, all frequencies are simultaneously excited. This fact justify *a posteriori* why one can limit oneself to adopt the Cauchy formula instead of the complete Sellmeier. On the counterpart, the presence of dispersion does not allow to discern a black body spectral shape, since only a window of frequencies is excited. Moreover, one has to take into account that, even in homogeneous transparent dielectrics, optical dispersion affects the spectral density of photons by introducing a phase space factor multiplying the standard Planckian distribution term which depends on both the refractive index $n(\omega)$ and on the group velocity $v_g(\omega)$ [Milonni (1995)]:

$$\rho(\omega) = \frac{\frac{\hbar\omega^3}{\pi^2 c^2}\frac{n^2(\omega)}{v_g(\omega)}}{\exp\left(\frac{\hbar\omega}{k_b T}\right) - 1}. \tag{9.150}$$

This formula is intended to hold true in the reference of the thermal bath. These optical dispersion contributions amounts to a sort of greybody factor arising from the interaction of photons with the dielectric material [Milonni

(1995)]. This is a sufficient reason for expecting deviations from the standard Planckian distribution. A very naive inclusion of dispersion in our case would also lead to $T = T(\omega)$, which would make even less plausible a pure Planckian spectrum.

The width of the window of spectral emission is determined by η. For example, consider a Gaussian pump pulse in fused silica, where $n_g(800 \text{ nm}) = 1.467$ so we find that for $\delta n = 10^{-3}$, equation (9.144) is satisfied over a $\sim$15 nm bandwidth at 435 nm. This window may become substantially large in lower dispersion regions, as shown in Fig. 9.13.

Let us now see what the phase velocity horizon condition is in the comoving frame: one has $v_\varphi = 0$ if $\omega = 0$ (for $k_x \neq 0$), so for $v > 0$

$$n\left(\gamma(\omega + vk_x), x\right)|_{\{\omega(k_x,x)=0, k_x\neq 0\}} = \frac{c}{v}. \tag{9.151}$$

In particular, $\omega = \omega(k_x, x)$ is solution of the two dimensional reduction of equation. (9.137). As in the dispersionless case, the above condition requires $n_0 < \frac{c}{v}$. If also $n_0 + \eta < \frac{1}{v}$ is satisfied, then one gets two disconnected regions:

$$-\frac{1}{\gamma v}\sqrt{\frac{\frac{c}{v} - n_0}{B_0}} < k_x \leq -\frac{1}{\gamma v}\sqrt{\frac{\frac{c}{v} - n_0 - \eta}{B_0}}$$
$$\frac{1}{\gamma v}\sqrt{\frac{\frac{c}{v} - n_0 - \eta}{B_0}} \leq k_x < \frac{1}{\gamma v}\sqrt{\frac{\frac{c}{v} - n_0}{B_0}}. \tag{9.152}$$

Only the latter region corresponds to $\omega_\ell > 0$, see Fig. 9.14.

For the case $n_0 + \eta \geq \frac{c}{v}$ one instead obtains a unique connected region, with

$$-\frac{1}{\gamma v}\sqrt{\frac{\frac{c}{v} - n_0}{B_0}} < k_x < \frac{1}{\gamma v}\sqrt{\frac{\frac{c}{v} - n_0}{B_0}}, \tag{9.153}$$

consistently with (9.147).

Let us now discuss the case when an horizon is defined by using the group velocity in place of the phase velocity. The horizon condition will be $v_{\ell,g} = v$ in the laboratory and $v_g = 0$ in the comoving frame. For the sake of simplicity we work in the two dimensional restriction defined by $k_\perp = 0$. Thus $v_{\ell,g} = v$ is equivalent to

$$\frac{c}{v} = n(\omega_\ell, u) + \frac{\partial n(\omega_\ell, u)}{\partial \omega_\ell}\omega_\ell. \tag{9.154}$$

In the Cauchy approximation, we find

$$\frac{c}{v} = n(\omega_\ell, u) + 2B_0\omega_\ell^2 \equiv n_g(\omega_\ell, u), \tag{9.155}$$

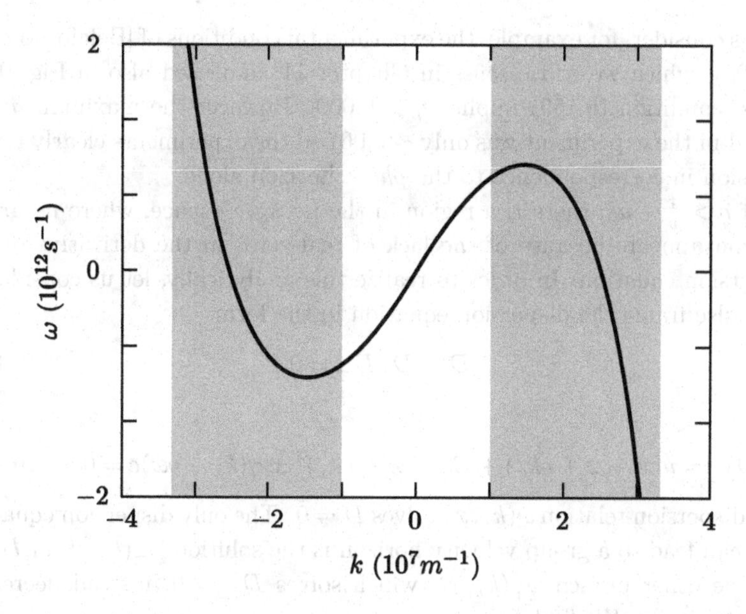

Fig. 9.14 The dispersion relation in the comoving frame for a 2D model at a fixed point x (only the branch relevant to horizon formation is shown). The shaded regions delimit the bands representing the values of k_x for which a phase horizon exists.

where n_g is the group refractive index. This way, in place of (9.144), we obtain

$$n_g(\omega_\ell) < \frac{c}{v} \leq n_g(\omega_\ell) + \eta, \tag{9.156}$$

so that the window is shifted w.r.t. the one corresponding to (9.144). Explicitly,

$$\sqrt{\frac{\frac{c}{v} - n_0 - \eta}{3B_0}} \leq \omega_l < \sqrt{\frac{\frac{c}{v} - n_0}{3B_0}}, \tag{9.157}$$

or, in terms of wavelength,

$$2\pi c\sqrt{\frac{3B_0}{\frac{c}{v} - n_0}} < \lambda \leq 2\pi c\sqrt{\frac{3B_0}{\frac{c}{v} - n_0 - \eta}}. \tag{9.158}$$

It follows that a phase horizon and a group horizon can coexist only if

$$\eta \geq \frac{2}{3}\left(\frac{c}{v} - n_0\right). \tag{9.159}$$

If this condition is not satisfied, an experiment should be able to distinguish between the two horizon conditions for phase velocity and group velocity.

Let us consider, for example, the experimental conditions of [Belgiorno *et al.* (2010a)] which we will discuss in Chapter 11, and used also in Fig. 9.13. Thus, condition (9.159) implies $\eta > 0.009$. However the maximum η obtained in the experiment was only ~ 0.001 so the experiments clearly reveal emission in correspondence to the *phase* horizon alone.

If $\eta > \frac{1}{v} - n_0$, there is a region in the (ω, k_x, x)-space, where no group horizon appears because of the lack of real zeros in the derivative of the dispersion equation. In order to realize this analytically, let us consider in the pulse frame the dispersion equation in the form

$$D = D_+ D_- = 0, \tag{9.160}$$

where

$$D_\pm := n(x)\gamma(\omega + vk_x) + B_0\gamma^3(\omega + vk_x)^3 \pm \gamma(k_x + v\omega)c = 0. \tag{9.161}$$

The dispersion relation $\omega(k_x, x)$ solves $D = 0$. The only dispersion equation that can lead to a group velocity horizon is the solution $\omega_-(k_x, x)$ of $D_- = 0$. The other branch $\omega_+(k_x, x)$, which solves $D_+ = 0$, instead decreases monotonically. We find

$$v_g = \frac{\partial \omega(k_x, x)}{\partial k_x} = \frac{\partial D}{\partial k_x}\left(\frac{\partial D}{\partial \omega}\right)^{-1}. \tag{9.162}$$

It follows that a group velocity horizon can emerge only by solving the system $D_- = 0$ and $\frac{\partial D_-}{\partial k_x} = 0$. But $\frac{\partial D_-}{\partial k_x} = 0$ leads to

$$k_x^{(-)} = -\frac{\omega}{v} \pm \frac{1}{\gamma v\sqrt{3B_0}}\sqrt{\frac{c}{v} - n(x)}. \tag{9.163}$$

Therefore, if $n_0 + \eta \geq \frac{c}{v}$, as for the case of a unique connected region for the phase velocity horizons (equation (9.153)), $k_x^{(-)}$ in (9.163) is complex valued for all x such that $n(x) > \frac{c}{v}$, so that the group velocity horizons disappear. We defer a further discussion on this topic in the next two chapters.

Chapter 10

Hawking radiation in a dispersive Kerr dielectric

In this chapter it will be presented a theoretical framework for studying Hawking radiation in homogeneous linearly dispersive media, where the dispersion is included mesoscopically from the beginning, from which the Sellmeier relations are deduced as a consequence.

The strategy will be to start from a model due to Hopfield [Hopfield (1958)], which describe the electromagnetic field linearly interacting with a "field of harmonic oscillators". Here the Hopfield mode will be extended in a Lorentz covariant form, a necessary step for our aims, since we need to be ready to pass from the laboratory frame to, eventually, the one co-moving with a pulse. We will also discuss its perturbative quantization before introducing the nonlinear Kerr effect. Asymptotic states, as well as generalized Manley-Rowe relations, are introduced.

Then, we will present semi-analytical and numerical results obtained from the study of Maxwell equations in the lab. This interlude concerns important achievements mainly discussed in [Rubino *et al.* (2012.); Petev *et al.* (2013)] directly by exploiting, from the analytical point of view, some approximations of the Maxwell equations, as well as numerical simulations which test both the aforementioned approximation and the feasibility of a thermal spectrum for a white hole configuration. Thermality can be safely found in presence of a blocking horizon. Also the subcritical case (absence of blocking horizon) is discussed.

Next, we will study the Hawking effect by phenomenologically introducing a RIP disturbance. In order to simplify technicalities the analysis will be performed for a scalar reduction of the problem, where the electric field E and the polarization field P are suitably replaced by two scalar fields φ and ψ respectively. This will provide first evidences of the effect in presence of *a priori* optical dispersion.

Finally, we will then add a quartic self interaction of the polariton field, modeling the Kerr effect. The analogue Hawking effect will be investigated in this setup.

10.1 The relativistic Hopfield model

Our aim to understand (analogue) Hawking radiation in situations where dispersion becomes a dominant phenomenon, as in opposite to the true gravitational case where far from the quantum gravity regime the dispersion relation is the one in vacuum. We already had some suggestion from the previous chapter, however, to be less speculative, instead of proceeding with a phenomenological quantization of the electromagnetic field in presence of a dielectric medium, see e.g. [Luks and Perinová (2009)] we would prefer to delve into a less phenomenological situation. This would be relevant non only for tackling the problem of analogue Hawking radiation, but also for more general investigations of phenomena involving photon pair creation associated with variations of the dielectric constant in a dielectric medium, see e.g. [Schwinger (1993a,d,b,c, 1994)].

The idea we adopt here is to adopt the Hopfield model: the electromagnetic field interacts linearly with a set of oscillators that reproduce sources for dispersive properties of the electromagnetic field in matter in such a way that the oscillators can be identified with the mesoscopic polarization fields [Hopfield (1958); Fano (1956); Kittel (1987); Davydov (1984)]. We will consider a more general situation where electric susceptibility, resonance frequencies and also the coupling between electromagnetic field and oscillators depend on spacetime variables, and we will refer to [Belgiorno *et al.* (2016a)]. Differently from other models generalizing the Hopfield model, see [Huttner and Barnett (1992); Suttorp and van Wonderen (2004); Suttorp (2007)], we will not include absorption in the present discussion. This is justified by the fact that we are interested in emission phenomena that take place far from the absorption region. Any framework including absorption would imply a much more tricky approach [Huttner and Barnett (1992); Suttorp and van Wonderen (2004); Suttorp (2007)].

The first aim of this generalization is to introduce the possibility to simulate in a perturbative way the presence of non linear effects, like the Kerr effect, by a phenomenological variation of the aforementioned parameters. The mean motion of a dielectric perturbation, resulting by the microscopical effects corresponding to non linear effects would be thus semi-phenomenologically reproduced by allowing a suitable spacetime dependence to the constants. In other words, we adopt a Lagrangian model which would be "first principle-based" apart for the appearance in the Lagrangian of a phenomenological contribution to the refractive index arising from the Kerr effect. Still, canonical quantization of this model leads to interesting results. Moreover, an advantage of this approach is that it allows to consider general situations where the RIP is uniformly moving or accelerating or also rotating and so on. Each of these situations, concretely realizable in laboratory, is a very interesting benchmark for photon pair creation in external fields or under changing external conditions. It is evident that in a more fundamental model, which will be considered later, such kind of phenomena should arise as solutions of a nonlinear theory, a fact that would make the task technically much more difficult.

Concretely, we can include even more general situations by allowing an almost completely arbitrary dependence on spacetime variables, for the susceptibility $\chi(t,x,y,z)$, the proper frequency $\omega_0(t,x,y,z)$ of the oscillators defining the polarization field, and the coupling "constant" $g(t,x,y,z)$ related to the linear interaction among the electric field and the oscillators. With "almost" arbitrary we mean that we need to ensure the existence of well defined IN and OUT particle states for the physical interpretation of the quantum theory. This is possible if asymptotically in time the aforementioned functions are constant.

In order to get rid of any possible ambiguity in actual calculations and respect the basic fundamental requirements of physics, we present our model in a full covariance form, together with its canonical quantization in a covariant gauge. On the other hand we must recall that there are several studies concerning electrodynamics of moving media, where covariance of the formalism plays a key-role. We limit ourselves to quote some seminal papers [Gordon (1923); Balasz (1955); Quan (1957); Synge (1956)], where phenomenological electrodynamics is adopted.

10.1.1 *The covariant generalization of the Hopfield model*

In the Hopfield model, the interaction between the electromagnetic field and a dielectric material is described by adding to the electromagnetic field another field, representing the polarization of the material, and that couples with the electric field via the Lagrangian

$$\mathcal{L}_{em} := \frac{1}{8\pi}\left(\frac{1}{c}\dot{\mathbf{A}} + \nabla\varphi\right)^2 - \frac{1}{8\pi}(\nabla\wedge\mathbf{A})^2 + \frac{1}{2\chi\omega_0^2}\left(\dot{\mathbf{P}}^2 - \omega_0^2\mathbf{P}^2\right)$$
$$- \frac{g}{2c}\left(\mathbf{P}\cdot\dot{\mathbf{A}} + \dot{\mathbf{A}}\cdot\mathbf{P}\right) - \frac{g}{2}\left(\mathbf{P}\cdot\nabla\varphi + \nabla\varphi\cdot\mathbf{P}\right), \tag{10.1}$$

where φ and $\mathbf{A}$ are the scalar potential and the vector potential respectively. A generic uniformly traveling perturbation can be described by assuming $\chi = \chi(x - vt, y, z)$, $\omega = \omega_0(x - vt, y, z)$, and $g = g(x - vt, y, z)$ [Belgiorno *et al.* (2015, 2016a)]. Only a first-principle introduction of the Kerr effect, like e.g. the one obtained by introducing a fourth power of $\mathbf{P}$ would require substantial modifications, but we will consider such situation only later.

In order to bring it in a covariant form, in particular, we need to tackle the problem of finding a covariant form for the polarization part of the Hopfield Lagrangian, which is not a so trivial task. To this aim, let us introduce v^μ as the 4-velocity of the bulk dielectric medium, that is the velocity of the dielectric sample, not to be confused with the velocity of the dielectric perturbation. Then, we propose the covariant Lagrangian density

$$\mathcal{L} = -\frac{1}{16\pi}F_{\mu\nu}F^{\mu\nu} - \frac{1}{2\chi\omega_0^2}\left[(v^\rho\partial_\rho P_\mu)(v^\sigma\partial_\sigma P^\mu)\right]$$
$$+ \frac{1}{2\chi}P_\mu P^\mu - \frac{g}{2c}(v_\mu P_\nu - v_\nu P_\mu)F^{\mu\nu}. \tag{10.2}$$

Here, the Minkowski metric $\eta_{\mu\nu}$ is chosen with the mostly minus signature, standard for quantum field theory, $\eta = \mathrm{diag}(+,-,-,-)$. This model looks to be the principal candidate in our consideration, since it provides a field equation for P_μ which, for constant χ, gives rise e.g. in the eikonal approximation to the correct covariant dispersion relation.

Covariance of the Lagrangian forces us to introduce a zeroth component of the polarization field, which now appears to be a four-vector. This component P^0 is absent in the rest frame and appears to have no physical meaning. Actually, however, this new component is not an independent one, but is a function of the spatial components, since it must satisfy the following constraint

$$v^\mu P_\mu = 0, \tag{10.3}$$

which in the rest frame, where $v^\mu = (c, 0, 0, 0)$, amounts just to $P_0 = 0$. It is worth mentioning that this condition is the correct one for our harmonic oscillator field P coupled to the electromagnetic field, regardless of its specific nature of polarization field.

The field equations are

$$-\frac{1}{4\pi}\partial_\nu F^{\nu\mu} - v^\nu \partial_\nu \left(\frac{g}{c} P^\mu\right) + v^\mu \partial_\nu \left(\frac{g}{c} P^\nu\right) = 0, \tag{10.4}$$

$$-v^\alpha \partial_\alpha \left(\frac{1}{\chi \omega_0^2}\right) v^\beta \partial_\beta P^\nu - \frac{1}{\chi} P^\nu + \frac{g}{c} v_\rho F^{\rho\nu} = 0. \tag{10.5}$$

By contracting the first equation with v_μ and taking into account the above constraint we get

$$\partial_\mu (E^\mu + 4\pi \frac{g}{c} P^\mu) = 0, \tag{10.6}$$

where $E^\mu := v_\nu F^{\nu\mu}$ is the **covariant electric field**. This is nothing but the Gauss law for the electric induction field $D^\mu := E^\mu + 4\pi \frac{g}{c} P^\mu$ and is the right condition to identify P^μ as a polarization field: it is required by compatibility among the transversality condition and the equations of motion.

More in general, the induction is described by a tensor $G^{\mu\nu}$, which is such that

$$D^\nu = v_\mu G^{\mu\nu}, \tag{10.7}$$

$$H_\mu = \frac{1}{2} v^\nu \epsilon_{\mu\nu\rho\sigma} G^{\rho\sigma} := \frac{1}{2} v^\nu \epsilon_{\mu\nu\rho\sigma} F^{\rho\sigma} = B_\mu. \tag{10.8}$$

In absence of free charges and currents, the laws of Gauss and Ampere are summarized by the equation $\partial_\mu G^{\mu\nu} = 0$, which, contracted with v_ν, gives $\partial_\mu (E^\mu + 4\pi \frac{g}{c} P^\mu) = 0$.

It may be useful to collect the fields A_μ and P_μ in an eight components field $\Phi_\mu := (A_\mu, P_\mu)$. Thus, we are dealing with a field theory which is quadratic in Φ_μ.

10.1.2 *Quantum covariant Hopfield model*

We now need to quantize the theory just obtained [Belgiorno *et al.* (2016a)]. This is not immediate since, first, we have to quantize a constrained system because of condition (10.3). This affects the quantization of the covariant Hopfield model, in the sense that the covariant form of the Heisenberg commutation relations has to be consistent with the constraints of the theory. More than this, the electromagnetic part of the theory should be treated adequately, being a gauge theory. We will follow a procedure that slightly departs from what could be considered as standard, t.i. the usual quantization under covariant gauge conditions like the Lorentz gauge $\partial_\mu A^\mu = 0$, and where one can adopt the Gupta-Bleuler formalism in order to get rid of spurious degrees of freedom, leaving only the transverse (physical) ones. We prefer to follow the Dirac approach in which all first-class constraints, which are associated with the gauge freedom, are first reduced to second-class ones by means of suitable explicit gauge-fixing terms in the Lagrange formalism, and then quantized [Dirac (1947, 1950)], so that the constraints are implemented operatorially, rather than in a weak sense, as in the Gupta-Bleuler approach [Gitman and Tyutin (1990)].

Adding all the constraints to the Lagrangian we get

$$\begin{aligned}\mathcal{L}_c := &-\frac{1}{16\pi}F_{\mu\nu}F^{\mu\nu} - \frac{1}{2\chi\omega_0^2}\left[(v^\rho\partial_\rho P_\mu)(v^\sigma\partial_\sigma P^\mu)\right] + \frac{1}{2\chi}P_\mu P^\mu \\ &- \frac{g}{2c}(v_\mu P_\nu - v_\nu P_\mu)F^{\mu\nu} + B(\partial_\mu A^\mu) + \frac{\xi}{2}B^2 + \lambda(v_\mu P^\mu), \qquad (10.9)\end{aligned}$$

where B is an auxiliary hermitian scalar field, usually called the B-field [Nakanishi and Ojima (1990); Gitman and Tyutin (1990)], and ξ is a constant reproducing various possible gauge conditions, known as R_ξ-gauges. The equations of motions are

$$-\frac{1}{4\pi}\partial_\nu F^{\nu\mu} - v^\nu\partial_\nu\left(\frac{g}{c}P^\mu\right) + v^\mu\partial_\nu\left(\frac{g}{c}P^\nu\right) + \partial^\mu B = 0, \qquad (10.10)$$

$$-v^\alpha\partial_\alpha\left(\frac{1}{\chi\omega_0^2}\right)v^\beta\partial_\beta P^\nu - \frac{1}{\chi}P^\nu + \frac{g}{c}v_\rho F^{\rho\nu} + \lambda v^\mu = 0, \qquad (10.11)$$

$$\partial_\nu A^\nu + \xi B = 0. \qquad (10.12)$$

In order to proceed with the canonical quantization we need to determine the canonically conjugate momenta:

$$\frac{\partial\mathcal{L}_c}{\partial\partial_t A_0} =: \Pi_A^0 = \frac{B}{c}, \qquad (10.13)$$

$$\frac{\partial\mathcal{L}_c}{\partial\partial_t A_i} =: \Pi_A^i = -\frac{1}{4\pi c}(\partial^0 A^i - \partial^i A^0) - \frac{g}{c^2}(v^0 P^i - v^i P^0), \qquad (10.14)$$

$$\frac{\partial\mathcal{L}_c}{\partial\partial_t B} =: \pi_B = 0, \qquad (10.15)$$

$$\frac{\partial\mathcal{L}_c}{\partial\partial_t P_\mu} =: \Pi_P^\mu = -\frac{1}{\chi\omega_0^2 c}v^0 v^\sigma\partial_\sigma P^\mu, \qquad (10.16)$$

$$\frac{\partial\mathcal{L}_c}{\partial\partial_t \lambda} =: \pi_\lambda = 0. \qquad (10.17)$$

Correspondingly, the classical Hamiltonian density is

$$\begin{aligned}
\mathcal{H} =&(\partial_t A^\mu)\Pi_{A\ \mu} + (\partial_t P^\mu)\Pi_{P\ \mu} - \mathcal{L}_c + u\pi_\lambda + y\pi_B + z\left(\Pi_A^0 - \frac{B}{c}\right) \\
=&2\pi c^2 (\Pi_A^i)^2 + \frac{1}{16\pi} F_{ij}F^{ij} + cA_0(\partial_i \Pi_{A\ i}) + 4\pi g(v_0 P_i - v_i P_0)\Pi_{A\ i} \\
&- c\frac{v^k}{v_0}(\partial_k P^\mu)\Pi_{P\ \mu} - \frac{\chi\omega_0^2 c^2}{2(v^0)^2}\Pi_{P\ \mu}\Pi_P^\mu - \frac{1}{2\chi}P_\mu P^\mu + \frac{2\pi g^2}{c^2}(v_0 P_i - v_i P_0)^2 \\
&+ \frac{g}{2c}(v_i P_j - v_j P_i)F^{ij} - B(\partial_i A^i) - \frac{\xi}{2}B^2 - \lambda(v_\mu P^\mu) \\
&+ u\pi_\lambda + y\pi_B + z\left(\Pi_A^0 - \frac{B}{c}\right).
\end{aligned} \tag{10.18}$$

There are three primary constraints $\pi_B, \Pi_A^0 - B/c, \pi_\lambda$: from the Poisson brackets of the primary constraints with the Hamiltonian we find the complete set of constraints:

$$\Gamma_1 =\pi_B, \tag{10.19}$$

$$\Gamma_2 =\Pi_A^0 - \frac{B}{c}, \tag{10.20}$$

$$\Gamma_3 =v_\mu P^\mu, \tag{10.21}$$

$$\Gamma_4 =v_\mu \Pi_P^\mu, \tag{10.22}$$

$$\Gamma_5 =\lambda, \tag{10.23}$$

$$\Gamma_6 =\pi_\lambda. \tag{10.24}$$

In order to get these expressions, we have taken into account that all functions of the constraints giving rise to the same submanifold $\Gamma_i = 0$, $i = 1, 2, 3, 4, 5, 6$ are to be considered equivalent. Notice that Γ_1, Γ_2, and Γ_6 represent primary second-class constraints of the theory. Indeed, we can be a little bit more explicit:

$$\{\pi_B, H\} =\partial_i A^i + \xi B + \frac{z}{c}, \tag{10.25}$$

$$\{\Pi_A^0 - B, H\} =- c\partial_i \Pi_{Ai} - y, \tag{10.26}$$

$$\{\pi_\lambda, H\} =- v_\mu P^\mu. \tag{10.27}$$

Equation (10.25) fixes z and is the same equation one obtains in QED [Gitman and Tyutin (1990)], so it implies exactly the same conditions. Equation (10.26) determines y, and the last equation (10.27) introduces a second-class second-stage constraint. If we take its Poisson bracket with the Hamiltonian

$$\{v_\mu P^\mu, H\} = -\frac{\chi\omega_0^2 c^2}{(v^0)^2} v_\mu \Pi_P^\mu - c\frac{v^k}{v^0}\partial_k(v_\mu P^\mu) \tag{10.28}$$

which, on the sub manifold defined by the above constraints amounts to requiring $v_\mu \Pi_P^\mu = 0$, which is then a second-class third-step constraint. Its Poisson bracket with the Hamiltonian is

$$\{v_\mu \Pi_P^\mu, H\} = \frac{1}{\chi} v_\mu P^\mu + c\frac{v^k}{v^0}(\partial_k \delta)(v_\mu \Pi_P^\mu) + v^\mu v_\mu \lambda, \tag{10.29}$$

which, on the submanifold of the previous constraints implies $\lambda = 0$. At this point we have a complete set of constraints, since

$$\{\lambda, H\} = u \tag{10.30}$$

simply fixes $u = 0$.

Following Dirac, we then define the matrix $\{C_{ij}\}$ whose entries are given by the Poisson brackets among all constraints at fixed time, $C_{ij} := \{\Gamma_i, \Gamma_j\}$. Its non-zero entries are

$$\{\Gamma_1(t,\mathbf{x}), \Gamma_2(t,\mathbf{y})\} = \delta^{(3)}(\mathbf{x}-\mathbf{y})/c, \tag{10.31}$$

$$\{\Gamma_3(t,\mathbf{x}), \Gamma_4(t,\mathbf{y})\} = v_\mu v^\mu \delta^{(3)}(\mathbf{x}-\mathbf{y}), \tag{10.32}$$

$$\{\Gamma_5(t,\mathbf{x}), \Gamma_6(t,\mathbf{y})\} = \delta^{(3)}(\mathbf{x}-\mathbf{y}). \tag{10.33}$$

Thus, the Dirac brackets, as provided from the theory of constrained systems, are

$$\{\mathcal{A}, \mathcal{B}\}_D = \{\mathcal{A}, \mathcal{B}\} - \{\mathcal{A}, \Gamma_i\} C_{ij}^{-1} \{\Gamma_j, \mathcal{B}\}, \tag{10.34}$$

that, by introducing collective symbols for phase-space variables

$$\{X_l\} = \{A^\mu, P^\mu, B, \lambda\}, \tag{10.35}$$

$$\{\bar{\Pi}_l\} = \{\Pi_A^\mu, \Pi_P^\mu, \pi_B, \pi_\lambda\}, \tag{10.36}$$

mean

$$\{\mathcal{A}, \mathcal{B}\} = \int d^3z \left(\frac{\delta \mathcal{A}}{\delta X_l(t,\mathbf{z})} \frac{\delta \mathcal{B}}{\delta \bar{\Pi}_l(t,\mathbf{z})} - \frac{\delta \mathcal{A}}{\delta \bar{\Pi}_l(t,\mathbf{z})} \frac{\delta \mathcal{B}}{\delta X_l(t,\mathbf{z})} \right),$$

$$\{\mathcal{A}, \Gamma_i\} C_{ij}^{-1} \{\Gamma_j, \mathcal{B}\} = \int d^3u\, d^3w \{\mathcal{A}, \Gamma_i(t,\mathbf{u})\} C_{ij}^{-1}(\mathbf{u},\mathbf{w}) \{\Gamma_j(t,\mathbf{w}), \mathcal{B}\}, \tag{10.37}$$

where repeated indices are summed. Finally, we get for the Dirac brackets

$$\{A^\mu(t,\mathbf{x}), \Pi_A^\nu(t,\mathbf{y})\}_D := \eta^{\mu\nu} \delta^{(3)}(\mathbf{x}-\mathbf{y}), \tag{10.38}$$

$$\{P^\mu(t,\mathbf{x}), \Pi_P^\nu(t,\mathbf{y})\}_D := \left(\eta^{\mu\nu} - \frac{1}{v_\rho v^\rho} v^\mu v^\nu\right) \delta^{(3)}(\mathbf{x}-\mathbf{y}), \tag{10.39}$$

$$\{B(t,\mathbf{x}), \pi_B(t,\mathbf{y})\}_D := 0, \tag{10.40}$$

$$\{\lambda(t,\mathbf{x}), \pi_\lambda(t,\mathbf{y})\}_D := 0. \tag{10.41}$$

According to the Dirac quantization scheme, for the quantum operators we must impose

$$[\hat{X}^l(t,\mathbf{x}),\hat{\bar{\Pi}}_k(t,\mathbf{y})] := i\hbar\{X^l(t,\mathbf{x}),\bar{\Pi}_k(t,\mathbf{y})\}_D, \tag{10.42}$$

whereas the constraints $\Gamma_i=0$, $i=1,\ldots,6$, are implemented as operators

$$\hat{\pi}_B = 0, \tag{10.43}$$

$$\hat{\Pi}^0_A = \frac{\hat{B}}{c}, \tag{10.44}$$

$$v_\mu \hat{P}^\mu = 0, \tag{10.45}$$

$$v_\mu \hat{\Pi}^\mu_P = 0, \tag{10.46}$$

$$\hat{\lambda} = 0, \tag{10.47}$$

$$\hat{\pi}_\lambda = 0. \tag{10.48}$$

The Hamiltonian is

$$\begin{aligned} H = \int d^3x \big[& 2\pi c^2(\Pi^i_A)^2 + \frac{1}{16\pi}F_{ij}F^{ij} + cA_0(\partial_i \Pi_{A\,i}) \\ & + 4\pi g(v_0 P_i - v_i P_0)\Pi_{A\,i} - c\frac{v^k}{v_0}(\partial_k P^\mu)\Pi_{P\,\mu} \\ & - \frac{\chi\omega_0^2 c^2}{2(v^0)^2}\Pi_{P\,\mu}\Pi^\mu_P - \frac{1}{2\chi}P_\mu P^\mu + \frac{2\pi g^2}{c^2}(v_0 P_i - v_i P_0)^2 \\ & + \frac{g}{2c}(v_i P_j - v_j P_i)F^{ij} - B(\partial_i A^i) - \frac{\xi}{2}B^2 \\ & - \lambda(v_\mu P^\mu) + u\pi_\lambda + y\pi_B + z(\Pi^0_A - \frac{B}{c})\big]. \end{aligned} \tag{10.49}$$

Since only second class primary constraints appear, we can explicit the Lagrange multipliers as functions of the canonical variables. We can limit our attention only to y, z, which display the only non-trivial behavior, and replace $y = -c\partial_i \Pi_{Ai}$ (cf. (10.26)), $z = -c(\partial_i A^i + \xi B)$ (cf. (10.25)) in (10.49).

From (10.47),(10.48), it follows that λ and its conjugate variable π_λ are "expelled" by constrained quantization, being reduced to the zero operator. The point is that (10.3) holds true as a consequence of the antisymmetric character of the induction tensor $G^{\mu\nu}$ defined at the end of the previous section. Indeed, $v_\nu D^\nu = v_\nu v_\mu G^{\mu\nu} = 0$, which, being $D^\mu := E^\mu + 4\pi\frac{g}{c}P^\mu$, and $v_\mu E^\mu := v_\mu v_\nu F^{\nu\mu} = 0$, necessarily it must be $v_\mu P^\mu = 0$. This condition is then preserved by quantization.

10.1.2.1 *The scalar product*

Having identified the quantum algebra of the fields, we need now to construct a representation on a Fock space [Belgiorno *et al.* (2016a)]. To this end, we first define a conserved scalar product. This can be easily obtained by employing the Hamiltonian structure of the theory as follows. Let us introduce the phase-space vectors $\Psi = (\{X^l\}, \{\bar{\Pi}_l\})$, to be intended as a direct sum, where $\{X^l\}$ is the ten-components vector

$$\{X^l\} = \begin{pmatrix} A^\mu \\ P^\mu \\ B \\ \lambda \end{pmatrix} \tag{10.50}$$

and P_l are the corresponding conjugate momenta. Ψ must be interpreted as a symplectic vectors w.r.t. the canonical symplectic form

$$\Omega := i \begin{bmatrix} 0 & 1_{10\times 10} \\ -1_{10\times 10} & 0 \end{bmatrix}, \tag{10.51}$$

where $1_{10\times 10}$ is the 10×10 identity matrix. Then the following scalar product

$$\langle (\{X^l\}, \{\bar{\Pi}_l\}), (\{\tilde{X}^l\}, \{\tilde{\bar{\Pi}}_l\}) \rangle := \int (\{X^{*l}, \bar{\Pi}_l^*\}) \cdot \Omega(\{\tilde{X}^l\}, \{\tilde{\bar{\Pi}}_l\}) d^3x, \tag{10.52}$$

where $\cdot$ stays for the usual Euclidean scalar product, is a conserved quantity, as one can easily check. By taking into account the definitions for $\bar{\Pi}_l$ it reduces to a scalar product

$$\begin{aligned} ((A_\mu, P_\mu, B, \lambda)|(\tilde{A}_\mu, \tilde{P}_\mu, \tilde{B}, \tilde{\lambda})) = \frac{i}{2} \int_{\Sigma_t} \Bigg[& \frac{1}{4\pi} F^{*0\nu} \tilde{A}_\nu + \frac{1}{\chi\omega_0^2} v^\rho \partial_\rho P^{*\sigma} \tilde{P}_\sigma v^0 \\ & - \frac{g}{c}(P^{*0} v^\rho - P^{*\rho} v^0) \tilde{A}_\rho - \frac{1}{4\pi} \tilde{F}^{0\nu} A_\nu^* - \frac{1}{\chi\omega_0^2} v^\rho \partial_\rho \tilde{P}^\sigma P_\sigma^* v^0 \\ & + \frac{g}{c}(\tilde{P}^0 v^\rho - \tilde{P}^\rho v^0) A_\rho^* - (B^* \tilde{A}^0 - \tilde{B} A^{0*}) \Bigg] d^3x. \end{aligned} \tag{10.53}$$

We can introduce the operator $\tilde{H}$ such that the Hamiltonian in (10.49) can be written as

$$\begin{aligned} H =& \frac{1}{2} ((\{X^l\}, \{\bar{\Pi}_l\}), \tilde{H}(\{X^l\}, \{\bar{\Pi}_l\})) \\ =& \frac{1}{2} \int (\{X^{*l}, \bar{\Pi}_l^*\}) \cdot \tilde{H}(\{\tilde{X}^l\}, \{\tilde{\bar{\Pi}}_l\}) d^3x. \end{aligned} \tag{10.54}$$

Thus, the Hamiltonian equations take the form

$$\dot{\Psi} = -i\Omega(\nabla_\Psi H) = -i\Omega \tilde{H} \Psi, \tag{10.55}$$

where $\nabla_\Psi = (\{\partial_{X^l}\}, \{\partial_{\bar{\Pi}_l}\})$. Notice that the trivialization of λ and π_λ, we can restrict our considerations to a 9×9 dimensions.

10.1.2.2 *Physical representation*

In constructing a representation of the algebra, we have to tackle the problem of selecting physical states working in a covariant gauge [Belgiorno *et al.* (2016a)]. It is well-known that for maintain covariance one must pay the price to face the presence of negative norm and zero norm states. This means that we have to work with a large Hilbert space that contains also nonphysical states, which, at the end must be eliminated to get the physical spectrum. Here, we will adopt a perturbative approach, implemented in the interaction representation, where the fields evolve as free fields. A delicate interplay of the interacting (renormalizable) theory with asymptotic free fields occurs. Asymptotic free fields (IN and OUT fields) satisfy canonical commutation relations and the corresponding creation and annihilation operators satisfy free-field commutation rules. In order to simplify our picture, we limit ourselves to the case of a homogeneous medium, without losses and without dielectric perturbations e.g. induced by means of the Kerr effect. This is not a so crude limitation, since in regions asymptotically far from the traveling dielectric perturbation, homogeneity can be assumed to be preserved in general. We will follow a strategy similar to the one in [Huttner *et al.* (1991)]: we quantize the fields as free, with the coupling $g = 0$, and then perform a Fano diagonalization for the physical part of the Hamiltonian, with $g \neq 0$. We will work in the Feynman gauge $\xi = 4\pi$ in order to avoid the problem of the dipole ghost.

Thus, we start by considering the fields separately, starting from the polarization field. Because of the transversality condition $v^\mu P_\mu = 0$, it carries only three degrees of freedom. Indeed, the set of eight fields $\{P_\mu, \partial_0 P_\mu\}$ over a Cauchy surface represents a complete but not independent set of initial conditions, $P_0, \partial_0 P_0$ are functions on the set $\{P_i, \partial_0 P_i\}$ that is both complete and independent. In particular, we can fix a set of polarization vectors e^μ_λ, $\lambda = 1, 2, 3$, such that

$$\sum_{\lambda=1}^{3} e^\mu_\lambda e^\nu_\lambda = \eta^{\mu\nu} - \frac{v^\mu v^\nu}{v_\rho v^\rho}. \tag{10.56}$$

We choose $\{e^\mu_1, e^\mu_2, \frac{v^\mu}{c} - p^\mu \frac{c}{\omega}\}$, where $\omega = v^\mu p_\mu$ and

$$e^\mu_\lambda = (0, \mathbf{e}_\lambda), \qquad \lambda = 1, 2, \tag{10.57}$$

are such that, for $\lambda = 1, 2$ it holds $\mathbf{e}_\lambda \cdot \mathbf{e}'_\lambda = \delta_{\lambda\lambda'}$, and

$$v_\mu e^\mu_\lambda = 0, \tag{10.58}$$

$$p_\mu e^\mu_\lambda = 0, \tag{10.59}$$

where p^μ is the usual wave-vector. The third polarization vector is orthogonal to e_1^μ, e_2^μ, and to v^μ.
The quantum polarization field is

$$P^\mu = \sum_{\lambda=1}^{3} \int d^4p \frac{1}{N_p} \delta(DR) \left[e_\lambda^\mu b_\lambda(\mathbf{p}) e^{ipx} + h.c. \right]. \tag{10.60}$$

Here, DR indicates the free field dispersion relation, and N_p is a suitable normalization factor. The operators $b_\lambda(\mathbf{p})$ satisfy the commutation relations

$$\left[b_\lambda(\mathbf{p}), b_{\lambda'}^\dagger(\mathbf{q}) \right] = \delta_{\lambda\lambda'} \delta^{(3)}(\mathbf{p} - \mathbf{q}). \tag{10.61}$$

Let us now consider the electromagnetic field together with the B-field. It is convenient to introduce a second set of polarization vectors, $\bar{e}_\lambda^\mu$, $\lambda = 0, 1, 2, 3$, and of operators $a_\lambda(\mathbf{p})$ such that

$$A^\mu = \sum_{\lambda=0}^{3} \int d^4p \frac{1}{W_p} \delta(DR) \left[\bar{e}_\lambda^\mu a_\lambda(\mathbf{p}) e^{ipx} + h.c. \right], \tag{10.62}$$

with a suitable normalization W_p. As usual, we impose

$$\sum_{\lambda=0}^{3} \bar{e}_\lambda^\mu \bar{e}_\lambda^\nu = \eta^{\mu\nu}. \tag{10.63}$$

In particular, we use the Gitman-Tyutin basis [Gitman and Tyutin (1990)]. Assume $v^\mu = \gamma(c, 0, 0, v)$, and choose e_λ^μ, $\lambda = 1, 2$ as in the case of the polarization field. The third polarization is chosen to be

$$e_3^\mu = -i \frac{1}{|\mathbf{p}|} p^\mu, \qquad p^0 = |\mathbf{p}|. \tag{10.64}$$

It is also useful to include

$$e_0^\mu = -i \frac{1}{2|\mathbf{p}|} (|\mathbf{p}|, -\mathbf{p}). \tag{10.65}$$

We shall indicate the basis

$$\{ -i \frac{1}{2|\mathbf{p}|} (|\mathbf{p}|, -\mathbf{p}), e_1^\mu, e_2^\mu, -i \frac{1}{|\mathbf{p}|} p^\mu \} \tag{10.66}$$

as the *Gitman-Tyutin basis*. In this basis

$$e_0^\mu e_0^\nu \eta_{\mu\nu} = 0 = e_3^\mu e_3^\nu \eta_{\mu\nu}, \tag{10.67}$$

and

$$e_0^\mu e_3^\nu \eta_{\mu\nu} = -1. \tag{10.68}$$

For the B-field we get

$$B = \int d^4p \frac{1}{S_p} \delta(DR) \left[\beta(\mathbf{p}) e^{ipx} + h.c.\right], \tag{10.69}$$

with a suitable normalization S_p. Again, DR is a free field dispersion relation. Because of equation (10.12), only four operators among a_λ, β, with $\lambda = 0, 1, 2, 3$, are independent. We choose

$$a_0(\mathbf{p}) = \beta(\mathbf{p}). \tag{10.70}$$

The commutation relations are

$$[\beta(\mathbf{p}), \beta^\dagger(\mathbf{q})] = 0, \tag{10.71}$$
$$[a_3(\mathbf{p}), \beta^\dagger(\mathbf{q})] = -\delta^{(3)}(\mathbf{p} - \mathbf{q}), \tag{10.72}$$
$$[a_\lambda(\mathbf{p}), \beta^\dagger(\mathbf{q})] = 0, \qquad \lambda = 1, 2, \tag{10.73}$$
$$\left[a_\lambda(\mathbf{p}), a^\dagger_{\lambda'}(\mathbf{q})\right] = \delta_{\lambda\lambda'} \delta^{(3)}(\mathbf{p} - \mathbf{q}), \qquad \lambda, \lambda' = 1, 2, 3. \tag{10.74}$$

All other commutation relations are zero. In particular,

$$\left[a_3(\mathbf{p}), a^\dagger_3(\mathbf{q})\right] = 0. \tag{10.75}$$

It is worth noticing that in our case the interaction is linear and, indeed, proportional to

$$E^\mu P_\mu. \tag{10.76}$$

The transversality of P^μ with respect to the velocity v^μ suggests to introduce the operator

$$\mathcal{P}^{\mu\nu} := \eta^{\mu\nu} - \frac{v^\mu v^\nu}{v^\rho v_\rho}, \tag{10.77}$$

which is such that $v_\mu \mathcal{P}^{\mu\nu} = 0$. It is evident that $\mathcal{P}^{\mu\nu} P_\nu = P^\mu$ and $\mathcal{P}^{\mu\nu} E_\nu = E^\mu$, so that the interaction term is transverse with respect to v^μ. Since we are considering isotropic dielectric media, $\mathbf{P} \propto \mathbf{E}$, and, then, $P^\mu \propto E^\mu$. Thus, the equation

$$\partial_\mu D^\mu = \partial_\mu \left(E^\mu + 4\pi \frac{g}{c} P^\mu \right) = 0. \tag{10.78}$$

implies that in the homogeneous case both $\partial_\mu E^\mu = 0$ and $\partial_\mu P^\mu = 0$. By passing to the Fourier representation, this amounts to

$$p_\mu E^\mu = 0, \tag{10.79}$$

i.e. also a transversality condition with respect to p^μ is implemented. Therefore, we can introduce a second projection operator

$$\bar{\mathcal{P}}^{\mu\nu} := \eta^{\mu\nu} - \frac{p^\mu p^\nu}{p^\rho p_\rho}, \tag{10.80}$$

so that

$$E^\mu = \bar{\mathcal{P}}^{\mu\nu} E_\nu, \tag{10.81}$$

and, for the interaction term

$$E^\mu P^\nu \eta_{\mu\nu} = \bar{\mathcal{P}}^{\mu\rho} E_\rho \mathcal{P}^{\nu\sigma} P_\sigma \eta_{\mu\nu}. \tag{10.82}$$

Thus the interaction term involves only polarizations that are transverse to both v^μ and p^μ: only physical transverse polarizations of the polarization field and of the electromagnetic field interact. Notice that the projection operators $\mathcal{P}^{\mu\nu}, \bar{\mathcal{P}}^{\mu\nu}$ commute. The polarizations corresponding to $\lambda = 0$ and $\lambda = 3$ correspond to interaction-free parts of the field, and are expected to completely decouple from the spectrum. In particular, the longitudinal component of the polarization field ($\lambda = 3$) does not participate to any physical process.

We can now discuss the construction of physical states. We will follow [Gitman and Tyutin (1990)]. We remark that we have very important simplifications to be taken into account, due to the fact that we are working with a free model, so that we can avoid discussing Ward identities, and also the in-field formalism.

If $|0\rangle$ is the vacuum state,

$$\beta(\mathbf{p})|0\rangle = a_\lambda(\mathbf{p})|0\rangle = b_\lambda(\mathbf{p})|0\rangle = 0, \qquad \lambda = 1, 2, 3, \tag{10.83}$$

then, the space of states $\mathcal{R}$ is spanned by elements of the form

$$(\beta^\dagger)^m (a_\lambda^\dagger)^n (b_\lambda^\dagger)^l |0\rangle, \qquad \lambda = 1, 2, 3, \tag{10.84}$$

where the dependence on momenta has been left implicit, and $l, m, n \in \mathbb{N}$. Since $a_3, a_3^\dagger, \beta, \beta^\dagger$ satisfy *non-canonical* commutation relations, $\mathcal{R}$ contains also negative and zero norm states. Indeed, define

$$d_0 := \frac{1}{\sqrt{2}}(a_3 + \beta), \tag{10.85}$$

$$d_3 := \frac{1}{\sqrt{2}}(a_3 - \beta), \tag{10.86}$$

which satisfy

$$[d_0(\mathbf{p}), d_0^\dagger(\mathbf{q})] = -\delta^{(3)}(\mathbf{p} - \mathbf{q}), \tag{10.87}$$

$$[d_3(\mathbf{p}), d_3^\dagger(\mathbf{q})] = \delta^{(3)}(\mathbf{p} - \mathbf{q}). \tag{10.88}$$

Then, the state $d_0^\dagger|0\rangle$ has negative norm (in the generalized sense)

$$\langle 0|d_0(\mathbf{p})d_0^\dagger(\mathbf{p})|0\rangle = -\delta^{(3)}(0). \tag{10.89}$$

So, the space $\mathcal{R}$ is a space with indefinite metric. This is the price to be paid in order to get an explicit Lorentz covariance. For example, we could consider only the subspace $\mathcal{R}_\perp$ of vectors of the form

$$(a_\lambda^\dagger)^n(b_{\lambda'}^\dagger)^l|0\rangle, \qquad \lambda = 1,2, \quad \lambda' = 1,2, \tag{10.90}$$

which has a positive definite metric, and thus it is an Hilbert space, but explicit covariance would be lost.

We define the *physical space* $\mathcal{R}_{ph}$, as the space spanned by vectors of the form

$$(\beta^\dagger)^m(a_\lambda^\dagger)^n(b_{\lambda'}^\dagger)^l|0\rangle, \qquad \lambda = 1,2, \quad \lambda' = 1,2. \tag{10.91}$$

$\mathcal{R}_{ph}$ is a proper subspace of $\mathcal{R}$, because id does not contain states generated by $a_3^\dagger$ and by $b_3^\dagger$. Let $\mathcal{R}_0$ be the subspace spanned by vectors (10.91) with $m \neq 0$, we have

$$\mathcal{R}_{ph} = \mathcal{R}_\perp \oplus \mathcal{R}_0. \tag{10.92}$$

$\mathcal{R}_0$ is made of states that have zero norm and are orthogonal to any state in $\mathcal{R}_{ph}$. The states in $\mathcal{R}_{ph}$ can be identified also by means of the condition

$$\beta|\Psi\rangle = 0, \qquad |\Psi\rangle \in \mathcal{R}_{ph}, \tag{10.93}$$

as β commutes with all other operators except for $a_3^\dagger$. This is essentially the usual Gupta-Bleuler condition, as it can be recast in the form

$$\hat{B}^{(+)}(t,\mathbf{x})|\Psi\rangle = 0, \tag{10.94}$$

where $\hat{B}^{(+)}(t,\mathbf{x})$ is the positive frequency part of the operator $\hat{B}(t,\mathbf{x})$, see [Gitman and Tyutin (1990)]. We then see that $\mathcal{R}_{ph}$ consists in the equivalence classes of vectors of positive norm which are equivalent if differ by zero norm states.

$$|\phi\rangle \simeq |\psi\rangle \Longleftrightarrow (|\phi\rangle - |\psi\rangle) \in \mathcal{R}_0. \tag{10.95}$$

Any physical operator $\hat{F}_{ph}$, i.e. any operator which is associated with a physical observable, should be such that [Gitman and Tyutin (1990)]

$$\hat{F}_{ph}\mathcal{R}_{ph} \subset \mathcal{R}_{ph}, \tag{10.96}$$

which means that $\mathcal{R}_{ph}$ must be invariant under the action of physical operators. In particular, the Hamiltonian $\hat{H}$ is a physical operator, and indeed it leaves the space of physical states invariant. Indeed, if $|\psi\rangle \in \mathcal{R}_{ph}$, then

$$\begin{aligned}\hat{B}^{(+)}(t,\mathbf{x})\hat{H}|\Psi\rangle =& [\hat{B}^{(+)}(t,\mathbf{x}),\hat{H}]|\Psi\rangle \\ =& i\hbar\partial_0\hat{B}^{(+)}(t,\mathbf{x})|\Psi\rangle = 0,\end{aligned} \tag{10.97}$$

which ensures that $\mathcal{R}_{ph}$ is left invariant under the action of the Hamiltonian $\hat{H}$.

10.1.2.3 *The interaction representation*

What we have done up to now refer to the interaction representation. In general, it is standard in quantum field theory as far as nontrivial interaction terms appear. In our case we do not need strictly to refer to this representation, as our theory can be dealt with exactly [Belgiorno *et al.* (2016b)] at least in the homogeneous case, also in the path integral formalism [Belgiorno *et al.* (2016c)]. Exact quantization for a scalar version of the Hopfield model can also be considered [Belgiorno *et al.* (2016d)].

Still, there are nontrivial subtleties related to the choice of the gauge. In standard QED the usual gauge the Feynman gauge, $\xi = 4\pi$, which avoids to be faced with the appearance of the so-called ghost pole in the theory [Nakanishi and Ojima (1990)], which represents a nontrivial and quite hard problem to be dealt with. In our model the choice $\xi = 4\pi$ is effective only in the interaction representation, whereas in the Heisenberg representation it is not available, [Belgiorno *et al.* (2016b)]. This is the reason here we adopt the interaction representation, so that field operators evolve freely, as dictated by the free Hamiltonian operator, whereas states evolve through the Dyson evolution operator defined the interaction term in the Hamiltonian. We have

$$H_0 = H_{em} + H_{pol}, \tag{10.98}$$

with

$$H_{em} = \int d^3x \left[2\pi c^2 (\Pi_A^i)^2 + \frac{1}{16\pi} F_{ij}F^{ij} + cA_0(\partial_i \Pi_{A\,i}) - B(\partial_i A^i) - \frac{\xi}{2} B^2 \right], \tag{10.99}$$

$$\begin{aligned} H_{pol} = \int d^3x \Big[&- \frac{\chi \omega_0^2 c^2}{2(v^0)^2} \Pi_{P\,\mu} \Pi_P^\mu - \frac{1}{2\chi} P_\mu P^\mu - c\frac{v^k}{v_0} (\partial_k P^\mu) \Pi_{P\,\mu} \\ &+ \frac{2\pi g^2}{c^2} (M_{0i})^2 \Big], \end{aligned} \tag{10.100}$$

where we have defined

$$M_{\alpha\beta} := v_\alpha P_\beta - v_\beta P_\alpha. \tag{10.101}$$

The interaction term is

$$\begin{aligned} H_{int} &= \int d^3x \left[\frac{g}{c} M_{0i} F^{0i} + \frac{g}{2c} M_{ij} F^{ij} \right] \\ &= \int d^3x \left[\frac{g}{2c} M_{\mu\nu} F^{\mu\nu} \right] = \int d^3x \left[\frac{g}{c} P_\mu E^\mu \right]. \end{aligned} \tag{10.102}$$

Some simplifications occur in this representation, since the momenta Π^i_A are considered at $g = 0$. However, notice that in (10.100) the last term has been considered as a contribution to the free Hamiltonian operator, despite it has coupling g^2. The reason that this term 'renormalizes' the proper frequency ω_0 of the polarization field rather than an improvement of the free Hamiltonian.

Also the Hamiltonian H_{em} can be simplified. By using the equations of motion for the electromagnetic field, the last three terms become

$$cA_0(\partial_i \Pi_{A\,i}) - B(\partial_i A^i) - \frac{\xi}{2}B^2 = B\partial_0 A_0 - A_0\partial_0 B + \frac{\xi^2}{2}B^2, \qquad (10.103)$$

both in the interaction representation, with $g = 0$, and in full interacting case. Then we obtain

$$H_{em} = \int d^3x \left[2\pi c^2 (\Pi^i_A)^2 + \frac{1}{16\pi} F_{ij}F^{ij} + B\partial_0 A_0 - A_0 \partial_0 B + \frac{\xi^2}{2} B^2 \right]. \qquad (10.104)$$

The last three terms in the free electromagnetic Hamiltonian do contribute only to the unphysical polarizations, and then are irrelevant for the physics at hand.

In order to specify the Hamiltonian contributions in terms of creators and annihilators, let us first specify the normalizations in the fields. After introducing the renormalized frequency

$$\Omega_0 = \omega_0 \sqrt{1 + 4\pi g^2 \chi}, \qquad (10.105)$$

we choose

$$P^\mu = \sum_{\lambda=1}^{3} \int \frac{d^3p}{(2\pi)^{3/2}} \sqrt{\frac{\chi\omega_0^2}{2\Omega_0}} \left[e^\mu_\lambda b_\lambda(\mathbf{p}) e^{-ipx} + h.c. \right], \qquad (10.106)$$

$$A^\mu = \sum_{\lambda=0}^{3} \int \frac{d^3p}{(2\pi)^{3/2}} \sqrt{\frac{(2\pi)}{p_0}} \left[e^\mu_\lambda a_\lambda(\mathbf{p}) e^{-ipx} + h.c. \right], \qquad (10.107)$$

$$B = \frac{1}{4\pi} \int \frac{d^3p}{(2\pi)^{3/2}} \sqrt{\frac{(2\pi)}{p_0}} p_0 \left[\beta(\mathbf{p}) e^{-ipx} + h.c. \right]. \qquad (10.108)$$

In the following, it will be convenient to distinguish the contributions to the Hamiltonian between transversal and non-transversal. $\hat{H}$ is at most quadratic in the creation and annihilation operators which span the space $\mathcal{R}$. We can write

$$\hat{H} = \hat{H}_{ph} + \hat{H}', \qquad (10.109)$$

where $\hat{H}_{ph}$ corresponds to the physical part of the Hamiltonian, which is the same as in the Coulomb gauge, and involves only transverse polarizations, whereas $\hat{H}'$ involves polarizations, which are decoupled from the physical spectrum.

Let us consider the specific case $v^\mu = (c, 0, 0, 0)$ for simplicity:

$$\begin{aligned} \hat{H}_{em,ph} &= \int d^3p\, p_0 \sum_{\lambda=1}^{2} a_\lambda^\dagger(\mathbf{p},t) a_\lambda(\mathbf{p},t), \\ \hat{H}_{pol,ph} &= \int d^3p\, \Omega_0 \sum_{\lambda=1}^{2} b_\lambda^\dagger(\mathbf{p},t) b_\lambda(\mathbf{p},t), \end{aligned} \tag{10.110}$$

for the free part, and

$$\begin{aligned} \hat{H}_{int,ph} =& i\frac{g}{c} \int d^3p \sqrt{\frac{\pi\chi\omega_0^2}{p_0\Omega}}\, c p_0 \sum_{\lambda=1}^{2} \Big[\Big(a_\lambda(\mathbf{p},t) b_\lambda(-\mathbf{p},t) - a_\lambda^\dagger(\mathbf{p},t) b_\lambda^\dagger(-\mathbf{p},t) \Big) \\ &+ \Big(a_\lambda(\mathbf{p},t) b_\lambda^\dagger(\mathbf{p},t) - a_\lambda^\dagger(\mathbf{p},t) b_\lambda(\mathbf{p},t) \Big) \Big], \end{aligned} \tag{10.111}$$

for the interaction term. For the unphysical part of the Hamiltonian (omitting the arguments when no ambiguity occurs) we get

$$\hat{H}'_{em} = \int d^3p \left[\left(-\frac{p_0}{4} \right) \beta^\dagger \beta \right] \tag{10.112}$$

for the electromagnetic field,

$$\hat{H}'_{pol} = \int d^3p\, \Omega_0 b_3^\dagger b_3 \tag{10.113}$$

for the polarization part, and

$$\begin{aligned} \hat{H}'_{int} =& \frac{g}{c} \int d^3p \sqrt{\frac{\pi\chi\omega_0^2}{p_0\Omega}} \left(\frac{1}{2} c p_0 \Big[\Big(\beta(\mathbf{p},t) b_3(-\mathbf{p},t) + \beta^\dagger(\mathbf{p},t) b_3^\dagger(-\mathbf{p},t) \Big) \right. \\ &\left. + \Big(\beta(\mathbf{p},t) b_3^\dagger(\mathbf{p},t) + \beta^\dagger(\mathbf{p},t) b_3(\mathbf{p},t) \Big) \Big] \right), \end{aligned} \tag{10.114}$$

for the interaction term. The matrix elements of the H' operators between physical states vanish:

$$\langle \Phi | H' | \Psi \rangle = 0, \qquad |\Phi\rangle, |\Psi\rangle \in \mathcal{R}_{ph}. \tag{10.115}$$

It is worth mentioning that in our construction, the condition

$$(\partial_\mu P^\mu)^{(+)} |\Psi\rangle = 0 \tag{10.116}$$

is implemented for any $|\Psi\rangle \in \mathcal{R}_{ph}$. This condition can be understood as a transversality condition in a weak sense.

For the physical part, we can find a Bogoliubov transformation, known as Fano transformation for this specific model [Hopfield (1958); Fano (1956)], which carries $\hat{H}_{ph}$ into a simplified form involving quasi-particle states $\alpha_\lambda(\mathbf{p})$, with $\lambda = 1, 2$. As in the original Hopfield paper [Hopfield (1958)], we impose for $\lambda = 1, 2$

$$[\alpha_\lambda(\mathbf{p}), \hat{H}_{ph}] = \omega(\mathbf{p})\alpha_\lambda(\mathbf{p}). \tag{10.117}$$

The corresponding commutation relations are

$$\left[\alpha_i(\mathbf{p}), \alpha_j^\dagger(\mathbf{q})\right] = \delta_{ij}\delta^{(3)}(\mathbf{p} - \mathbf{q}). \tag{10.118}$$

In this way we recover the same dispersion relation as for the full model, as it is easy to verify.

10.1.2.4 *Multi-resonant case*

It is interesting to see how our model can be generalized to the case of $N > 1$ material polarization fields coupled with the electromagnetic field. We can simply perform substitutions $P^\mu \mapsto P^\mu_{(k)}$, with $k = 1, \ldots, N$, $\omega_0 \mapsto \omega_{0(k)}$, $\chi \mapsto \chi_{(k)}$ and $g \mapsto g_{(k)}$, to get the desired form of the Lagrangian:

$$\begin{aligned}\mathcal{L}_c := &-\frac{1}{16\pi}F_{\mu\nu}F^{\mu\nu} - \sum_{k=1}^{N}\left[\frac{1}{2\chi_{(k)}\omega_{0(k)}^2}\left[(v^\rho\partial_\rho P_{(k)\mu})(v^\sigma\partial_\sigma P^\mu_{(k)})\right]\right.\\ &\left.-\frac{1}{2\chi_{(k)}}P_{(k)\mu}P^\mu_{(k)} + \frac{g_{(k)}}{2c}(v_\mu P_{(k)\nu} - v_\nu P_{(k)\mu})F^{\mu\nu}\right]\\ &+ B(\partial_\mu A^\mu) + \frac{\xi}{2}B^2 + \sum_{k=1}^{N}\lambda_{(k)}(v_\mu P^\mu_{(k)}).\end{aligned} \tag{10.119}$$

Similarly, for the constraints, in place of Γ_3, Γ_4 we get 2N constraints $\Gamma_{3(k)}, \Gamma_{4(k)}$, and, analogously, 2N constraints $\Gamma_{5(k)}, \Gamma_{6(k)}$. The fields $P^\mu_{(k)}$ satisfy:

$$\{P^\mu_{(k)}, \Pi^\nu_{P_{(l)}}\}_D := \delta_{(k)(l)}\left(\eta^{\mu\nu} - \frac{1}{v_\rho v^\rho}v^\mu v^\nu\right)\delta^{(3)}(\mathbf{x} - \mathbf{y}). \tag{10.120}$$

The reason for considering such extension of the theory is not just for sake of generality, but is due to the fact that it correctly reproduce the Sellmeier dispersion relations with more resonances. For example, typically, a dielectric material made of fused silica is described by a Sellmeier relation with three resonances. For a realistic theoretical modelization we would need to use the model with $N = 3$ resonances.

10.2 Uniformly moving RIP

Because of our interest in the specific case of analogue Hawking emission in a dielectric material, in particular in the simplest case of a stationary analogue horizon, we want now to shortly discuss here the case where the dielectric perturbation induced by means of the Kerr effect is traveling with constant velocity v in the laboratory frame. Moreover, we will assume independence on transverse coordinates y, z, in order to ensure staticity, as discussed in the previous chapter. Thus, in the reference frame comoving with the perturbation, in particular, one gets a static dependence of the parameters χ, ω_0, g on x'/γ, where $x' = \gamma(x - vt)$, $t' = \gamma(t - \frac{v}{c^2}x)$ represent the Lorentz boost connecting laboratory and comoving frame. Then, energy is conserved in the comoving frame, i.e. it is possible to perform a variable separation involving the time coordinate with a conserved frequency ω'. This result is very important and helpful in interpreting scattering in presence of the perturbation, and amply corroborates the relevance of a covariant approach and a consistent quantization. In particular, it is possible to find out a scattering basis for the quantum fields in a straightforward way, without any problem arising because of a possible time-dependence of the perturbation. Notice that, as far as the full model with the uniformly traveling perturbation is concerned, in the laboratory frame one can still implement quantization by means of standard canonical commutation relations: the electromagnetic part can be quantized in the Coulomb gauge, whereas the polarization field part does not require a particular care in the definition of the variable conjugate to $\mathbf{P}$. So one gets the equal time commutation relations

$$\left[P_i(t, \mathbf{x}), \frac{1}{\chi\omega_0^2(t, \mathbf{y})}\partial_t P_j(t, \mathbf{y})\right] = i\hbar\delta_{ij}\delta^{(3)}(\mathbf{x} - \mathbf{y}). \tag{10.121}$$

In the laboratory frame, the presence of the traveling perturbation induces an explicit time dependence dependence in the total Hamiltonian, which implies that energy is not conserved in this frame. Still one can expect that some sort of conservation occurs.

We will enter much more into details in the next section, where, in order to simplify calculations and several technical points, we will work with a scalar version. We limit ourselves here to discuss the general behavior of the solutions in the background of a RIP with constant velocity, in the comoving frame. The Hamiltonian (10.18) allows variables separation for the solutions: the ansatz

$$A^\mu(x, y, z, t) = e^{-i\omega t + ik_y y + ik_z z} a^\mu(x),$$

$$\begin{aligned} P^{\mu}(x,y,z,t) &= e^{-i\omega t+ik_y y+ik_z z}p^{\mu}(x), \\ B(x,y,z,t) &= e^{-i\omega t+ik_y y+ik_z z}b(x), \\ \lambda(x,y,z,t) &= e^{-i\omega t+ik_y y+ik_z z}l(x) \end{aligned} \tag{10.122}$$

leads to a second order system of ordinary differential equations for the variables $a^{\mu}(x)$, $p^{\mu}(x)$, $b(x)$ and $l(x)$. The equation for $l(x)$ is $l(x) = 0$ so we can omit this variable from the system. The equation involving $b(x)$ is algebraic and can be used to simplify the remaining equations. As usual, we can reduce the second order system to an equivalent first order system by introducing

$$\begin{aligned} &\alpha^{\mu}(x) := \partial_x a^{\mu}(x), \quad \pi^{\mu}(x) := \partial_x p^{\mu}(x), \\ &\beta(x) := \partial_x b(x). \end{aligned} \tag{10.123}$$

Then if $W(x) := (a^{\mu}(x), \alpha^{\mu}(x), p^{\mu}(x), \pi^{\mu}(x))$, we obtain the system

$$W'(x) = K_{16}W(x), \tag{10.124}$$

where K_{16} is a suitable 16×16 operator. This can be written as $K_{16} = \mathcal{C} + \mathcal{R}(x)$, where $\mathcal{C}$ is a constant 16×16 matrix and $\mathcal{R}(x)$ contains the non constant part. For the sake of simplicity, we consider the case where only the dielectric susceptibility varies; then $\mathcal{R}(x)$ has the form:

$$\mathcal{R}(x) := \begin{pmatrix} 0_4 & 0_4 & 0_4 & 0_4 \\ A_4 & B_4 & C_4 & D_4 \\ 0_4 & 0_4 & 0_4 & 0_4 \\ 0_4 & 0_4 & 0_4 & 0_4 \end{pmatrix}, \tag{10.125}$$

where 0_4 is the 4×4 identity matrix, and

$$A_4 = -\frac{iv^0\omega\omega_0^2\chi(x)}{c(v^1)^2}I_4, \tag{10.126}$$

$$B_4 = \frac{\omega_0^2\chi(x)}{cv^1}I_4, \tag{10.127}$$

$$C_4 = -\frac{iv^0\omega\chi'(x)}{v^1\chi(x)}I_4, \tag{10.128}$$

$$D_4 = \frac{\chi'(x)}{\chi(x)}I_4, \tag{10.129}$$

with I_4 the 4×4 identity matrix. Then, the condition

$$\int_a^{\infty} dx|\mathcal{R}(x)| < \infty, \tag{10.130}$$

which physically can match very well the nature of traveling perturbation of $\delta\chi$, Implies that the asymptotic behavior of solutions is governed by the eigenvalues of $\mathcal{C}$, both as $x \to \infty$ and as $x \to -\infty$, see e.g. [Eastham (1989)]. This implies that the basis obtained for $\delta\chi = 0$ is asymptotically a good scattering basis also for the perturbed problem, which is what we wanted to prove. To be more precise, solutions of the full equations asymptotically behave as the asymptotic region solutions, which is then a good scattering basis.

10.3 Hawking radiation in the perturbative formulation

We want now to tackle the problem of providing an analytical *nonperturbative* deduction of thermality of the analogue Hawking emission in dielectric media. It is worth to explain the contrast between the present statement and the title of this section: the Hawking effect is intrinsically nonperturbative in the sense that it cannot recovered by means of standard perturbative techniques; instead, the "perturbative" in the title refers to the fact that the RIP source of the radiation is introduced at hand as a perturbation and does not comes out from first principles. A more complete formulation where the RIP is a solution of the equation of motion will be considered in the subsequent sections. However, we also mention that this fact is not secondary in the sense that because of the expected universality of the Hawking effect, it should not depend on the exact origin of the RIP.

Our reference model will be the covariant Hopfield model just discussed, which represents our basic tool for analyzing the conceptual issues characterizing the physics at hand. Nevertheless, we will concretely work with a technical simplified model, which is still covariant, maintains the main physical characteristics and in particular provides the same dispersion relation, but has the great advantage to be not involved in a constrained quantization procedure. We will call it the $\varphi\psi$-model, from the names of the scalar fields that replace the (four-potential of the) electromagnetic and the polarization field respectively. Their scalar nature is a second important simplification. Still, its physical content is nontrivial, and can allow for a number of very interesting physical situations, which can be also experimentally tested. It is important to mention that also the full Hopfield model can be solved exactly in the same physical situations, with the difference that it is much more difficult to be handled. Since it is not particularly instructive to include such technical difficulties here, we will not discuss the vector case in the present book.

The main focus of this section is the analytical deduction of the thermality for dielectric black holes. This will obviously obtained for the case of weak dispersion and it will be related to a peculiar Fuchsian singularity structure of the field equations near the horizon. We will also derive the so-called generalized Manley-Rowe relations, which play an important role in presence of pair-creation.

10.3.1 *The $\varphi\psi$–model*

The electromagnetic Lagrangian for the full Hopfield model is quite involved. It is not much manageable for testing quantum effects, so, in order to gain insights into the real situation and carry out analytical calculations as far as possible, we now introduce a simplified model where the electromagnetic field and the polarization field are simulated by a pair of scalar fields, φ, and ψ respectively. Obviously we are interested in constructing a model in such a way to maintain exactly the same dispersion relation and to simulate the same coupling as in the full case. Our model is related to the two dimensional reduction of the Hopfield model adopted in [Finazzi and Carusotto (2013)].

Let us introduce

$$\mathcal{L}_{\varphi\psi} = \frac{1}{2}(\partial_\mu \varphi)(\partial^\mu \varphi) + \frac{1}{2\chi\omega_0^2}\left[(v^\alpha \partial_\alpha \psi)^2 - \omega_0^2 \psi^2\right] + \frac{g}{c}(v^\alpha \partial_\alpha \psi)\varphi, \quad (10.131)$$

where χ plays the role of the dielectric susceptibility, v^μ is the usual four-velocity vector of the dielectric, ω_0 is the proper frequency of the medium, and g is the coupling constant between the fields. As discussed in the previous section for the full Hopfield model, we work with a phenomenological model where we can leave room for a spacetime dependence of the microscopic parameters χ, ω_0, g. We will also interested in extending the model in such a way to include also $N > 1$ polarization fields ψ_i, each one with its own resonance frequency ω_{0i}, susceptibility χ_i, and coupling parameter g_i. However, we shall not use the latter freedom till the end of the section where the generalization will be discussed, and focus our attention on the single-resonance mode.

Omitting spacetime arguments, the equations of motion are

$$\Box\varphi - \frac{g}{c}(v^\alpha \partial_\alpha \psi) = 0, \quad (10.132)$$

$$(v^\alpha \partial_\alpha)\frac{1}{\chi\omega_0^2}(v^\alpha \partial_\alpha \psi) + \frac{1}{\chi}\psi + \frac{1}{c}v^\alpha \partial_\alpha (g\varphi) = 0. \quad (10.133)$$

These can be also rewritten as follows. Let G_ψ be the Green function for ψ:

$$\left[(v^\alpha\partial_\alpha)\frac{1}{\chi\omega_0^2}(v^\alpha\partial_\alpha)+\frac{1}{\chi}\right]G_\psi=\delta, \tag{10.134}$$

where δ is the Dirac delta function. Denoting $\tilde{x}=(t,x,y,z)$, we then get the following system

$$\psi(\tilde{x})=\frac{1}{c}\int d^4w G_\psi(\tilde{x}-w)(v^\alpha\partial_\alpha g\varphi)(w), \tag{10.135}$$

$$\Box\varphi(\tilde{x})-\frac{g}{c^2}(v^\alpha\partial_\alpha)\int d^4w G_\psi(\tilde{x}-w)(v^\beta\partial_\beta g\varphi)(w)=0, \tag{10.136}$$

which represent a simplified model equations set, simulating dispersive effects in optics.

It is also useful to introduce the corresponding Hamiltonian equations. The conjugate momenta are

$$\pi_\varphi:=\frac{\partial\mathcal{L}}{\partial\partial_0\varphi}=\partial_0\varphi, \tag{10.137}$$

$$\pi_\psi:=\frac{\partial\mathcal{L}}{\partial\partial_0\psi}=\frac{1}{\chi\omega_0^2}v^0v^\alpha\partial_\alpha\psi+\frac{g}{c}v^0\varphi, \tag{10.138}$$

so that the corresponding the Hamiltonian density $\mathcal{H}$ is

$$\begin{aligned}\mathcal{H}=&\frac{1}{2}\pi_\varphi^2+\frac{\chi\omega_0^2}{2(v^0)^2}\pi_\psi^2-\frac{v^k}{v^0}(\partial_k\psi)\pi_\psi-\frac{g}{c}\frac{\chi\omega_0^2}{v^0}\pi_\psi\varphi\\&+\frac{1}{2c^2}\chi\omega_0^2g^2\varphi^2+\frac{1}{2\chi}\psi^2+\frac{1}{2}(\partial_k\varphi)^2.\end{aligned} \tag{10.139}$$

Let us also introduce $H:=\int dx\mathcal{H}$. Then, we get

$$\partial_0\varphi=\{\varphi,H\}, \tag{10.140}$$

$$\partial_0\psi=\{\psi,H\}, \tag{10.141}$$

$$\partial_0\pi_\varphi=\{\pi_\varphi,H\}, \tag{10.142}$$

$$\partial_0\pi_\psi=\{\pi_\psi,H\}. \tag{10.143}$$

Notice that the following non trivial Poisson brackets hold true

$$\{\varphi,\pi_\varphi\}=\delta, \tag{10.144}$$

$$\{\psi,\pi_\psi\}=\delta, \tag{10.145}$$

which will play a role in the quantization of the fields, as in the Hopfield model.

As we have done in the electromagnetic case, we can extract a conserved scalar product from the complexified Hamiltonian structure of the model.

However, we want here to follow another line of thought, in order to illustrate the origin of the product from a particular symmetry. Assume the fields are complex and consider the following (global) phase transformation

$$\varphi \mapsto e^{ia}\varphi; \qquad \varphi^* \mapsto e^{-ia}\varphi^*, \tag{10.146}$$

$$\psi \mapsto e^{ia}\psi; \qquad \psi^* \mapsto e^{-ia}\psi^*. \tag{10.147}$$

This is a symmetry for the complexified Lagrangian density

$$\begin{aligned} \mathcal{L}^{complex}_{\varphi\psi} =& \frac{1}{2}(\partial_\mu\varphi^*)(\partial^\mu\varphi) + \frac{1}{2\chi\omega_0^2}\left[(v^\alpha\partial_\alpha\psi^*)(v^\alpha\partial_\alpha\psi) + \omega_0^2\psi^2\right] \\ &+ \frac{g}{2c}(v^\alpha\partial_\alpha\psi^*)\varphi + \frac{g}{2c}(v^\alpha\partial_\alpha\psi)\varphi^*. \end{aligned} \tag{10.148}$$

Then, we can apply Noether's theorem in order to determine the current

$$\begin{aligned} J^\mu =& \frac{i}{2}\left[\varphi^*\partial^\mu\varphi - (\partial^\mu\varphi^*)\varphi + \frac{1}{\chi\omega_0^2}v^\mu\psi^* v^\alpha\partial_\alpha\psi - \right. \\ &\left. \frac{1}{\chi\omega_0^2}v^\mu\psi v^\alpha\partial_\alpha\psi^* + \frac{g}{c}v^\mu(\psi^*\varphi - \psi\varphi^*)\right], \end{aligned} \tag{10.149}$$

which is conserved along the equation of motion: $\partial_\mu J^\mu = 0$. This provides the conserved charge

$$\int_{\Sigma_t} dx J^0, \tag{10.150}$$

which is what we need in order to define a conserved scalar product. In particular, we obtain

$$\begin{aligned} ((\varphi\ \psi)|(\tilde\varphi\ \tilde\psi)) =& \frac{i}{2}\int_{\Sigma_t}\left[\varphi^*\partial^0\tilde\varphi - (\partial^0\varphi^*)\tilde\varphi + \frac{1}{\chi\omega_0^2}v^0\psi^* v^\alpha\partial_\alpha\tilde\psi\right. \\ &\left. - \frac{1}{\chi\omega_0^2}v^0\tilde\psi v^\alpha\partial_\alpha\psi^* + \frac{1}{c}g v^0(\psi^*\tilde\varphi - \tilde\psi\varphi^*)\right], \end{aligned} \tag{10.151}$$

which will obviously play a particularly important in the definition of the quantum states for the model at hand.

This is exactly the same scalar product one obtains by analyzing the symplectic structure of the classical Hamiltonian equations: the starting point is still the complexified Lagrangian; one first defines

$$\pi_\varphi := \frac{\partial\mathcal{L}_c}{\partial\partial_0\varphi} = \frac{1}{2}\partial_0\varphi, \tag{10.152}$$

$$\pi^*_\varphi := \frac{\partial\mathcal{L}_c}{\partial\partial_0\varphi^*} = \frac{1}{2}\partial_0\varphi^*, \tag{10.153}$$

$$\pi_\psi := \frac{\partial\mathcal{L}_c}{\partial\partial_0\psi} = \frac{1}{2\chi\omega_0^2}v^0 v^\alpha\partial_\alpha\psi + \frac{g}{2c}v^0\varphi, \tag{10.154}$$

$$\pi^*_\psi := \frac{\partial\mathcal{L}_c}{\partial\partial_0\psi^*} = \frac{1}{2\chi\omega_0^2}v^0 v^\alpha\partial_\alpha\psi^* + \frac{g}{2c}\varphi^*. \tag{10.155}$$

Then, introducing the canonical matrix

$$\Omega := -i \begin{bmatrix} 0 & 1 \\ -1 & 0 \end{bmatrix}, \tag{10.156}$$

where the square matrix represent the canonical symplectic form of standard Hamiltonian classical mechanics, we can rewrite the conserved scalar product as follows

$$< \Psi_1, \Psi_2 >:= (\Psi_1, \Omega\Psi_2) =: \int dx \Psi_1^* \cdot \Omega\Psi_2, \tag{10.157}$$

where $\cdot$ is the ordinary Euclidean scalar product in $\mathbb{R}^4$, and

$$\Psi := \begin{pmatrix} \psi \\ \varphi \\ \pi_\psi \\ \pi_\varphi \end{pmatrix}. \tag{10.158}$$

10.3.2 *Separation of variables*

What we said up to now is quite general. Now we specialize to the case when a RIP moves with constant velocity v, and we work in the comoving frame. From the system (10.135,10.136) it is not immediately evident that in such a situation it is possible to separate variables, even though it is quite obvious for the transversal spatial variables y, z. In the case of the time variable t, we can perform the separation after writing the system (10.135), (10.136) as a first order system in t, which amounts to rewrite the aforementioned system of equations of motion in Hamiltonian form. This gives rise to the following Hamiltonian form

$$i\partial_0 \begin{pmatrix} \varphi \\ \psi \\ \pi_\varphi \\ \pi_\psi \end{pmatrix} = i \begin{bmatrix} 0 & 0 & 1 & 0 \\ -\frac{g}{c}\chi\omega_0^2 & -\frac{v^k}{v^0}\partial_k & 0 & -\frac{\chi\omega_0^2}{v^0} \\ (\nabla^2 - \frac{g^2}{c^2}\chi\omega_0^2) & 0 & 0 & \frac{g}{c}\chi\omega_0^2 \\ 0 & -\frac{1}{\chi} & 0 & -\frac{v^k}{v^0}\partial_k \end{bmatrix} \begin{pmatrix} \varphi \\ \psi \\ \pi_\varphi \\ \pi_\psi \end{pmatrix}. \tag{10.159}$$

In a more concise form

$$i\partial_t \Psi = \hat{H}\Psi, \tag{10.160}$$

where $\Psi := (\varphi, \psi, \pi_\varphi, \pi_\psi,)^T$ and where $\hat{H}$ is the matrix operator displayed in (10.159). Notice that $\hat{H}$ is formally self-adjoint in the scalar product $<,>$ defined above. Indeed, it is not difficult to show that the hermitian conjugate $\hat{H}_c = (\Omega H \Omega)^\dagger$ of $\hat{H}$, has the same form as the operator $\hat{H}$.

Since $\hat{H}$ is independent on t, we can find stationary solutions in the form

$$\Psi = \exp(-i\omega t)F(x, y, z), \tag{10.161}$$

this way separating the time variable t. Notice that ω is a conserved quantity. It corresponds to the following conservation law in the laboratory frame

$$\omega_{lab} - vk_{lab,x} = \text{const.} \tag{10.162}$$

This has been previously obtained in a perturbative approach and is now confirmed in an exact model. This conservation law is nothing but the conservation of energy in the comoving frame.

10.3.3 *Quantization of the fields*

The conserved scalar product induces the norm of the states as follows

$$\begin{aligned} ||\Psi||^2 :=< \Psi, \Psi >= \frac{i}{2}\int_{\Sigma_t} \Big[& \varphi^* \partial^0 \varphi - (\partial^0 \varphi^*)\varphi + \frac{1}{\chi\omega_0^2} v^0 \psi^* v^\alpha \partial_\alpha \psi \\ & - \frac{1}{\chi\omega_0^2} v^0 \psi v^\alpha \partial_\alpha \psi^* + \frac{g}{c} v^0 (\psi^* \varphi - \psi \varphi^*) \Big]. \end{aligned} \tag{10.163}$$

Particle states correspond to positive norm states (which is a notion which remains invariant under Lorentz group), whereas negative norm states correspond to antiparticles. This can be exemplified in the homogeneous case, i.e. in absence of the perturbation. For plane wave solutions, one finds that

$$||\Psi||^2 :=< \Psi, \Psi > \propto \omega_{lab}, \tag{10.164}$$

so that $\text{sign}||\Psi||^2 = \text{sign}\omega_{lab}$.

This result matches well what happens in the electromagnetic case. It shows that particles in the laboratory frame are defined by the condition $\omega_{lab} > 0$. Correspondingly, in the frame the particle states are defined by the condition $\omega > -vk$. These conditions represent a good indication for particle and antiparticle states also for the full problem. In order to clarify this point, let us consider the asymptotic behavior in x in the case of the two dimensional model, since in the four dimensional case, separation of variables also on transverse variables allows to show that only a little and not substantial modification of the calculations displayed below occurs. Working on stationary solutions, we get the following second order system of ordinary differential equations

$$\psi'' =((\log \chi\omega_0^2)' + 2i\frac{\omega}{v})\psi' + \left[-i\frac{\omega}{v}(\log \chi\omega_0^2)' - \frac{\omega_0^2}{\gamma^2 v^2}) + \frac{\omega^2}{v^2}\right]\psi$$
$$- i\frac{g}{c}\chi\omega_0^2\frac{\omega}{\gamma v^2}\varphi - \frac{1}{\gamma v c}\chi\omega_0^2\partial_x(g\varphi), \tag{10.165}$$

$$\varphi'' =\frac{g}{c}\gamma v\psi' - i\frac{g}{c}\gamma\omega\psi - \frac{\omega^2}{c^2}\varphi. \tag{10.166}$$

It is transformed in a first order system by introducing

$$p := \psi', \qquad q := \varphi'. \tag{10.167}$$

Thus, if $W(x) := (\psi(x), \varphi(x), p(x), q(x))^T$, we obtain the following first order system

$$W'(x) = K_4 W(x), \tag{10.168}$$

where the 4×4 matrix operator $K_4(x)$ has the structure

$$K_4 := \begin{bmatrix} 0_2 & 1_2 \\ A_2 & B_2 \end{bmatrix}, \tag{10.169}$$

where $0_2, 1_2$ are 2×2 matrices, the first one having all entries equal to zero, and the second one is the identity. The matrices A_2 and B_2 are

$$A_2 := \begin{bmatrix} \frac{\omega^2}{v^2} - i\frac{\omega}{v}(\log \chi\omega_0^2)' - \frac{\omega_0^2}{\gamma^2 v^2} & -i\frac{g}{c}\frac{\omega}{v}\frac{\chi\omega_0^2}{\gamma v} - \frac{\chi\omega_0^2}{\gamma v c}g' \\ -i\frac{g}{c}\gamma\omega & -\frac{\omega^2}{c^2} \end{bmatrix}, \tag{10.170}$$

and

$$B_2 := \begin{bmatrix} 2i\frac{\omega}{v} + (\log \chi\omega_0^2)' & -\frac{g}{c}\frac{\chi\omega_0^2}{\gamma v} \\ \frac{g}{c}\gamma v & 0 \end{bmatrix}. \tag{10.171}$$

Let us write

$$K_4 = \mathcal{C} + \mathcal{R}, \tag{10.172}$$

where $\mathcal{C}$ is a constant matrix, and $\mathcal{R} = \mathcal{R}(x)$:

$$\mathcal{C} := \begin{bmatrix} 0_2 & 1_2 \\ A_c & B_c \end{bmatrix}, \tag{10.173}$$

with

$$A_c := \begin{bmatrix} \frac{\omega^2}{v^2} - \frac{\omega_0^2}{\gamma^2 v^2} & 0 \\ 0 & -\frac{\omega^2}{c^2} \end{bmatrix}, \tag{10.174}$$

$$B_c := \begin{bmatrix} 2i\frac{\omega}{v} & 0 \\ 0 & 0 \end{bmatrix}; \tag{10.175}$$

whereas,

$$\mathcal{R} := \begin{bmatrix} 0_2 & 0_2 \\ A_r & B_r \end{bmatrix}, \tag{10.176}$$

with

$$A_r := \begin{bmatrix} -i\frac{\omega}{v}(\log \chi\omega_0^2)' & -i\frac{g}{c}\frac{\omega}{v}\frac{\chi\omega_0^2}{\gamma v} - \frac{\chi\omega_0^2}{\gamma v c}g' \\ -i\frac{g}{c}\gamma\omega & 0 \end{bmatrix}, \tag{10.177}$$

$$B_r := \begin{bmatrix} (\log \chi\omega_0^2)' & -\frac{g}{c}\frac{\chi\omega_0^2}{\gamma v} \\ \frac{g}{c}\gamma v & 0 \end{bmatrix}. \tag{10.178}$$

If we make the hypothesis

$$\int_a^\infty dx|\mathcal{R}(x)| < \infty, \tag{10.179}$$

as we have discussed in the previous section, we can thus infer that, both as $x \to \infty$ and as $x \to -\infty$, the asymptotic behavior of solutions is governed by the eigenvalues of $\mathcal{C}$. This implies that the basis for the homogeneous case with g, ω_0, and χ asymptotically constants, is asymptotically a good scattering basis also for the perturbed problem. Notice that this condition physically can match very well the nature of traveling perturbation of δn. This can be implemented by means of a suitable choice of the microscopic parameters g, ω_0, and χ. In general, we are interested in localized wave-packets, which have finite support.

This discussion corroborates the common use of the asymptotic dispersion relation in order to identify particle and antiparticle states. We will refer to these ad (DR)-asymptotic in the following, Fig. 10.1 is an example. In the Cauchy approximation we have

$$n(\omega_{lab}) = n(0) + B\omega_{lab}^2 + \delta n(x - vt), \tag{10.180}$$

where $n(0)$ is $n(\omega_{lab} = 0)$, B is a constant, and the Kerr effect induces the right-moving perturbation $\delta n(x - vt)$. In the Cauchy approximation, and in the comoving frame, the two branches are given by

$$G_\pm := \gamma(\omega + vp)(B\gamma^2(\omega + vp)^2 + n(x)) \pm \frac{c}{v}\left(\gamma(\omega + vp) - \frac{\omega}{\gamma}\right) = 0. \tag{10.181}$$

$G_+ = 0$ corresponds to the monotone branch, whereas $G_- = 0$ is the branch involved in the Hawking effect. The (DR)-asymptotic displays three states on the same branch G_- defied by equation (10.181), all having the same ω in the comoving frame: the positive group velocity particle state, which we

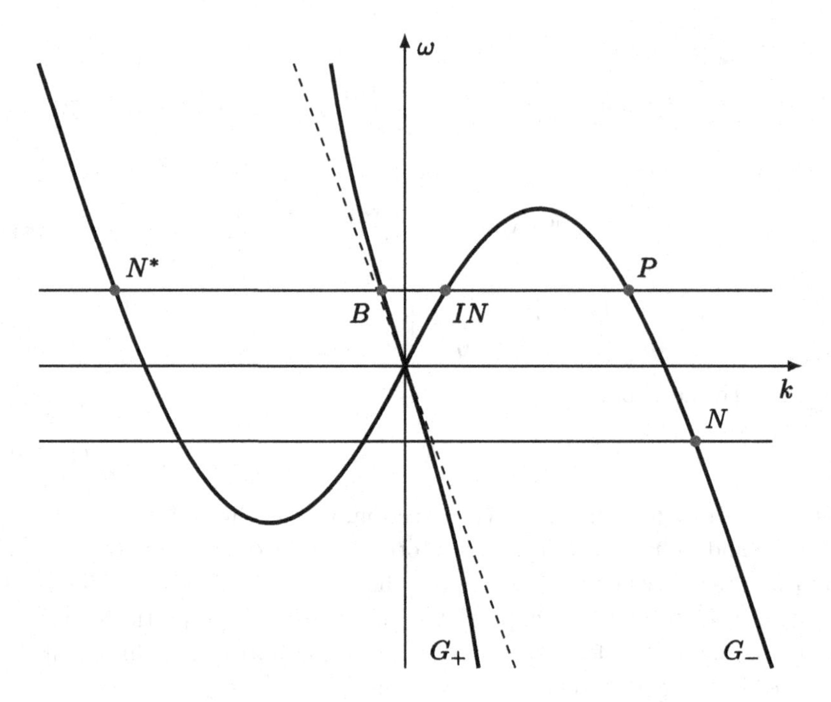

Fig. 10.1 Asymptotic dispersion relation for the Cauchy case in the comoving frame. The monotone branch is G_+, the non-monotone one is G_-. The dashed line divides antiparticle states (below it) and particle ones (above it). Two lines at ω =const and at $-\omega$ =const are also drawn, and relevant states introduced in the text are explicitly indicated.

call IN state, the negative group velocity particle state, which we call the P state, and a negative group velocity antiparticle state N^*. Notice that each state specified by a pair (ω, k) is conjugated to a state corresponding to $(-\omega, -k)$, so, in particular, we can consider the state N, which is the conjugate of N^*, in place of N^*.

There is also a second branch G_+, defined by equation (10.181), which for the same ω displays only a single negative group velocity particle state B. However, we will consider only former branch G_- as involved in the Hawking process.

We also stress that, with some abuse of language, we maintain also for the black hole states the denomination we use for white hole ones. In particular, IN is an incoming state in the case of scattering on a white hole horizon (which is relevant for the experiment [Belgiorno *et al.* (2010a)]), and states P, N, B are late time states in a scattering framework. In the black hole case, we have a time-reversed situation, where P, N, B are early time scattering states, and IN is the emergent one (the Hawking state, sometimes also indicated with H [Linder *et al.* (2016); Belgiorno *et al.* (2017b)]).

Now, the quantization of the fields follows the general lines we have illustrated in the section devoted to the full electromagnetic case, with the important simplification deriving by the absence of constraints, which makes much easier and standard the whole treatment. In particular, in the static comoving frame, we have

$$\Psi(t, \mathbf{x}) = \int \frac{d\omega}{2\pi} \frac{d\mathbf{k}}{(2\pi)^3} \frac{1}{N_\Psi} \delta(DR) \left(a(\omega, \mathbf{k}) U(\omega, \mathbf{k}; t, \mathbf{x}) + a^\dagger(\omega, \mathbf{k}) V(\omega, \mathbf{k}; t, \mathbf{x}) \right), \tag{10.182}$$

where N_Ψ is a normalization factor, $\delta(DR)$ indicates that we are considering "on shell" solutions, and U and V are positive norm solutions and negative norm solutions of the field equations respectively. More precisely, we are interested in the field

$$\Psi_{(IN)}(t, \mathbf{x}), \tag{10.183}$$

which represents the field in the asymptotic past, as $t \to -\infty$, and the corresponding field in the asymptotic future,

$$\Psi_{(OUT)}(t, \mathbf{x}). \tag{10.184}$$

We then consider a single transition from a region with a given set of parameters ω_0, χ, and g to a region with parameters ω_0', χ', and g', and interpolate the above asymptotic field by the field (10.182).

Notice that this on shell quantization, in a two-dimensional model, can give rise both to an ω-representation, where only the integration in $d\omega$ is left, or also to a k-representation, where the dispersion relation is used to leave only the integral in dk. It is clear that in a four-dimensional model, the latter choice is to be preferred.

In order to compute amplitudes for pair-creation, let us expand in plane waves both the IN state and the P, N, B states emerging from the scattering process. Then consider

$$|N|^2 := \frac{|J_x^N|}{|J_x^{IN}|}, \tag{10.185}$$

where J_x is the conserved current (10.149), and the indexes N, IN indicate that one is considering the N-particle states and the IN-particle states respectively, and similar for P and B. The quantity $|N|^2$ defined above is the ratio between the outgoing flux along x of negative energy particle states and the flux along x of ingoing particle states: it coincides with the mean number per unit time and unit volume of created particles [Damour (1975); Nikishov (1969); Gavrilov *et al.* (2008)].

Indeed, we now provide a more systematic account of pair-creation amplitudes, which can be used both for analytical and for numerical calculations. First of all, we will focus on the two-dimensional problem and fix a scattering basis in the asymptotic regions. In the region $x \to -\infty$, at a fixed value of ω we have a number of states which is equal to the number of intersections between the horizontal line ω =const and the asymptotic dispersion relation $\omega = f(k)$, represented in figure 10.1. Some of them can have positive group velocity $v_g > 0$, thus representing right moving packets, and some others can have negative group velocity $v_g < 0$, which thus are left moving. In situations where a blocking horizon is present, only states that belong to the asymptotic region on the left (with $\delta n = 0$) are involved. Instead, when blocking is absent or if the frequencies involved do not admit blocking, then also (transmitted) scattering states that belong to the asymptotic region on the right (with $\delta n \neq 0$) are involved.

For what regards the Hawking effect, it is worth to remark that in the present section we concentrate on the scattering process involved for a white hole, whereas in 10.5 we will discuss thermality for a black hole. There is no contradiction, since white hole is the time-reversal of a black hole, but, at the level of scattering this is not a trivial statement and it is better to investigate what happens. Time reversal implies $(\omega, v) \mapsto (-\omega, -v)$. Notice that the original Hopfield model is invariant under time reversal, and the same holds for the generalization with varying χ, ω_0, g, as the Lagrangian of the model is. Also for the $\varphi\psi$-model the equations of motion are invariant under time-reversal provided that $g \mapsto -g$. This freedom for the scalar model can be assumed without any problem, provided that the correct branch for the relation between microscopic parameters and macroscopic ones is chosen.

In general, we expect to deal with several branches of the dispersion relation. In that case, it may looks more useful to consider the asymptotic dispersion relations in the laboratory frame. An analogous reasonings leads to a number of states as $x \to \infty$, for which, again, the group velocity may be positive or negative. We then obtain a complete scattering basis by considering both a scattering with one initial right moving state, which can give rise to a transmitted state and several reflected states, and an analogous scattering which starts with an initial left moving state. This is particularly relevant for the actual computation of the pair creation amplitude in the case of spontaneous emission.

There are different possible ways of performing the calculation of amplitudes: the traditional framework, by means of Bogoliubov transformations, or by means of the conserved fluxes in the scattering process in the comoving frame. We will adopt the latter frame (see also [Kravtsov *et al.* (1974); Ostrovskii (1972); Sorokin (1972)] for former calculations leading to generalized Manley-Rowe relations in optics). In Section 10.3.1, we have shown that in the comoving frame it exists a conserved current J_x. In particular, we can consider J_x as a bilinear form

$$J_x(\Psi_1, \Psi_2), \tag{10.186}$$

for a pair of asymptotic plane wave solutions of the equations of motion Ψ_1, Ψ_2. In particular, we will indicate with $\Psi_{\rightarrow}$ the scattering solutions for a process with a single initial state which is right moving. An obviously specular definition is given for the left scattering, whose scattered states will be denoted with $\Psi_{\leftarrow}$. For independent solutions we can determine various "wronskian relations". For example we can consider

$$J_x(\Psi^*_{\rightarrow}, \Psi_{\rightarrow}). \tag{10.187}$$

Let us consider the process $IN \rightarrow P + N^* + B + T$, where T stays for a possible transmitted state. Consider the state

$$\Psi_{\rightarrow} = N_{\Psi} e^{-i\omega t} \begin{cases} T^{\rightarrow}_{IN} W_{IN} e^{ik_{IN}x} + T^{\rightarrow}_{P} W_P e^{-ik_P x} & \\ +T^{\rightarrow}_{N^*} W_{N^*} e^{-ik_{N^*}x} + T^{\rightarrow}_{B} W_B e^{-ik_B x} & \text{for } x \rightarrow -\infty, \\ T^{\rightarrow}_{T} W_T e^{ik_T x} & \text{for } x \rightarrow \infty, \end{cases} \tag{10.188}$$

where W_{IN} etc. are vector Fourier components of the considered plane wave, and $T^{\rightarrow}_{IN}$ etc. are the usual scattering coefficients with the additional label $\rightarrow$ according to the notation introduced above. Then, (10.187) takes the form

$$1 - |P|^2 - |B|^2 - |T|^2 + |N|^2 = 0. \tag{10.189}$$

Now, as usual, for long time scales with respect to the interaction scale, we expect that photon packets for the various mode will separate. Indeed, this emerges from current conservation

$$J_x^{left} = J_x^{right}, \tag{10.190}$$

where "left" and "right" indicate states on the left and on the right of the step-like potential as $x \to -\infty$ and $x \to \infty$ respectively. For well distinct packets

$$J_x^{IN} + J_x^P + J_x^{N^*} + J_x^B = J_x^T. \tag{10.191}$$

The sign of these quantities is determined by

$$J_x^U = \text{sign}(v_g^U)|J_x^U|, \tag{10.192}$$

where U is intended to be an asymptotic solution with positive norm. Furthermore, we have that the antiparticle state current is opposite to the corresponding particle state current, t.i.

$$J_x^{N^*} = -J_x^N. \tag{10.193}$$

Then, let us introduce the following quantities:

$$\frac{J_x^U}{J_x^{IN}} =: \text{sign}\left(\frac{v_g^U}{v_g^{IN}}\right)|U|^2, \qquad U = P, B, T, N, \tag{10.194}$$

which satisfy

$$|U|^2 = F(\omega, k_U)|T_{\vec{U}}|^2, \qquad U = P, B, T, N \tag{10.195}$$

where $F(\omega, k_U)$ is a positive kinematic coefficient which depends explicitly on the current structure. The extension of this analysis to the case of an arbitrary number of states is straightforward and is left to the reader.

In the spontaneous case, we have to consider

$$< J_x >:=< 0|J_x|0 >= \sum_{\rightarrow}\sum_{\leftarrow}\int \frac{d\omega}{(2\pi)}\frac{dk}{(2\pi)}\frac{1}{N_\Psi}\delta(\text{dispersion relation})V^*V, \tag{10.196}$$

where the sum is extended to both the initial left-moving states and the initial right moving ones. Remarkably, this leads to the same particle creation amplitudes associated with (10.189).

One can also conclude that the same amplitudes as above would be obtained starting from the definition of the Poynting vector for the theory at hand to determine the conservation of the fluxes.

10.4 An interlude: Semi-analytical and numerical calculations from Maxwell equations in the lab

The analysis in [Rubino *et al.* (2012.); Petev *et al.* (2013)] achieved important results involving both numerical and semi-analytical arguments, relative to white holes in dielectric media, which our analytical calculations to be discussed below corroborate, to the extent they are able to cover. The model is more phenomenological with respect to the Hopfield model which represent the topic of the present chapter, still Maxwell equations and their approximations in the lab frame are able to provide a valid benchmark for the analogue Hawking effect in dielectric media. There is a different naming of states, which is more appropriate in relation with studies on the electromagnetic field. See in particular [Rubino *et al.* (2012)]. The positive norm state which is indicated previously with P is therein indicated as RR state, which stays for 'resonant radiation', whereas the negative norm state N is called NRR, which stays for 'negative resonant radiation'. The ratio

$$r := \frac{|NRR|^2}{|RR|^2} \tag{10.197}$$

is equivalent to eq. (10.224). To sum up the content of the aforementioned studies, from the one hand, a perturbative (Born approximation) semi-analytical method involving the so-called unidirectional optical pulse propagation equation (UPPE) approximation [Kolesik *et al.* (2002); Kolesik and Moloney (2004)] was developed, and applied in [Rubino *et al.* (2012.); Petev *et al.* (2013); Rubino (2013)]. On the other hand, numerical simulations involving both the UPPE approximation and, independently, the finite difference time domain (FDTD) [Taflove and Hagness (2005)], confirmed the presence of a blackbody spectrum in presence of a blocking horizon, and the presence of a grey-body spectrum also in the nonblocking case, where an additional transmitted mode F appears [Petev *et al.* (2013)]. The counter-propagating mode B in all cases gives negligible contributions. We provide below a more detailed description.

10.4.1 *Born approximation*

We evaluate the elements *in the lab frame*, of the S-matrix in the first Born approximation, giving an analytical formulation for the spectral amplitude of the scattered wave. We also generalize the treatment in order to include the possibility of a third output scattered mode, i.e. the NRR mode that, as argued above may also be generated during the scattering process. In this section the flight direction is indicated with z_ℓ. The role of the scattering potential, $R(z_\ell, t_\ell)$, is played by the moving relativistic inhomogeneity (RI), which we assume to be stationary along propagation, i.e. $R(z_\ell, t_\ell) = R(t_\ell - z_\ell/v) = n(t_\ell - z_\ell/v)$. In detail, let us consider the laser pulse evolution 'unidirectional pulse propagation equation' (UPPE) in the laboratory reference frame:

$$\partial_{z_\ell} E_{\omega_\ell} - ik_{z_\ell}(\omega_\ell) E_{\omega_\ell} = i \frac{\omega_\ell^2}{2\epsilon_0 c^2 k_{z_\ell}(\omega_\ell)} P_{\omega_\ell}, \tag{10.198}$$

where

$$\begin{aligned} P(z_\ell, t_\ell) &= \epsilon_0 \left\{ [n_0 + \delta n(t_\ell - z_\ell/v)]^2 n_0^2 \right\} E(z_\ell, t_\ell) \\ &\simeq 2\epsilon_0 n_0 \delta n(t_\ell - z_\ell/v) E(z_\ell, t_\ell), \end{aligned} \tag{10.199}$$

so that

$$P_{\omega_\ell} = \frac{1}{2\pi} \int dt_\ell 2\epsilon_0 n_0 \delta n(t_\ell - z_\ell/v) E(z_\ell, t_\ell) e^{i\omega_\ell t_\ell}. \tag{10.200}$$

Equation (10.198) is the so-called unidirectional pulse propagation equation described in the supplementary information of [Rubino *et al.* (2012.)]. We want to solve this equation perturbatively by writing

$$E_{\omega_\ell}(z_\ell) = E_{\omega_\ell}^{(0)}(z_\ell) + E_{\omega_\ell}^{(1)}(z_\ell) + E_{\omega_\ell}^{(2)}(z_\ell) + \ldots, \tag{10.201}$$

where $E_{\omega_\ell}^{(p)}(z_\ell)$ is of order $(\delta n)^p$. Then we get the equations

$$\partial_{z_\ell} E_{\omega_\ell}^{(0)} - ik_{z_\ell}(\omega_\ell) E_{\omega_\ell}^{(0)} = 0, \tag{10.202}$$

$$\begin{aligned} \partial_{z_\ell} E_{\omega_\ell}^{(1)} - ik_{z_\ell}(\omega_\ell) E_{\omega_\ell}^{(1)} &= i \frac{\omega_\ell^2 n_0}{c^2 k_{z_\ell}(\omega_\ell)} \frac{1}{2\pi} \int dt_\ell \delta n(t_\ell - z_\ell/v) \\ &\quad \times \int d\bar{\omega} E_{\bar{\omega}}^{(0)}(z_\ell) e^{-i\bar{\omega} t_\ell},, \\ &\vdots \end{aligned} \tag{10.203}$$

From (10.202) we obtain

$$E_{\omega_\ell}^{(0)}(z_\ell) = e^{ik_{z_\ell}(\omega_\ell) z_\ell} E_{\omega_\ell}^0, \tag{10.204}$$

where $E_{\omega_\ell}^0$ is constant. Assuming that for $z_\ell \to -\infty$ there is only the unperturbed (zeroth order) solution, we also find

$$E_{\omega_\ell}^{(1)}(z_\ell) = \int d\bar{\omega} \sigma(z_\ell, \omega_\ell, \bar{\omega}) E_{\bar{\omega}}^0(z_\ell), \tag{10.205}$$

where we have indicated with σ the S-matrix density:

$$\begin{aligned} \sigma(z_\ell, \omega_\ell, \omega_{\ell IN}) : &= i \frac{\omega_\ell^2 n_0}{2\pi c^2 k_{z_\ell}(\omega_\ell)} e^{i[k_{z_\ell}(\omega_\ell) - k_{z_\ell}(\omega_{\ell IN})]} \int dt_\ell \int_{-\infty}^{z_\ell} d\xi \\ &\quad \times e^{i[\omega_\ell t_\ell - k_{z_\ell}(\omega_\ell)\xi]} \delta n(t_\ell \xi/v) e^{-i[\omega_{\ell IN} t_\ell - k_{z_\ell}(\omega_{\ell IN})\xi]}. \end{aligned} \tag{10.206}$$

By introducing $u := t_\ell - \xi/v$ and $w := t_\ell + \xi/v$ we can evaluate the integrals in the last expression, to obtain (as $z \to \infty$)

$$\sigma(z_\ell,\omega_\ell,\omega_{\ell IN}) \sim i\frac{v\omega_\ell^2 n_0}{2c^2 k_{z_\ell}(\omega_\ell)} e^{i[k_{z_\ell}(\omega_\ell)-k_{z_\ell}(\omega_{\ell IN})]}$$
$$\times\delta\left[(\omega_\ell - vk_{z_\ell}(\omega_\ell)) - (\omega_{\ell IN} - vk_{z_\ell}(\omega_{\ell IN}))\right]$$
$$\times\hat{R}\left(\frac{1}{2}\left[\omega_\ell - \omega_{\ell IN} + v[k_{z_\ell}(\omega_\ell) - k_{z_\ell}(\omega_{\ell IN})]\right]\right), \quad (10.207)$$

where $\hat{R}$ is the Fourier transform of the scattering potential $\delta n(u)$:

$$\hat{R}(\bar{\omega}) := \int du\delta n(u)e^{i\bar{\omega}u}. \quad (10.208)$$

Then we obtain

$$E^{(1)}_{\omega_\ell}(z_\ell) = S(z_\ell,\omega_\ell,\omega_{\ell IN})E^{(0)}_{\omega_{\ell IN}}(z_\ell), \quad (10.209)$$

where the S-matrix element is

$$S(z_\ell,\omega_\ell,\omega_{\ell IN}) = i\frac{v\omega_\ell^2 n_0}{2c^2 k_{z_\ell}(\omega_\ell)} e^{i(\omega_\ell-\omega_{\ell IN})z_\ell/v}\hat{R}(\omega_\ell - \omega_{\ell IN}), \quad (10.210)$$

and it holds (cf. the δ-function in (10.207))

$$\omega_\ell - vk_{z_\ell}(\omega_\ell) = \omega_{\ell IN} - vk_{z_\ell}(\omega_{\ell IN}), \quad (10.211)$$

which amounts to frequency conservation in the comoving frame.

10.4.2 *Thermality for a Gaussian perturbation*

In the supplementary information relative to [Petev *et al.* (2013)] we have shown that thermality can be achieved also analytically also in this framework, at least in the limit as $\omega \to 0$.

In order to simplify the notation, we use $\omega_{NRR}, \omega_{RR}, \omega_{IN}$ for frequencies in the lab frame and ω, k_{IN}, k_P, k_N for frequencies and momenta in the comoving frame.

The scattering amplitudes of the modes RR and NRR are given by

$$S(\omega_{\rm RR}) = i\frac{v\omega_{\rm RR}^2 n_0}{2c^2 k_z(\omega_{\rm RR})} e^{i\frac{(\omega_{\rm RR}-\omega_{\rm IN})z}{v}}\hat{R}(\omega_{\rm RR} - \omega_{\rm IN}) \quad (10.212)$$

$$S(\omega_{\rm NRR}) = i\frac{v\omega_{\rm NRR}^2 n_0}{2c^2 k_z(\omega_{\rm NRR})} e^{i\frac{(\omega_{\rm NRR}+\omega_{\rm IN})z}{v}}\hat{R}(\omega_{\rm NRR} + \omega_{\rm IN}). \quad (10.213)$$

These amplitudes can be used for calculating the following ratio. In the limit as $\omega \to 0$ in the comoving frame, the ratio $|N|^2/|P|^2$ corresponds to a thermal spectrum. In the lab frame we have to compute

$$\frac{\omega_{NRR}^2}{k_{NRR}^2}\frac{k_{RR}^2}{\omega_{RR}^2}\left|\frac{\hat{R}(\omega_{NRR}+\omega_{IN})}{\hat{R}(\omega_{RR}-\omega_{IN})}\right|^2. \quad (10.214)$$

We choose

$$\delta n = \delta n_{max}e^{-u^2/\sigma^2}. \quad (10.215)$$

Then

$$\hat{R}(s) = \sqrt{\frac{\sigma^2}{2}} e^{-1/4s^2\sigma^2}, \tag{10.216}$$

which implies

$$\frac{\hat{R}(\omega_{NRR} + \omega_{IN})}{\hat{R}(\omega_{RR} - \omega_{IN})} = e^{-1/4(\omega_{NRR}+\omega_{RR})(\omega_{NRR}-\omega_{RR}+2\omega_{IN})\sigma^2}. \tag{10.217}$$

In the comoving frame, the latter quantity becomes

$$e^{-1/4\gamma^2 v^2 (k_N+k_P)(k_N-k_P+2k_{IN})\sigma^2}. \tag{10.218}$$

In the limit as $\omega \to 0$, both k_P and k_N tend to a same positive value k_0 (as can be seen from the dispersion curve in the comoving frame), whereas k_{IN} tends to zero linearly in ω: $k_{IN} \sim \zeta\omega$, where ζ is a constant.

As to the factor

$$\frac{\omega_{NRR}^2}{k_{NRR}^2} \frac{k_{RR}^2}{\omega_{RR}^2}, \tag{10.219}$$

it is easy to show that it tends to 1 as $\omega \to 0$.

As a consequence, we get

$$|N|^2/|P|^2 \sim e^{-\alpha\omega} \qquad \text{as } \omega \to 0, \tag{10.220}$$

with

$$\alpha = 2\gamma^2 v^2 k_0 \sigma^2 \zeta. \tag{10.221}$$

Thermality can be recovered also for other perturbations (step-like ones are also discussed in the supplementary information associated with [Petev *et al.* (2013)]). As to the cases of super-Gaussian profiles with $m > 1$ discussed in [Petev *et al.* (2013)], we can point out that, as m increases, the trailing profile of the pulse more and more looks like a step-like function (for $l \gg 1$ and $q \gg 1$ respectively) of the kind we studied in the supplementary information associated with [Petev *et al.* (2013)], so we can safely expect an analogous behavior as $\omega \to 0$. We also recall that the photon numbers, denoted e.g. with $|RR|^2$ and $|NRR|^2$ where calculated as

$$\int_{\mathrm{RR,NRR}} \frac{|E(\omega_\ell)|^2}{\omega_\ell} d\omega_\ell \tag{10.222}$$

where $\int_{\mathrm{RR,NRR}}$ indicates integration over the extent of the RR or NRR spectral peaks. These quantities are of course only proportional to the actual photon numbers as they neglect some constants that in any case cancel out when normalizing to the input photon number and in any case do not change the functional dependence on ω_ℓ that is the main focus of the present analysis. The normalized photon number is thus taken as

$$|RR|^2, |NRR|^2 = \frac{\int_{\text{RR,NRR}} \frac{|E(\omega_\ell)|^2}{\omega_\ell} d\omega_\ell}{\int_{\text{IN}} \frac{|E(\omega_\ell)|^2}{\omega_\ell} d\omega_\ell}. \tag{10.223}$$

We note that in a non-dispersive medium the energy density is simply $u \propto |E|^2$ but in a dispersive medium it is corrected by a frequency dependent term, i.e. $u \propto |E|^2(\varepsilon + \omega_\ell d\varepsilon/d\omega_\ell)$, where $\varepsilon = \varepsilon(\omega_\ell)$ is the frequency dependent dielectric constant of the medium. In simulations occurring in [Petev *et al.* (2013)] this corrective factor does actually modify in any way the results and can be safely neglected.

10.4.3 *A sample of numerical results*

A series of configurations and profiles were tested numerically in [Petev *et al.* (2013)]. Results obtained by means of the aforementioned Born approximation of the UPPE have been confirmed by other independent numerical methods which are not unidirectional: *a priori*, this limitation could be jeopardized by the fact that the fourth counter-propagating mode B cannot be simulated. Still, at least in the range of the explored configurations the approximation has been shown to be robust, and B is substantially negligible. It is also interesting to point out that in the nonblocking case (i.e. the subcritical case, which is occurring e.g. in the Bessel filament experiment described in [Belgiorno *et al.* (2010a); Rubino *et al.* (2011)]) it is still possible to find out thermality, albeit with a grey body factor which is associated with the transmitted mode F we mentioned above.

However, we note that in crossing over into the non-blocking case the appearance of the additional transmitted mode ($|F|^2 > 0$) significantly distorts the blackbody spectrum. Indeed, if we plot the NRR photon number for a non-blocking case [dotted line in Fig. 10.2(a)] we see that at small frequencies the mode number drops to zero rather than diverging as would be expected for a 1D blackbody emission. This can be understood by noting that for high enough frequencies a blocking horizon is still present [region between the dispersion curve maxima indicated with stars in Fig. 10.2(c)] but low frequencies are not blocked at all, implying that the interaction time with the δn trailing edge is significantly reduced and any scattering

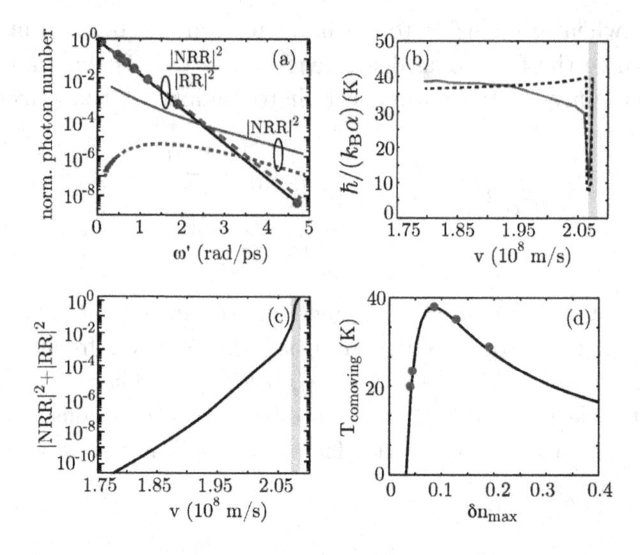

Fig. 10.2 (a) Numerical simulation for $v = 1.9 \times 10^8$ m/s of $r = |NRR|^2/|RR|^2 = \exp(-\alpha\omega')$ (cf. 10.224) for varying input frequency (dots) and best fit with exponential function (solid line). ω' stays for the comoving frame frequency (indicated as ω in the text). Dashed line: Born approximation calculation. Also shown is the normalized photon number $|NRR|^2$ for the blocking case (solid line) and the nonblocking one (dotted line). (b) $\hbar/k_B\alpha$ (derived from graphs as in (a)) for varying δn speed-simulations (solid line) and Born approximation model (dashed line). The shaded area indicates speeds for which a blocking horizon is formed. (c) Total normalized photon count for varying speed. (d) Comparison between numerically estimated temperature $T = \hbar/k_B\alpha$ (dots) and theoretical Hawking emission temperature estimated from the δn gradient (solid line) for varying maximum index change δn_{max} (blocking case).

process is suppressed. This suppression is also observed in Fig. 10.2(c) that shows the total photon number $|NRR|^2 + |RR|^2$ evaluated from the numerical simulations. As can be seen, this number varies by several orders of magnitude with varying v and is dramatically enhanced only in the presence of a blocking horizon.

Furthermore, a confirmation of the signature of the Hawking emission as a peak in the lab at a frequency which in the comoving frame corresponds to the zero frequency is also obtained.

10.5 Calculation of thermality in the $\phi\psi$ model

We will now set up an analytical approach allowing us to infer thermality for our model. We will refer principally to [Corley (1998); Coutant *et al.*

(2012b)], and for the specific case of dielectric media to [Belgiorno *et al.* (2015); Linder *et al.* (2016); Belgiorno *et al.* (2017b)]. More precisely, we will identify a common mechanism for thermality, which includes both our model and the general fluid model discussed in [Coutant *et al.* (2012b)], which is a refinement of the seminal calculations by Corley [Corley (1998)]. Our strategy will be as follows: one starts from a WKB approximation, introduced at the level of the calculations for asymptotic states far from the horizon. Then one matches this expansion with the near-horizon expansion, which is instead treated in Fourier space, in the approximation up to the linear order in x. If one assumes a not very strong gradient of the refractive index, it is not difficult to show that the state B (one of the four asymptotic states, which belongs to the monotone branch of the dispersion relation), decouples and gives rise to a scattering phenomenon which is an almost negligible fraction of the dominant phenomenon represented by the Hawking effect. On the other hand, as a consequence of the construction of the states in the near-horizon region, the states which match with the asymptotic states P and N, which are positive and negative norm states emerging from the scattering and lying on the non-monotone branch of the dispersion relation, are such that the ratio

$$\frac{|N|^2}{|P|^2} = \exp\left(-\beta_h \hbar\omega\right), \tag{10.224}$$

where β_h is the black hole temperature. Thus, it leads to the standard thermal spectrum as far as the fourth state is negligibly coupled, t.i. when $|B|^2$ is negligible.

The model which is discussed in [Belgiorno *et al.* (2015)] is based on the assumption that only the coupling constant is perturbed, so that we can get $g \mapsto g_0 + \delta g(x)$, where g_0 is constant. In this chapter, we present a different choice, where we allow a dependence on x of the susceptibility $\chi \mapsto \chi_0 + \delta\chi(x)$, as well as $\omega_0^2 \mapsto \omega_0^2 + \delta\omega_0^2(x)$, and which is inspired by our study for the full so-called Hopfield-Kerr model appearing in [Belgiorno *et al.* (2017b)]. We improve our previous model by introducing a nonlinear term in the Lagrangian and also by our different choice for the origin of thermality. We add to the Lagrangian (10.131) the fourth order term

$$\mathcal{L}_{nonlinear} := -\frac{\lambda}{4!}\psi^4. \tag{10.225}$$

We linearize around a solitonic solution $\psi_0(x)$ of the exact equations of motion inherited from the above action (for explicit calculations, see [Belgiorno *et al.* (2017b)]). As a consequence, we obtain the linearized equations of motion

$$\Box\phi - \frac{g}{c}v^\mu\partial_\mu\psi = 0, \tag{10.226}$$

$$\frac{1}{\chi\omega_0^2}(v^\mu\partial_\mu)^2\psi + \frac{g}{c}v^\mu\partial_\mu\phi + \frac{1}{\chi}\psi + \frac{\lambda}{2}\psi_0^2\psi = 0. \tag{10.227}$$

We observe that this amounts to the following shifts in χ and in ω_0^2 with respect to the linear case $\lambda = 0$:

$$\frac{1}{\chi} \mapsto \frac{1}{\chi}\left(1 + \chi\frac{\lambda}{2}\psi_0^2(x)\right), \tag{10.228}$$

$$\omega_0^2 \mapsto \omega_0^2\left(1 + \chi\frac{\lambda}{2}\psi_0^2(x)\right), \tag{10.229}$$

in such a way that $\chi\omega_0^2$ remains invariant: $\chi\omega_0^2 \mapsto \chi\omega_0^2$.
If we linearize near the horizon $x = 0$

$$\frac{1}{\chi}(1 + \chi\frac{\lambda}{2}\psi_0^2(x)) \sim \frac{1}{\chi}\left(1 + \chi\frac{\lambda}{2}\psi_0^2(0) + \chi\frac{\lambda}{2}(\psi_0^2)'(0)x\right), \tag{10.230}$$

where $(\psi_0^2)'(0)$ stays for the derivative of $\psi_0^2(x)$ at $x = 0$, and pass to the Fourier space, by taking into account that $x \mapsto i\partial_k$, we obtain the following system:

$$\tilde{\phi} = \frac{-i\frac{g}{c}v^\mu k_\mu}{\frac{\omega^2}{c^2} - k^2}\tilde{\psi}, \tag{10.231}$$

$$\left[i\frac{\lambda}{2}(\psi_0^2)'(0)\partial_k - \frac{1}{\chi\omega_0^2}(v^\mu k_\mu)^2 + \frac{1}{\chi}\left(1 + \chi\frac{\lambda}{2}\psi_0^2(0)\right) \right.$$
$$\left. + \frac{g^2}{c^2}(v^\mu k_\mu)^2\frac{1}{\frac{\omega^2}{c^2} - k^2}\right]\tilde{\psi} = 0. \tag{10.232}$$

10.5.1 *Determination of the microscopic parameters in terms of the physical ones*

In the eikonal approximation (WKB) we find

$$n^2(s, x) = 1 + \frac{g^2\chi\omega_0^2}{\omega_0^2(x) - s^2}, \tag{10.233}$$

where

$$s = \gamma(\omega + vk), \tag{10.234}$$

and

$$\omega_0^2(x) = \omega_0^2\left(1 + \chi\frac{\lambda}{2}\psi_0^2(x)\right). \tag{10.235}$$

In the Cauchy approximation we have

$$n(s,x) = n_0 + Bs^2 + \delta n(x), \tag{10.236}$$

so that, by comparison and in the limit as $s^2 \ll \omega_0^2$, we obtain

$$n_0^2 = 1 + g^2\chi, \tag{10.237}$$

$$2n_0 B = g^2\chi\frac{1}{\omega_0^2}, \tag{10.238}$$

$$\chi\frac{\lambda}{2}\psi_0^2(x) = -2n_0\frac{1}{g^2\chi}\delta n(x), \tag{10.239}$$

so that

$$g^2\chi = n_0^2 - 1 \sim \frac{c^2}{v^2}\frac{1}{\gamma^2}, \tag{10.240}$$

$$\omega_0^2 = g^2\chi\frac{1}{2n_0 B} \sim \frac{1}{2B}\frac{c}{v}\frac{1}{\gamma^2}, \tag{10.241}$$

$$\chi\frac{\lambda}{2}\psi_0^2(x) = -\frac{2n_0}{n_0^2 - 1}\delta n(x). \tag{10.242}$$

It is interesting to note that at the horizon, it holds

$$\chi\frac{\lambda}{2}(\psi_0^2)'(0) \sim -2\frac{v}{c}\gamma^2 n'(0). \tag{10.243}$$

Note also that $n'(x) = (\delta n(x))'$.

10.5.2 *Near horizon approximation: Solutions of equation (10.232)*

Let us define

$$\alpha := -2n_0 n'(0)\frac{1}{g^2\chi} \sim -2\frac{v}{c}\gamma^2 n'(0). \tag{10.244}$$

We note that, at a black hole horizon, $n'(0) < 0$, and then we get $\alpha = |\alpha| > 0$. Then (10.232) becomes

$$i\alpha\partial_k\tilde{\psi} = \left[\frac{\gamma^2(\omega + vk)^2}{\omega_0^2} - \frac{g^2}{c^2}\chi\gamma^2(\omega + vk)^2\frac{1}{\frac{\omega^2}{c^2} - k^2} - C_0\right]\tilde{\psi}, \tag{10.245}$$

where

$$C_0 = 1 - 2n_0\delta n(0)\frac{1}{g^2\chi} \tag{10.246}$$

is a constant.

We obtain

$$i\alpha \log \tilde{\psi} = -(C_0 - \frac{g^2}{c^2}\chi\gamma^2 v^2)k + a_+\omega \log(kc-\omega) - a_-\omega \log(kc+\omega) + \frac{\gamma^2(\omega+vk)^3}{3v\omega_0^2} + C_1 \tag{10.247}$$

where C_1 is an integration constant, and

$$a_\pm := \frac{g^2}{c^2}\chi\gamma^2\frac{c}{2}(1\pm\frac{v}{c})^2 \sim \frac{1}{v^2}\frac{c}{2}(1\pm\frac{v}{c})^2. \tag{10.248}$$

Then, we obtain

$$\tilde{\psi} = C_2 e^{\frac{i}{\alpha}\left[-\frac{\gamma^2(\omega+vk)^3}{3v\omega_0^2}+(C_0-\frac{g^2}{c^2}\chi\gamma^2 v^2)k\right]}(kc-\omega)^{-i\omega a_+/\alpha}(kc+\omega)^{i\omega a_-/\alpha}, \tag{10.249}$$

where $C_2 := e^{-iC_1/\alpha}$. We notice that

$$C_0 - \frac{g^2}{c^2}\chi\gamma^2 v^2 \sim 1 - \frac{g^2}{c^2}\chi\gamma^2 v^2 \sim 0, \tag{10.250}$$

i.e. this linear term is very small and can be neglected in the saddle point approximation, where it would amount to a small shift in the saddle point equation.

10.5.3 *Steepest descent approximation*

We shall refer to [Miller (2006); Wong (1989); Temme (2014)] as general reference on asymptotic expansion of integrals. We have

$$\psi(x) = \int_\Gamma dk e^{ikx}\tilde{\psi}(k), \tag{10.251}$$

where $\tilde{\psi}(k)$ is given by (10.249) and Γ is a suitable path in the complex k-plane. We also have

$$\psi(x) = \int_\Gamma dk e^{p(k;x)}h(k), \tag{10.252}$$

where

$$p(k;x) := i\left(kx - \frac{\gamma^2(\omega+vk)^3}{3\alpha v\omega_0^2}\right), \tag{10.253}$$

(we neglect the terms $\propto k$), and

$$h(k) := (kc-\omega)^{-i\omega a_+/\alpha}(kc+\omega)^{i\omega a_-/\alpha}. \tag{10.254}$$

We also define

$$\alpha_\pm = \frac{a_\pm}{\alpha}\omega = \frac{\omega}{\alpha}\frac{g^2}{c^2}\chi\gamma^2\frac{c}{2}\left(1\pm\frac{v}{c}\right)^2, \tag{10.255}$$

which is useful in the following.

We pass to variable s (cf. (10.234)) and we also define

$$u := \frac{s}{\sqrt{|\alpha|\omega_0^2}}, \tag{10.256}$$

in such a way that, by defining

$$\eta := \frac{\sqrt{|\alpha|\omega_0^2}}{\gamma v}, \tag{10.257}$$

we obtain

$$k = \eta u - \frac{\omega}{v}, \tag{10.258}$$

and then

$$\psi(x) = \eta e^{-i\frac{\omega}{v}x} \int_\Gamma du e^{f(u;x)} h(u) \tag{10.259}$$

where

$$f(u;x) := i\eta\left(ux - \frac{u^3}{3}\right), \tag{10.260}$$

and

$$h(u) = c^{-i(\alpha_+ - \alpha_-)} \left(\eta u - \frac{\omega}{v} - \frac{\omega}{c}\right)^{-i\alpha_+} \left(\eta u + \frac{\omega}{c} - \frac{\omega}{v}\right)^{i\alpha_-}. \tag{10.261}$$

η in (10.257) is to be considered the big parameter in the saddle point approximation: $\eta \to \infty$. Indeed we have

$$\eta \propto \frac{1}{\sqrt{B}}, \tag{10.262}$$

where B is very small, as usual in the Cauchy approximation. Note that g is order of 1 (we could put $g = 1$ everywhere, as we are exploiting a model whose nonlinearity involves only ψ).

Saddle points are obtained by solving

$$f'(u;x) = 0 \tag{10.263}$$

where the derivative is taken with respect to u. Then we find

$$u_\pm = \begin{bmatrix} \pm\sqrt{x} & x \geq 0, \\ \pm i\sqrt{|x|} & x < 0. \end{bmatrix} \tag{10.264}$$

It is interesting to point out that, as $x \to 0$, saddle points coalesce, i.e. we get

$$u_\pm = 0. \tag{10.265}$$

Coalescence of saddle points has to be handled with care [Temme (2014)]. As to the branch points

$$k_b^{\pm} := \pm \frac{\omega}{c}, \tag{10.266}$$

we obtain

$$u_b^{\pm} = \frac{\omega}{v}\left(1 \pm \frac{v}{c}\right)\frac{1}{\eta}. \tag{10.267}$$

We note that $u_b^{\pm} > 0$, and also that

$$\lim_{\eta \to \infty} u_b^{\pm} = 0, \tag{10.268}$$

i.e. there is a coalescence of branch points in the limit as $\eta \to \infty$, which is the typical limit where the saddle point approximation holds. In [Belgiorno *et al.* (2015)], we adopted implicitly the approximation where branch points coalesce.

There is also another interesting possibility for the coalescence of branch points, which consists in the following limit:

$$\lim_{\omega \to 0} u_b^{\pm} = 0, \tag{10.269}$$

which is also to be considered. We recall that ω is the frequency in the comoving frame, is a conserved quantity in the process and $\omega = 0$ is an allowed value for such a frequency. In this case, coalescence of branch points is exact. We shall discuss this point in the following (cf. sec. 10.5.14).

In the complex u-plane, we can find various possible configurations when x varies (we recall that x appears as an external parameter in such a framework):

$$u_- < u_b^- < u_b^+ < u_+, \tag{10.270}$$

$$u_- < u_b^- < u_+ < u_b^+, \tag{10.271}$$

$$u_- < u_+ < u_b^- < u_b^+, \tag{10.272}$$

according to possible different values of $x > 0$. The first configuration above could be designed as standard, but it is by no means clear a priori if it should be considered as the only relevant one. Indeed, as $x \to 0$, the other two configurations definitely play a role.

Note also that

$$h(u) = (c\eta)^{-i(\alpha_+ - \alpha_-)}(u - u_b^+)^{-i\alpha_+}(u - u_b^-)^{i\alpha_-}. \tag{10.273}$$

10.5.4 *Convergence regions*

In order to perform the saddle point approximation, we have to fix the regions where the integral (10.252) is convergent as $k \to \infty$. We need in particular to check where the phase $i\eta(ux - \frac{u^3}{3})$ gives rise to a negative real contribution (convergence regions). We find the dominant contribution from the cubic term, and we have

$$-i\frac{u^3}{3} = -i\rho^3 \cos(3\theta) + \rho^3 \sin(3\theta), \tag{10.274}$$

where $\rho = |u|$ and θ is the argument of u. We impose $\sin(3\theta) < 0$, which means

$$\frac{\pi}{3} < \theta < \frac{2\pi}{3}, \tag{10.275}$$

$$\pi < \theta < \frac{4\pi}{3}, \tag{10.276}$$

$$\frac{5\pi}{3} < \theta < 2\pi. \tag{10.277}$$

Convergence regions amount to valleys, and paths in the approximation of steepest descent start at a valley and end in another valley, as usual [Miller (2006); Wong (1989); Temme (2014)].

10.5.5 *Decreasing mode inside the black hole $x < 0$*

The saddle point $u_- = -i\sqrt{x}$ gives rise to a decreasing mode in the black hole region, which decays exponentially inside the black hole. See also [Belgiorno *et al.* (2015); Linder *et al.* (2016); Belgiorno *et al.* (2017b)]. Its contribution corresponds to the Fig. 10.3.

10.5.6 *Possible diagrams in the external region $x > 0$*

Explicit calculations can be performed e.g. as suggested in [Miller (2006)], by putting branch cuts along steepest descent lines. We draw qualitatively the possible diagrams. See Fig. 10.4 for the configuration corresponding to (10.270), Fig. 10.5 for the configuration corresponding to (10.271), and Fig. 10.6 for the configuration corresponding to (10.272). It is to be noted that the last configuration does not involve any branch cut, and then thermality is not expected. Even if nonstandard configurations like (10.271), (10.272) a priori could give contributions, in the saddle point approximation where $\eta \to \infty$, due to the coalescence to zero of the branch point, we expect that they provide negligible contributions as $\eta \to \infty$.

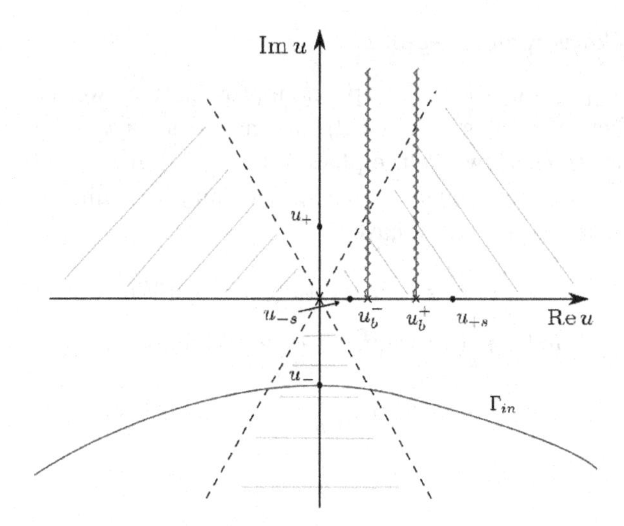

Fig. 10.3 Path corresponding to the decaying mode inside the horizon. Vertical branch cuts are displayed.

10.5.7 *Special configurations and thermality of pair-creation*

We consider the saddle point contribution to the integral in (10.259). Let us introduce

$$F(\eta) = \int_{\Gamma} du e^{f(u;x)} h(u), \tag{10.278}$$

where $h(u)$ is given by (10.273), and

$$g(u;x) := \frac{f(u;x)}{\eta} = i\left(ux - \frac{u^3}{3}\right). \tag{10.279}$$

We have

$$g(u^s_{\pm}) = \pm i\frac{2}{3}x^{3/2}, \tag{10.280}$$

$$g''(u^s_{\pm}) = \mp 2ix^{1/2}, \tag{10.281}$$

where for brevity we discuss only the case $x \geq 0$.

We start from the standard configuration where saddle points lie outside the interval (u_b^-, u_b^+). The path through the saddle point u^s_+ provides a factor

$$|u^s_+ - u_b^+|^{-i\alpha_+}|u^s_+ - u_b^+|^{i\alpha_-}. \tag{10.282}$$

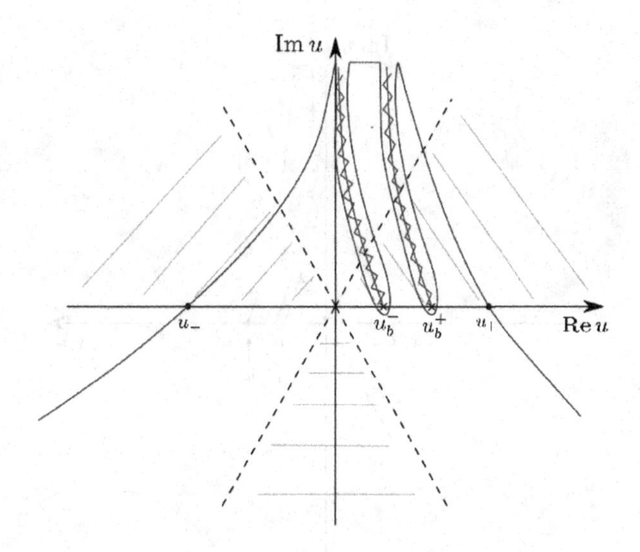

Fig. 10.4 Standard configuration (10.270) with branch cuts along steepest descent lines (according to Ref. [Miller (2006)]).

On the other hand, the path through the saddle point u^s_- is associated with a factor

$$|u^s_- - u^-_b|^{-i\alpha_+} e^{-\pi\alpha_+} |u^s_- - u^-_b|^{i\alpha_-} e^{\pi\alpha_-}. \tag{10.283}$$

In the limit as $\eta \to \infty$, we have

$$F(u^s_+) \sim \frac{1}{\sqrt{\eta}} e^{i\eta\frac{2}{3}x^{3/2}} e^{i\theta_+} \frac{\sqrt{\pi}}{x^{1/4}} |u^s_+ - u^+_b|^{-i\alpha_+} |u^s_+ - u^+_b|^{i\alpha_-}, \tag{10.284}$$

$$F(u^s_-) \sim \frac{1}{\sqrt{\eta}} e^{-i\eta\frac{2}{3}x^{3/2}} e^{i\theta_-} \frac{\sqrt{\pi}}{x^{1/4}}$$
$$\times |u^s_- - u^-_b|^{-i\alpha_+} |u^s_- - u^-_b|^{i\alpha_-} e^{-\pi\alpha_+} e^{\pi\alpha_-}, \tag{10.285}$$

where $\theta_\pm$ are pure phase factors we are not interested in for our aims. As a consequence, we obtain

$$\frac{|F(u^s_-)|^2}{|F(u^s_+)|^2} = e^{-2\pi(\alpha_+ - \alpha_-)} = e^{-\beta\omega}, \tag{10.286}$$

where

$$\beta = \frac{2\pi c}{\gamma^2 v^2 \kappa} \tag{10.287}$$

is the inverse of the expected Hawking temperature.

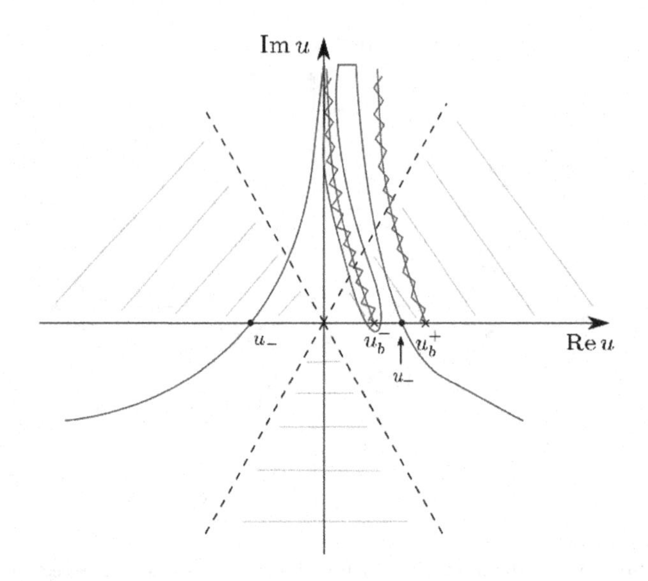

Fig. 10.5 Configuration (10.271) with branch cuts along steepest descent lines. Only a branch remains involved.

The intermediate configuration where $u_b^- < u_+^s < u_b^+$ is such that the branch point u_b^+ is not involved in the calculation. We obtain

$$|u_+^s - u_b^+|^{-i\alpha_+} e^{\pi\alpha_+} |u_+^s - u_b^+|^{i\alpha_-} \tag{10.288}$$

for the saddle point u_+^s. The other saddle point contributes a factor

$$|u_-^s - u_b^+|^{-i\alpha_+} e^{\pi\alpha_+} |u_-^s - u_b^-|^{i\alpha_-} e^{\pi\alpha_-}. \tag{10.289}$$

Then we obtain

$$\frac{|F(u_-^s)|^2}{|F(u_+^s)|^2} = e^{2\pi\alpha_-}. \tag{10.290}$$

This result is puzzling, in the sense that it is still possible to interpret it as a thermal contribution but only at the price of assuming a negative temperature, as $\alpha_- > 0$.

The last configuration we can consider is the one where $u_+^s < u_b^-$, and then no branch cut is involved. In this situation, we have

$$u_s^\pm - u_b^+ < 0, \qquad u_s^\pm - u_b^- < 0, \tag{10.291}$$

so we gain the same factors for both the saddle points:

$$|u_+^s - u_b^+|^{-i\alpha_+} e^{\pi\alpha_+} |u_+^s - u_b^+|^{i\alpha_-} e^{-\pi\alpha_-} \tag{10.292}$$

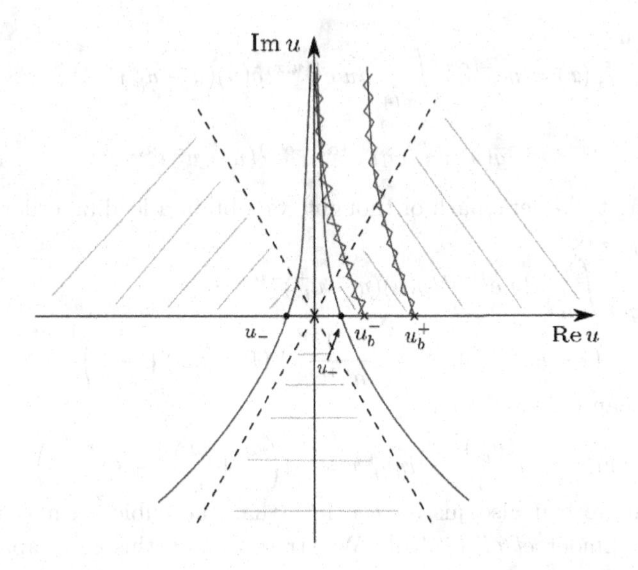

Fig. 10.6 Configuration (10.270) with branch cuts along steepest descent lines. No branch is involved, and then no thermality is expected.

for the saddle point u_+^s, and

$$|u_-^s - u_b^+|^{-i\alpha_+} e^{\pi\alpha_+} |u_-^s - u_b^+|^{i\alpha_-} e^{-\pi\alpha_-}, \tag{10.293}$$

and then

$$\frac{|F(u_-^s)|^2}{|F(u_+^s)|^2} = 1. \tag{10.294}$$

Evidently, this contribution is non-thermal.

As a consequence of the above discussion, if all the configurations described above contribute to the physics at hand, it is hard to find out a really thermal spectrum. The spectrum superimposes a genuinely thermal contribution from the standard configuration, and two non-thermal contributions from the exceptional configurations. A discussion on the exceptional contributions is found in the following. We point out that nothing similar can occur in the case of fluid models like the ones discussed in [Corley (1998)] and [Coutant *et al.* (2012b)], as therein a single branch cut appears.

10.5.8 *Branch cuts along steepest descent paths*

Let us assume an approach as suggested by [Miller (2006)], i.e. by adopting branch cuts along steepest descent paths. See the figures above. Then, we can obtain

$$\psi_{cut+}(x) = \eta e^{-i\frac{\omega}{v}x} \int_{cut+} du\, e^{f(u;x)} \tilde{g}(u)(u - u_b^+)^{-i\alpha_+}, \qquad (10.295)$$

where

$$\tilde{g}(u) := (c\eta)^{-i(\alpha_+ - \alpha_-)} (u - u_b^-)^{i\alpha_-}. \qquad (10.296)$$

According to Miller's path of thought, we obtain a leading order contribution as $\eta \to \infty$

$$\int_{cut+} du\, e^{f(u;x)} \tilde{g}(u)(u - u_b^+)^{-i\alpha_+} \sim$$
$$(1 - e^{-2\pi\alpha_+}) e^{f(u_b^+;x)} \frac{1}{\eta^{1-i\alpha_+}} \Gamma(1 - i\alpha_+) \left(\frac{1}{i\eta x}\right)^{1-i\alpha_+}, \qquad (10.297)$$

where it can be noted that

$$f(u_b^+;x) = -i\eta \frac{(u_b^+)^3}{3} - i\eta u_b^+ x = -i\left(\frac{\omega}{v} + \frac{\omega}{c}\right) x + O\left(\frac{1}{\eta^2}\right), \qquad (10.298)$$

and then we can also justify the fact that the cubic term is negligible. Cf. also [Linder *et al.* (2016)]. We stress that in this case, apart for the saddle point approximation, no other approximations are introduced. It is interesting to note that the same behavior in x of $\psi_{cut+}(x)$ is obtained. For the second cut, calculations are analogous. It is hard to identify these contributions with states in the asymptotic spectrum in our case. See also the following subsection.

10.5.9 *Vertical branch cuts*

Vertical branch cuts could be chosen as well e.g. as in [Linder *et al.* (2016)], and then one should perform the integrals around the branch cuts in some other approximation.

We start by showing that along the cut one obtains the same contribution as $\eta \to \infty$. Indeed, one has at $u = u_b^+$ and at $u = u_b^-$

$$\psi_{cut+}(x) = \eta e^{-i\frac{\omega}{v}x} \int_{cut+} du\, e^{f(u;x)} h(u), \qquad (10.299)$$

$$\psi_{cut-}(x) = \eta e^{-i\frac{\omega}{v}x} \int_{cut-} du\, e^{f(u;x)} h(u). \qquad (10.300)$$

One finds

$$\psi_{cut+}(x) = e^{-\frac{\pi}{2}\alpha_+} \eta (c\eta)^{-i(\alpha_+ - \alpha_-)} e^{-i\frac{\omega}{v}x} 2i \sinh(\pi\alpha_+)$$
$$\times \int_0^\infty dy\, e^{f(u_b^+ + iy;x)} y^{-i\alpha_+} (u_b^+ - u_b^- + iy)^{i\alpha_-}, \qquad (10.301)$$

and

$$\psi_{cut-}(x) = -e^{\frac{\pi}{2}\alpha_-}\eta(c\eta)^{-i(\alpha_+-\alpha_-)}e^{-i\frac{\omega}{v}x}2i\sinh(\pi\alpha_-) \times \int_0^\infty dy\, e^{f(u_b^-+iy;x)}y^{i\alpha_-}(u_b^- - u_b^+ + iy)^{-i\alpha_+}. \tag{10.302}$$

It is easily shown that, by neglecting the cubic term in the phase $f(u,x)$, one obtains a contribution which can be expressed in terms of the Gamma function [Linder *et al.* (2016)]. We can also give a better approximation for $x \geq 0$ for

$$I_+ := \int_0^\infty dy\, e^{-\eta y x}y^{-i\alpha_+}(u_b^+ - u_b^- + iy)^{i\alpha_-}, \tag{10.303}$$

$$I_- := \int_0^\infty dy\, e^{-\eta y x}y^{i\alpha_-}(u_b^- - u_b^+ + iy)^{-i\alpha_+}. \tag{10.304}$$

Let us define

$$\delta := u_b^+ - u_b^- = 2\frac{\omega}{c}\frac{1}{\eta} > 0. \tag{10.305}$$

Recalling that

$$\psi(a;c;z) = \frac{1}{\Gamma(a)}\int_0^\infty dt\, e^{-zt}t^{a-1}(1+t)^{c-a-1} \tag{10.306}$$

is an integral representation for the confluent hypergeometric Tricomi function

$$\psi(a;c;z) = \frac{\Gamma(1-c)}{\Gamma(a-c-1)}{}_1F_1(a,c,z) + \frac{\Gamma(c-1)}{\Gamma(a)}z^{1-c}{}_1F_1(a+1-c,2-c,z), \tag{10.307}$$

we then get

$$\begin{aligned} I_+ = &-\frac{1}{\Gamma(-i\alpha_-)}\alpha_+e^{-\frac{\pi}{2}\alpha_+}\delta^{1-i(\alpha_+-\alpha_-)}\Gamma(-i\alpha_+)\Gamma(-1+i(\alpha_+-\alpha_-)) \\ &\times{}_1F_1(1-i\alpha_+,2-i(\alpha_+-\alpha_-),-i\delta\eta x) \\ &+e^{-\frac{\pi}{2}\alpha_-}(\eta x)^{-1+i(\alpha_+-\alpha_-)}\Gamma(1-i(\alpha_+-\alpha_-)) \\ &\times{}_1F_1(-i\alpha_-,i(\alpha_+-\alpha_-),-i\delta\eta x). \end{aligned} \tag{10.308}$$

Analogously, we obtain

$$\begin{aligned} e^{-\pi\alpha_+}I_- = &-\frac{1}{\Gamma(i\alpha_+)}\alpha_-e^{-\frac{\pi}{2}\alpha_-}\delta^{1-i(\alpha_+-\alpha_-)}\Gamma(i\alpha_-)\Gamma(-1+i(\alpha_+-\alpha_-)) \\ &\times{}_1F_1(1+i\alpha_-,2-i(\alpha_+-\alpha_-),i\delta\eta x) \\ &+e^{\frac{-\pi}{2}\alpha_+}(\eta x)^{-1+i(\alpha_+-\alpha_-)}\Gamma(1-i(\alpha_+-\alpha_-)) \\ &\times{}_1F_1(i\alpha_+,i(\alpha_+-\alpha_-),i\delta\eta x). \end{aligned} \tag{10.309}$$

One could check the behavior of the above expressions. Note that

$$\delta\,\eta\,x = 2\frac{\omega}{c}x =: z, \tag{10.310}$$

where $\pm iz$ is the argument of the Kummer functions above. Expansions for $z \to \infty$ are well-known. Instead of performing such expansions, we note that a large z limit amounts to a large x limit. The latter should be taken with care: indeed, from a physical point of view, we have

$$n(x) = n_0 + \delta n(x), \tag{10.311}$$

where $\delta n(x) \ll n_0$, and as usually $\delta n(x) \sim 10^{-3}$ for the so-called χ_3 materials. Then, in the linear region

$$n(x) \sim n_0 + n'(0)x =: n_0 - \kappa x, \tag{10.312}$$

with $\kappa = |n'(0)|$, we must respect the following bound:

$$\kappa|x| \leq \sup(\delta n(x)) \ll n_0, \tag{10.313}$$

and then

$$\frac{\kappa}{n_0}|x| \ll 1. \tag{10.314}$$

In the following estimates, as n_0 is order of 1, we limit ourselves to consider for simplicity the bound

$$\kappa|x| \ll 1. \tag{10.315}$$

We notice that, as known, Hawking radiation is mostly populated for $\omega \sim c\kappa$, and the region $0 < \omega \leq c\kappa$ receives the leading contribution to Hawking radiation. Then we have also the bound

$$z \ll 1. \tag{10.316}$$

But then the large z expansion is inhibited by the above bound. x may be large, still with

$$|x| \ll \frac{1}{\kappa}. \tag{10.317}$$

A large $z \gg 1$ implies to violate the condition as $\delta n \ll n_0$. For a coherent picture as $z \to \infty$, one should allow a linear behavior of $\delta n(x)$ and also for $\omega_0^2(x)$, in such a way to allow consistently $n \to 1$ and $k_{\pm s} \to \pm\frac{\omega}{c}$, as is assumed in [Linder *et al.* (2016)].

10.5.10 *Near horizon: A different saddle point approximation*

Let us start again from the original relevant integral, that this time is written as follows:

$$F(\eta) = \int_\Gamma du e^{\eta s(u;x)}, \tag{10.318}$$

where, by defining

$$\epsilon = \frac{1}{\eta}, \tag{10.319}$$

and

$$\omega_\pm = \frac{\omega}{c} \pm \frac{\omega}{v}, \tag{10.320}$$

we have the following phase factor:

$$s(u;x) := i\left(ux - \frac{u^3}{3}\right) - i\epsilon\alpha_+ \log(u - \epsilon\omega_+) + i\epsilon\alpha_- \log(u + \epsilon\omega_-). \tag{10.321}$$

Saddle points are solutions of the following equation:

$$\frac{ds}{du} = 0 \Longleftrightarrow i\left(x - u^2 - \epsilon\alpha_+ \frac{1}{u - \epsilon\omega_+} + \epsilon\alpha_- \frac{1}{u + \epsilon\omega_-}\right) = 0. \tag{10.322}$$

It is evident that, for $\epsilon = 0$, we recover the well-known saddle points we calculated above. The former equation corresponds for $\epsilon \neq 0$ to a quartic equation:

$$(x - u^2)(u - \epsilon\omega_+)(u + \epsilon\omega_-) - \epsilon\alpha_+(u + \epsilon\omega_-) + \epsilon\alpha_-(u - \epsilon\omega_+) = 0, \tag{10.323}$$

which, in line of principle, can be solved exactly (one can do it with Mathematica). Still, very long and involved expressions are obtained. Then we adopt a different route: we expand (10.323) in powers of ϵ, for $\epsilon \to 0$, and study order by order the solutions, which are obtained as series in ϵ too:

$$u = u^{(0)} + \epsilon u^{(1)} + \epsilon^2 u^{(2)} + \dots \tag{10.324}$$

We expect that two solutions, at the zeroth order, correspond to the 'standard' saddle points

$$u_\pm^{(0)} = \pm\sqrt{x}, \tag{10.325}$$

and obviously $u_s^\pm = u_\pm^{(0)}$, whereas the other two solutions vanish at the zeroth order as $\epsilon \to 0$, as it can be also verified by manipulating the full solutions provided by Mathematica. These vanishing solutions correspond to the double root $u^{(0)} = 0$. Corrections to the standard saddle points are of limited interest, still we provide them at the first order:

$$u_{\pm}^{(1)} = -\frac{\alpha_+ - \alpha_-}{2x} = -\frac{c\omega}{2\gamma^2 v^2 |k| x}. \tag{10.326}$$

We are more interested in checking the nature of the further saddle points emerging from the above series expansion in ϵ. We have to consider the $O(\epsilon^2)$ equation in order to find the first corrections $u_{\pm s}^{(1)}$ to the double root $u^{(0)} = 0$. Let us define

$$A_{+-}(x) := (\alpha_+ - \alpha_-) + (\omega_+ - \omega_-)x, \tag{10.327}$$

$$B_{+-}(x) := 4x(\alpha_+\omega_- + \alpha_-\omega_+ + \omega_+\omega_- x). \tag{10.328}$$

We get

$$u_{\pm s}^{(1)} = \frac{A_{+-}(x) \pm \sqrt{(A_{+-}(x))^2 + B_{+-}(x)}}{2x}, \tag{10.329}$$

which amounts to

$$u_{\pm s}^{(1)} = \frac{\omega}{v} \frac{1}{\alpha x} \left(1 + \alpha x \pm \sqrt{(1 + \alpha x)\left(1 + \alpha x \frac{v^2}{c^2}\right)} \right). \tag{10.330}$$

In the linear region where $0 \leq \alpha x \ll 1$ one obtains

$$u_{+s}^{(1)} \sim \frac{2\omega}{\alpha v x} + \frac{\omega}{2v}\left(3 + \frac{v^2}{c^2}\right) \sim \frac{\omega c}{\gamma^2 v^2 |\kappa| x}, \tag{10.331}$$

$$u_{-s}^{(1)} \sim \frac{\omega}{2\gamma^2 v}. \tag{10.332}$$

In the linear region, we notice that it holds

$$u_{+s}^{(1)} \gg u_{-s}^{(1)}. \tag{10.333}$$

As a consistency condition for our expansion, we should expect that corrections to the main saddle points are really very smaller than the main saddle points themselves. We also should ensure that the perturbative saddle points are much smaller than the main ones. The latter requirement can be implemented as follows:

$$u_{+}^{(0)} \gg u_{+s}^{(1)} \Longleftrightarrow \sqrt{x} \gg \frac{1}{\eta} \frac{\omega c}{\gamma^2 v^2 |\kappa| x}, \tag{10.334}$$

i.e.

$$|\kappa| x \gg \left(\frac{c^2}{\gamma^2 v^2} B \omega^2 \right)^{1/3}. \tag{10.335}$$

This requirement identifies a subset of the linear region such that

$$\left(\frac{c^2}{\gamma^2 v^2} B\omega^2\right)^{1/3} \ll |\kappa| x \ll 1, \tag{10.336}$$

which defines our consistency region, where the expansion holds true, as also the perturbations of the main saddle points are suitably small therein. This region is the same as in the WKB expansion, see the following subsection. It is also worth noting that, in this region, we get

$$\sqrt{x} \gg \frac{\omega}{\eta v}. \tag{10.337}$$

Note also that in this region the main saddle points falls externally with respect to the branch points, and then we get automatically the standard configuration ensuring thermality. A priori, this is not mandatory, and it is just a condition which is implicit in our approximation scheme.

States in the near horizon approximation can be obtained from considering

$$e^{-i\frac{\omega}{v}x} F(\eta), \tag{10.338}$$

and nothing substantially changes for standard saddle points $u^s_\pm$. It is instead interesting to check what happens for the new saddle point contributions: we start from $u^{(1)}_{-s}$, for which the phase contribution is

$$\eta s(u^{(1)}_{-s}, x) \sim i\eta\epsilon u^{(1)}_{-s} x + \ldots = i\frac{\omega}{2\gamma^2 v} x + \ldots, \tag{10.339}$$

where in the neglected terms we find constant (i.e. x-independent) contributions. Then we obtain

$$e^{-i\frac{\omega}{v}x} F(\eta) \sim e^{-i\frac{\omega}{2v}(1+\frac{v^2}{c^2})x}, \tag{10.340}$$

which matches the behavior which can be found for k_{s-} in the WKB approximation. As to the amplitude, we easily find that the amplitude is constant. Then, we have a strong link with the linear region behavior of the k_{-s} state, with a good matching.

As to the saddle point $u^{(1)}_{+s}$, there are much more subtleties, as expected. We obtain

$$\eta s(u^{(1)}_{+s}, x) \sim i(\alpha_+ - \alpha_-) + i\frac{\omega_+\alpha_+ + \omega_-\alpha_-}{\alpha_+ - \alpha_-} x + i(\alpha_+ - \alpha_-) \log x + \ldots, \tag{10.341}$$

and we obtain the right behavior for the leading logarithmic term $\propto \log x$, which confirms that the state at hand is the best candidate to be identified with the Hawking mode, which is characterized by a logarithmic divergence

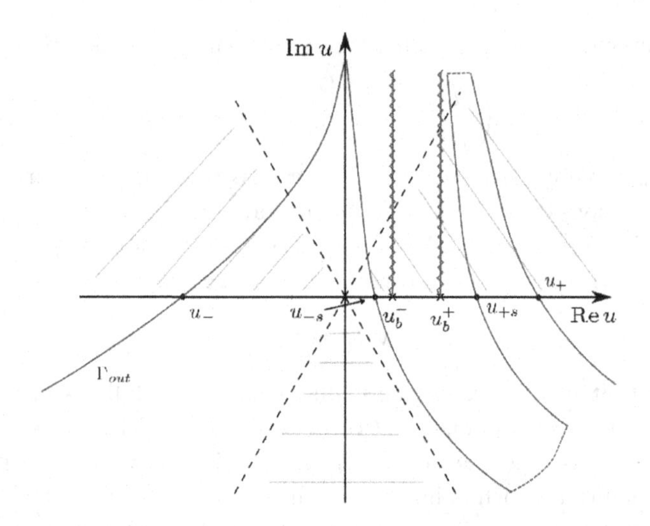

Fig. 10.7 Standard configuration (10.270). The dashed parts of the contour are taken at Re [s(u, z)] constant and asymptotically in the allowed regions, such that their contribution is negligible.

at the horizon in the nondispersive approximation, with the right coefficient. For the linear term in x, due to the subleading contribution $\propto x$, we do not match the WKB approximation, which is discussed in the following. Still, we point out that, at the level of the amplitude, we obtain $B_0 \propto \frac{1}{x}$ in the linear region, as for the WKB approximation. Summarizing, we are able to reproduce almost all the characteristics for the matching, and indicate the extreme sensibility to accuracy in the calculation one needs in order to manage properly the Hawking state k_{+s} as a possible reason for the remaining mismatching.

In Fig. 10.7 we display the contour for the standard configuration in the external region in our latest calculation scheme. As to the configuration inside the black hole, nothing changes with respect to Fig. 10.3.

10.5.11 *WKB solutions*

We adopt the following expansion for the WKB approximation:

$$\begin{pmatrix} \varphi \\ \psi \end{pmatrix} = \sum_{k \geq 0} \left(\frac{\hbar}{i} \right)^k \exp \left(i \frac{\omega}{\hbar} t + i \frac{S(x)}{\hbar} \right) \begin{pmatrix} A_k \\ B_k \end{pmatrix}. \tag{10.342}$$

WKB solutions are deduced by introducing the following trick. One starts from the equations of motion involving both the fields ϕ, ψ. The zeroth order contribution to the WKB approximation is represented by the solutions of the equation

$$\det M_0 = 0, \tag{10.343}$$

where

$$M_0 = \begin{bmatrix} -\frac{\omega^2}{c^2} + k^2 & i\frac{g}{c}\gamma(\omega + vk) \\ -i\frac{g}{c}\gamma(\omega + vk) & \frac{1}{\chi\omega_0^2}\left(\omega_0^2 - \gamma^2(\omega + vk)^2\right) \end{bmatrix}. \tag{10.344}$$

These solutions provide us the phases $k(\omega, x)$ of the physical states:

$$i\frac{\omega}{\hbar}t + i\frac{1}{\hbar}\int^x dyk(y), \tag{10.345}$$

where $k(x)$ is a solution of $\det M_0 = 0$. In order to obtain explicit solutions, in [Belgiorno *et al.* (2015)] we assumed that ω is such that $\omega \ll ck$. In the present case, we adopt an analogous but not strictly equivalent scheme, where is assumed that $\omega_0 \gg \omega$. In particular, by passing from k to u by means of the equation (10.258), and by taking into account that $\omega_0 = \frac{\gamma v}{\sqrt{\alpha}}\eta \sim \sqrt{\frac{cv}{2\kappa}}\eta$ (cf. eq. (10.257)), and by defining $\bar{\omega}_0 := \sqrt{\frac{cv}{2\kappa}}$, we obtain

$$\left(u^2 - 2\epsilon\frac{\omega}{v}u + \epsilon^2\frac{\omega^2}{\gamma^2 v^2}\right)\left(1 - \frac{v^2\gamma^2}{\bar{\omega}_0^2}u^2\right) - \frac{g^2}{c^2}\chi\gamma^2 v^2 u^2 = 0, \tag{10.346}$$

where

$$\epsilon = \frac{1}{\eta}. \tag{10.347}$$

We expand equation (10.346) in powers of ϵ, for $\epsilon \to 0$, and study order by order the solutions, which are obtained as series in ϵ too:

$$u = u^{(0)} + \epsilon u^{(1)} + \epsilon^2 u^{(2)} + \dots \tag{10.348}$$

We obtain a vanishing solution $u^{(0)}$ (with multiplicity two), and two other solutions

$$u_\pm^{(0)} := \pm\frac{\bar{\omega}_0}{\gamma v}\sqrt{1 - g^2\gamma^2\chi\frac{v^2}{c^2}}; \tag{10.349}$$

in the linear region, where it holds

$$0 \leq |\kappa x| \ll 1, \tag{10.350}$$

they are such that

$$u_\pm^{(0)} \sim \pm\sqrt{x}. \tag{10.351}$$

First order corrections for these solutions are

$$u_\pm^{(1)} = -\frac{g^2\gamma^2\chi v\omega}{c^2(1 - g^2\gamma^2\chi\frac{v^2}{c^2})} \sim -\frac{\omega}{v} + \frac{c\omega}{2\gamma^2 v^2|\kappa|x}. \tag{10.352}$$

We are interested to the corrections to the vanishing solutions, which are obtained by considering the $O(\epsilon^2)$ contribution. We obtain

$$\begin{aligned} u^{(1)}_{\pm s} &= \frac{1}{1-g^2\gamma^2\chi\frac{v^2}{c^2}}\frac{\omega}{v}\left(1\pm\sqrt{1-\left(1-g^2\gamma^2\chi\frac{v^2}{c^2}\right)\frac{1}{\gamma^2}}\right) \\ &\sim \frac{c\omega\left(1\pm\left(1-\frac{v}{c}|\kappa|x\right)\right)}{2\gamma^2v^2|\kappa|x}. \end{aligned} \tag{10.353}$$

So, in the linear region we obtain

$$u^{(1)}_{+s}\sim\frac{c\omega}{\gamma^2v^2|\kappa|x}-\frac{\omega}{2\gamma^2v}, \tag{10.354}$$

and

$$u^{(1)}_{-s}\sim\frac{\omega}{2\gamma^2v}. \tag{10.355}$$

We can come back to k, and then we get

$$k^{(0)}_{\pm}=\pm\eta\sqrt{x}-\frac{\omega}{v}\sim\pm\sqrt{\frac{|\kappa|x}{B}\frac{1}{\gamma v}}-\frac{\omega}{v}, \tag{10.356}$$

and

$$k_{+s}=-\frac{\omega}{2v}\left(3-\frac{v^2}{c^2}\right)+\frac{c\omega}{\gamma^2v^2|\kappa|}\frac{1}{x}+\dots, \tag{10.357}$$

$$k_{-s}=-\frac{\omega}{2v}\left(1+\frac{v^2}{c^2}\right)+\dots \tag{10.358}$$

It must be noticed that, in the linear region, it holds

$$k_{+s}\sim\frac{c}{\gamma^2v^2}\frac{\omega}{|\kappa|x}, \tag{10.359}$$

as $\omega/(|\kappa|x)\gg\omega$ in the linear region.

There is a consistency condition for our series expansion to which we can appeal: we should expect that corrections to the zeroth order nonvanishing solutions above are much smaller than their subleading corrections, as well as the perturbative solutions $u^{(1)}_{\pm s}$ are much smaller than the main nonvanishing ones $u^{(0)}_{\pm}$. The aforementioned consistency requirement can be implemented as follows:

$$\sqrt{x}\gg\frac{c\omega}{\gamma^2v^2|\kappa|x}, \tag{10.360}$$

which requires

$$|\kappa|x\gg\left(\frac{c^2}{\gamma^2v^2}B\omega^2\right)^{1/3}. \tag{10.361}$$

This defines the same consistency region as in the near horizon approximation, as we have shown.

In order to match solutions with the ones obtained in the near horizon region, we need to investigate also the amplitude part of the WKB solutions. We need to look for the first order equation. It is

$$M_{(1)} \begin{pmatrix} A_0 \\ B_0 \end{pmatrix} + M_{(0)} \begin{pmatrix} A_1 \\ B_1 \end{pmatrix} = 0, \tag{10.362}$$

where

$$M_{(1)} = \begin{bmatrix} -i(\partial_x k) - 2ik\partial_x & -\frac{1}{c}\gamma v g \partial_x \\ \frac{1}{c}\gamma v g \partial_x & i\frac{\gamma v}{\chi \omega_0^2}((\partial_x s) + 2s\partial_x) \end{bmatrix}, \tag{10.363}$$

and where we used (10.234). Amplitudes can be obtained by the first order corrections in $\hbar$ and we need to appeal at the theory of the multicomponent WKB, see [Ehlers *et al.* (1987); Fedoryuk (1993)]. See also [Belgiorno *et al.* (2015)]. Due to (10.343), we have

$$B_0 = ig\gamma(\omega + vk)\frac{\chi\omega_0^2}{\omega_0^2(x) - \omega^2}A_0. \tag{10.364}$$

As to the amplitudes, we obtain for states $k_\pm$

$$\begin{aligned} A_0 &\propto \frac{1}{x^{3/4}}, \\ B_0 &\propto \frac{1}{x^{1/4}}. \end{aligned} \tag{10.365}$$

As to the states $k_{\pm s}$, they are involved in more subtle calculations. For k_{-s} we find

$$\begin{aligned} A_0 &\propto \text{const}, \\ B_0 &\propto \text{const}. \end{aligned} \tag{10.366}$$

The most tricky case involves the Hawking mode k_{+s}, for which in the linear region we find

$$\begin{aligned} A_0 &\propto \text{const}, \\ B_0 &\propto \frac{1}{x}. \end{aligned} \tag{10.367}$$

10.5.12 *A dimensionless parameter for the saddle point approximation*

The parameter η (10.257) we have previously introduced is not dimensionless. Indeed, a simple inspection reveals that

$$[\eta] = [L]^{-3/2}. \tag{10.368}$$

It is not difficult to see that we can equally well define a dimensionless parameter, by rescaling in the following way:

$$\eta \mapsto \frac{\eta}{\kappa^{3/2}}, \tag{10.369}$$

where $\kappa = |n'(0)|$ is associated with a natural length scale for the physics at hand. This new parameter, which we indicate with η_d, is such that

$$\eta_d \simeq \frac{1}{\sqrt{B}\gamma v \kappa}. \tag{10.370}$$

We can estimate it as follows. Let us consider for example fused silica as a dielectric medium. We have for the Cauchy approximation

$$n(\lambda) = n_0 + \frac{B}{\lambda_l^2}, \tag{10.371}$$

where λ_l is the wavelength in the lab frame. That approximation is good in the visible spectrum $\lambda_l \in [0.4\mu m, 0.75\mu m]$, and $n_0 = 1.458$, and $B = 0.00354\mu m^2$. As an example, let us consider $\lambda_l = 0.6\mu m$. Then we obtain $B\lambda_l^2 \sim 10^{-2}$.

In η_d, let us consider $v \sim 0.685c$ (we should obtain $c/v \sim n_0$). Then we obtain $\gamma \sim 1.37$. For the order of magnitude, we can limit ourselves to consider the order of magnitude for $\sqrt{B}c\kappa$. We notice that $c\kappa$ is related to the peak frequency $\omega_{Hawking}$ (in the comoving frame) of the Hawking emission. We have

$$\omega_{Hawking} = \kappa_{sg}\frac{1}{c} = \gamma^2 v^2 \kappa \frac{1}{c} = \frac{\beta^2}{1-\beta^2} c\kappa, \tag{10.372}$$

where κ_{sg} is the surface gravity [Belgiorno *et al.* (2011b)] and $\beta = \frac{v}{c}$, as usual. Furthermore, we have

$$B\omega_{Hawking}^2 = B(\gamma(\omega_l - v k_l))^2, \tag{10.373}$$

where in the lab frame

$$k_l = \frac{1}{c} n(\omega_l)\omega_l, \tag{10.374}$$

so we get

$$B\omega_{Hawking}^2 = B\gamma^2\omega_l^2 \left(1 - \frac{v}{c} n(\omega_l)\right)^2, \tag{10.375}$$

where we have considered the measured (asymptotic) frequency, and then

$$n(\omega_l) = n_0 + B\omega_l^2, \tag{10.376}$$

without the correction $\delta n(x)$. With the above estimates, we get

$$(1 - \frac{v}{c} n(\omega_l))^2 \sim \delta n^2 \sim 10^{-6}, \tag{10.377}$$

so we obtain

$$\eta_d \sim \frac{1}{\sqrt{B\omega_{Hawking}^2}} \sim 10^4. \tag{10.378}$$

One may also consider $\lambda_l \sim 0.8\mu m$, and similar conclusions are reached.

10.5.13 *A further rescaling and insights for thermality*

We can note that u is such that

$$[u] = [L]^{1/2}. \tag{10.379}$$

In order to obtain a dimensionless u, we define

$$\bar{u} := \kappa^{1/2} u. \tag{10.380}$$

Then we obtain for $x \geq 0$

$$\bar{u}^s_\pm := \kappa^{1/2} u^s_\pm = \pm\sqrt{\bar{x}}, \tag{10.381}$$

where

$$\bar{x} := \kappa x \tag{10.382}$$

is dimensionless. Furthermore, we get

$$\bar{u}^\pm_b = \frac{1}{\eta_d \kappa} \frac{\omega}{v} \left(1 \pm \frac{v}{c}\right). \tag{10.383}$$

For the typical frequency $\omega_{Hawking}$, we obtain

$$\bar{u}^\pm_b \sim 10^{-4}, \tag{10.384}$$

whereas an upper bound for $\bar{x}$ is provided by

$$\bar{x} \leq \sup \delta n(x) \sim 10^{-3}, \tag{10.385}$$

so we can meet the condition $|\bar{u}^s_\pm| \sim 3 \cdot 10^{-2} \gg |\bar{u}^\pm_b|$, which is associated with the standard diagram where saddle points are external with respect to the interval of branch points.

The choice of the length scale is a subtle problem. We could also proceed as in [Coutant and Parentani (2014a)]: we identify the appropriate scale length by considering the integral (10.252), in particular by selecting the leading term in k in (10.253): this term is

$$\frac{\gamma^2 v^2}{3\alpha\omega_0^2} k^3, \tag{10.386}$$

and then we define

$$d_{br} := \left(\frac{\gamma^2 v^2}{\alpha\omega_0^2}\right)^{1/3} = \eta^{-2/3}. \tag{10.387}$$

We notice that

$$\eta = (d_{br})^{-3/2}, \tag{10.388}$$

and that

$$\eta_d = \frac{1}{(\kappa d_{br})^{3/2}}. \tag{10.389}$$

The ansatz is that this scale dominates the behavior of the process. We can also rescale both x and k:

$$\hat{x} := \frac{x}{d_{br}}, \tag{10.390}$$

$$\hat{k} := k d_{br}. \tag{10.391}$$

In order to give an order of magnitude evaluation for the scale d_{br}, we can find out the order of magnitude of κ in a typical configuration for a Gaussian perturbation:

$$\delta n(x) = l_0 e^{-\frac{x^2}{2\sigma^2}}, \tag{10.392}$$

where $l_0 \sim 10^{-3}$ and $\sigma \sim 10^{-5} m$. We get

$$\kappa = (\frac{c}{v} - n_0)\frac{1}{\sigma}\sqrt{-2\log\left[\left(\frac{c}{v} - n_0\right)\frac{1}{l_0}\right]}, \tag{10.393}$$

and for fused silica and $v = 0.6855c$ we obtain $\kappa \sim 54$. The corresponding temperature is order of $10^{-2} K$ (cf. [Belgiorno *et al.* (2011b)]). So, we can infer that $d_{br} \sim 4 \cdot 10^{-5} m$.

We assume that, as in [Coutant and Parentani (2014a)], the length scale d_{br} is such that the physical system is not able to resolve distances shorter than the scale itself. It seems to be legitimate to consider then the following lower bound for x:

$$x \geq d_{br}. \tag{10.394}$$

As a consequence, we must assume

$$|u^s_{\pm}| \geq |u^s_{\pm}|_{min} := \sqrt{d_{br}}. \tag{10.395}$$

In order to understand which configuration gives the leading contribution, we have to compare the aforementioned lower bound with

$$(u^{\pm}_b)_H := d_{br}^{3/2}\frac{1}{v}\omega_{Hawking}(1 \pm \beta) = d_{br}^{3/2}\frac{\beta}{1 \mp \beta}\kappa. \tag{10.396}$$

We obtain easily from $u^s_+ > u^+_b$

$$1 > d_{br}\kappa\frac{\beta}{1-\beta} = \eta_d^{-2/3}\frac{\beta}{1-\beta} \sim 10^{-3}, \tag{10.397}$$

which is obtained under the above hypotheses on the values of the parameters. This inequality is also automatically satisfied for any $\omega \in$

$(0, \omega_{Hawking})$. As a consequence of this reasoning, if β is not very near 1, and if ω is in the range of the Hawking effect, the dominant contribution to the amplitude of pair-creation comes from the standard configuration, and thermality is recovered.

There is a further possible way to infer when the standard configuration is the one to be considered. In presence of a group horizon, we have a turning point which can occur on the left of the horizon $x = 0$. Indeed, the equation to be satisfied is [Belgiorno *et al.* (2015)]

$$\frac{c}{v} - n(x_{GH}) = c_0 (B\omega^2)^{1/3}, \tag{10.398}$$

where

$$c_0 := \frac{3}{2^{2/3}} \gamma^{-5/3} \left(\frac{c}{v}\right)^{2/3}. \tag{10.399}$$

The (ω-dependent) position of the group horizon is such that

$$n(x_{GH}) = \frac{c}{v} - c_0 (B\omega^2)^{1/3} < \frac{c}{v}, \tag{10.400}$$

and, being n a decreasing function of x in a neighborhood of the horizon $x = 0$, we have

$$x_{GH}(\omega) \geq 0, \tag{10.401}$$

and, in particular, $x_{GH}(\omega = 0) = 0$ and $x_{GH}(\omega > 0) > 0$. Apart for ω very near to $\omega = 0$, we obtain a turning point on the right of $x = 0$, and then we can expect that all $x < x_{GH}(\omega > 0)$ eventually do not play a relevant role in the scattering process at $\omega > 0$ fixed. In other terms, our expectation is that the presence of the turning point makes it possible to stay away from $x = 0$. As a consequence, even if in the spontaneous process it is hard to justify a leading thermal contribution, in the stimulated process, with a suitable choice of the frequencies ω, and with a suitable enhancement of the stimulated contribution with respect to the spontaneous one, it should be still possible to recover a thermal spectrum and thermality of the Hawking radiation. Still, it is remarkable that the mechanism contributing to the particle production is horizon-generated in all cases.

10.5.14 *Coalescence of branch points as $\omega \to 0$*

As discussed above, in the limit as $\omega \to 0$ we have coalescence of branch cuts at $u = 0$. In line of principles, this coalescence would require an uniform asymptotic expansion, in order to reach an agreement between the limit for

$\omega \to 0$ of the asymptotic approximation and the asymptotic approximation taken at $\omega = 0$ (which should be a legitimate asymptotic expansion). In the following, we show that no discontinuous behavior occurs, and that the same result is obtained form taking the limits of the integrals and calculating the integral at $\omega = 0$. The main point is that a quite mild behavior actually takes place: indeed, for $\omega = 0$ no branch cut occurs in the equation for ψ. Then, at $\omega = 0$ no cut contribution arises, and this is perfectly coherent with the fact that cut contributions vanish as $\omega \to 0$.

It is worthwhile stressing that $\omega = 0$ is not only an allowed parameter in the physics at hand, but also it corresponds to the main contribution to particle creation in the experimental situation, as verified by the group leaded by D. Faccio.

Let us start from the original system as $\omega = 0$:

$$\tilde{\phi} = \frac{i\frac{g}{c}\gamma v}{k}\tilde{\psi}, \tag{10.402}$$

$$\left[i\frac{\lambda}{2}(\psi_0^2)'(0)\partial_k - \frac{1}{\chi\omega_0^2}(\gamma v k)^2 + \frac{1}{\chi}(1 + \chi\frac{\lambda}{2}\psi_0^2(0)) - \frac{g^2}{c^2}(\gamma^2 v^2)\right]\tilde{\psi} = 0. \tag{10.403}$$

As it is evident, there is no more any branch cut in the differential equation for $\tilde{\psi}$. The solution is

$$\tilde{\psi} = C_3 e^{-i\frac{\gamma^2 v^2}{3\alpha\omega_0^2}k^3}, \tag{10.404}$$

where again a linear term is suppressed (see above). We can perform a saddle point approximation for the integral corresponding to the Fourier transform $\psi(x)$, and, as in the previous section, we get

$$F(u_{\pm}^s) \sim \frac{1}{\sqrt{\eta}} e^{\pm i\eta\frac{2}{3}x^{3/2}} e^{i\theta_{\pm}} \frac{\sqrt{\pi}}{x^{1/4}}, \tag{10.405}$$

and then we find

$$\frac{|F(u_{-}^s)|^2}{|F(u_{+}^s)|^2} = 1, \tag{10.406}$$

as expected from the above results in any configuration. Indeed, we recall that

$$\lim_{\omega \to 0} \alpha_{\pm} = 0, \tag{10.407}$$

and, moreover, from the above analysis it is easily verified that

$$\lim_{\omega \to 0} \psi_{cut\pm} = 0, \tag{10.408}$$

due to the vanishing of $\sinh(\pi\alpha_{\pm})$ in the same limit. This confirms that there is no need of any sort of uniform asymptotic expansion.

10.5.15 *Horizons and dispersion*

Some comments are in order now. We have discussed the thermality assuming to identify the horizon with the geometrical one, i.e. the one defined in absence of dispersion, which we call *non-dispersive horizon* NDH. However this choice is not the only possible, since we know there are at list two other horizons which could be considered: the *group horizon*, call it GH, and the *phase horizon* PH. These are defined respectively as the point where the group velocity of the wave packet vanishes, and the locus where the phase velocity of the waves composing the wave packet vanishes. At a first sight, from a physical point of view there is no doubt that GH is more appealing and meaningful than the PH, and should play the role of NDH in the dispersive case. Pure thermality of the spectrum is supposed to rely on the presence of a GH. However, we prefer to keep both concepts, as they could play a role which is not yet made as evident from our previous calculations. Before continuing, let us thus recall some fact about geometrical optics.

An usual tool for analyzing solutions in the framework of analogue gravity is eikonal approximation. We explore also this conceptual frame, in order to get useful suggestions and analytical tools for a better comprehension of the phenomenon at hand. It is remarkable that thermality of the Hawking radiation arises as associated with the presence, in the comoving frame, of a GH, which is a turning point (TP) for the waves which reach the perturbation, at least for frequencies in a given interval. In a WKB approximation, a turning point is to be handled with care, since it violates the requirements of the approximation itself. Similarly, a TP in geometrical optics represents a caustic for rays, so, again, the eikonal approximation fails there. Since all the phenomenology which we are interested in arises near such a TP, it reveals necessary to point out immediately the limits of the given approximation. Near a TP some other analytical tool are expected to be necessary. Despite this, in what follows, we point out that even in presence of dispersion, the eikonal approximation gives useful suggestions, and the method of characteristics can be used in order to explore solutions. Moreover, the problems of the group horizon, and of the phase horizon, can be exactly solved in the Cauchy approximation.

Let us consider the two-dimensional eikonal equation

$$\omega_{lab} n(\omega_{lab}, x - vt) = \pm c k_{lab}, \tag{10.409}$$

and assume the Cauchy approximation (10.180). In the comoving frame we get

$$G = 0 \Longleftrightarrow (\omega + vk)(n(x) + B\gamma^2(\omega + vk)^2) - ck - \frac{v}{c}\omega = 0. \tag{10.410}$$

The group horizon (if any), is obtained by solving the system

$$G = 0, \tag{10.411}$$

$$\partial_k G = 0. \tag{10.412}$$

From the latter

$$\partial_k G = 0 \Longleftrightarrow 3B\gamma^2 v(\omega + vk)^2 - v\left(\frac{c}{v} - n(x)\right) = 0, \tag{10.413}$$

which can be solved explicitly

$$(\omega + vk) = \pm \left(\frac{\frac{c}{v} - n(x)}{3B\gamma^2} \right)^{1/2}. \tag{10.414}$$

As we mean to get the group horizon for positive norm waves, by substituting the positive root in $G = 0$, we get an equation for $n(x)$ which from which we obtain explicitly the group horizon

$$\frac{c}{v} - n(x) = 3B\gamma^2 \left(\frac{1}{2B\gamma^4} \frac{c}{v} \right)^{2/3} \omega^{2/3} =: \zeta_B \omega^{2/3}, \tag{10.415}$$

where $\zeta_B \propto B^{1/3}$. This determines $x_{GH}(\omega)$, which is a function of the frequency ω, as expected.

On the other hand, in the comoving frame the PH corresponds to $\omega = 0$. By taking into account the dispersion relation (10.410), we find

$$\frac{c}{v} - n(x) = B\gamma^2 v^2 k_0^2, \tag{10.416}$$

where k_0 is the value at which the dispersion relation $G = 0$ intersect the axis $\omega = 0$. Thus, $x_{PH}(k_0)$ is a function of the aforementioned parameter, which is independent from ω.

Now, let us turn back to our thermal analysis, where the horizon surface has been picked at a generic locus simply indicated as "horizon" and shifted to $x = 0$, and only at the end of calculations our microscopic parameters where transformed into the macroscopic ones. Equation for the black hole temperature remains unaltered by the exact choice of an horizon, and changes occurs only at the level of the subsequent approximation (cf. e.g. [Belgiorno *et al.* (2015)]). We also point out that in fluid models the above distinction between different kind of horizons, and in particular between non dispersive horizon and group horizon, is a *posteriori* irrelevant, in the sense that conditions can be given such that the wave function of the modes is not able to distinguish between the aforementioned horizons, and any correction to thermality is washed out, see [Coutant and Parentani (2014a)]. Since our framework has several analogies with those one, we expect that a similar conclusion is reasonable also in our case, despite the formal dependence we found above on the expansion point. However, we will not consider further this point here.

It is interesting to notice that the $\varphi\psi$ model is at least in principle exactly solvable, and could provide hints on what happens in a strongly dispersive regime, where the approximations we done are not valid. Whereas at the moment the last goal is not achievable, it is remarkable that the model can provide us the exact dispersion relation of the theory. Indeed, let us consider the case where $\chi = \chi(x - vt)$ in the lab frame. In the full four dimensional case for $k = (\frac{\omega}{c}, \vec{k})$, from equation (10.136) we get

$$k^\mu k_\mu - \frac{g^2}{c^2}(v^\mu k_\mu)^2 \tilde{G}_\psi(k_\nu) = 0, \tag{10.417}$$

which represent the *exact dispersion relation* for the theory at hand. Notice that this relation holds for a generic spacetime dependence of the susceptibility χ, and even of the proper frequency ω_0. It contains all the necessary information in order to explore the problem of pair-creation [Belgiorno *et al.* (2015)].

10.6 Recapitulation

Given the length of the present chapter, it may be useful, before introducing the next argument, stopping for a while and recapitulate what we have done up to now. We considered the Hawking effect in dispersive dielectric media, and in particular looking for an analytical proof of thermally in the spectrum of emitted photons, focusing on a phenomenological model. Our reference framework was the Hopfield model, suitably modified in order to account of some characteristics of the Kerr effect. We have taken into account several properties of the model, obtaining a number of interesting results:

(1) We started from a model apt to the phenomenological description of the process involved in thermal pair creation previously identified with the Hopfield model. We have developed a model where microscopic parameters are left free to vary smoothly in spacetime coordinates. This model allows extensions to multi-resonances situations for the polarization field. We have also introduced a simpler model involving a scalar field doublet in place of the full electromagnetic field and the polarization field, in order to get a simpler and more manageable model. The scalar model, we called it the $\varphi\psi$ model preserves the same dispersion

relation as the Hopfield one. We have identified a conserved scalar product providing a norm for identifying particle and anti-particles states. In particular, from Wronskian relations we obtained the generalized Manley-Rowe relations in a scattering framework;

(2) We have obtained the thermality in an analytic by matching the asymptotic solutions obtained in the WKB approximation and the solutions of the differential equation in Fourier space where an expansion up to the linear order of the refractive index has been performed near "the horizon". In particular, we have shown that calculations can be developed in a nice parallel way with respect to known calculations concerning fluid models [Corley (1998); Coutant *et al.* (2012b)]. A first order expansion near the horizon in Fourier space put in evidence a Fuchsian singularity structure which can be associated with Hawking effect. In all cases, thermality is involved with a substantially ternary process, since we have shown that a fourth state decouples from the spectrum.

(3) The aforementioned picture extends to the multi-resonances situation, see e.g. [Belgiorno *et al.* (2015)].

(4) We have also tried to discuss the problem concerns what should be meant by "horizon" in the near horizon expansion. In our model, the non-dispersive horizon, the group horizon and also the phase horizon could all have chances to be the right places. GH looks the strongest candidate, but our analysis cannot yet be conclusive. In particular, the expected role of a blocking horizon does not emerge in a neat way from the present. Indeed, a detailed discussion, in comparison with the analysis of the paper [Coutant and Parentani (2014a)], where it is shown that in the case of fluids the wave function associated with the modes involved in the scattering process is not able to distinguish between different kinds of horizons, suggests that the same could happen for the dielectric system, but this would require further analysis than the one presented here.

10.7 Further readings

It is worth referring to a series of papers which have been devoted to the analysis of Hawking radiation in dielectrics. The studies contained in [Finazzi and Carusotto (2013, 2014)] are dedicated to the numerical analysis of pair creation in white hole and in black hole configurations in dielectrics, described through a 2D reduction of the Hopfield model. A non-smooth step-like profile for the dielectric perturbation is chosen, and also the non-trivial situation involving fused silica is discussed, with the aim to compare theoretical results with the white hole experiment discussed in [Belgiorno *et al.* (2010a); Rubino *et al.* (2011)]. The analysis takes into account both spontaneous emission and stimulated one, and also both the presence and the absence of a group horizon, showing that also in absence of an horizon a pair creation process can still occur. In the white hole case, a very marked sensitivity on the pulse velocity is stressed [Finazzi and Carusotto (2014)].

We also signal [Bermudez and Leonhardt (2016)] concerning the analogous Hawking effect in fiber optics, and a further study on a quantum emission from a moving dielectric perturbation [Jacquet and König (2015)]. Optical black hole lasers have been taken into account in [Faccio *et al.* (2012)]. See also [Faccio (2012)]. Pair creation from a moving dielectric perturbation has been taken into account also in [Belgiorno *et al.* (2014)], and a discussion on the stimulated Hawking effect, with reference also to the white hole scattering case discussed above is found in [Belgiorno and Cacciatori (2016)]. A study of a 2D model with a half-line filling dielectric medium and solitonic solutions when implementing a nonlinear contribution to the polarization field are considered in [Belgiorno *et al.* (2017c)].

10.8 Hawking radiation in a dispersive nonlinear dielectric

We have seen how in certain limits a RIP moving with constant velocity can give rise to a thermalized Hawking like spectrum, whose temperature is compatible with what predicted by the non dispersive limit. We have done this in a phenomenological model, where the desired dependence of the RIP is substituted by an at hand suitably chosen spacetime dependence of the parameters. Nevertheless, we have also already stated that if one would like to go beyond the range of validity of the phenomenological model, then it is necessary to consider a more "first principles" based model. Here we have proposed a first step in this direction, by introducing a modification of the $\varphi\psi$ model including a nonlinear modification of the Lagrangian aimed to simulate the Kerr effect. Indeed, such a modification can be realized in the full Hopfield model, but, for sake of simplicity, we will not discuss it here, referring to the literature for the more physical situation [Belgiorno *et al.* (2017b)]. This model is more fundamental than the previous one in the sense that it is completely self consistent and no external "at hand" modifications, like the spacetime dependent variation of the constants, must be considered. Nevertheless, it cannot be considered as a definitive and fully microscopic description of the analogue Hawking effect for at least two reasons: first, the model contains dispersion but not dissipation, so that we may have a correct answer only far from the spectral regions where dissipative phenomena become important (e.g. near the resonances); second, the polarization field does not obviously provide a microscopic description of the matter but only a mesoscopic one, where the dynamics of matter is absent and only elementary properties are recorded in a finite number of independent oscillating fields.

Nevertheless, we expect that this simple model should capture the main features of analogue Hawking radiation in dielectric media, and that more sophisticated models including dissipation, matter dynamics and so on should only provide minor corrections. But confirming or rejecting such conjecture, which surely is an important task, will require much harder efforts which we are not yet ready to include here, as we are not yet ready to deal with analytical calculations for the nonblocking (subcritical) case, which is still the one occurring for Bessel filaments in [Belgiorno *et al.* (2010a); Rubino *et al.* (2011)]. See also the following chapter.

Chapter 11

Hawking radiation in the lab

This chapter is dedicated to experiments involved with the revelation of analogue Hawking radiation in labs. We start from the Como experiment, where a dielectric perturbation created by means of a intense laser pulse in fused silica plays the role of white hole in the comoving frame. Then we discuss the Vancouver experiment, in which analogue Hawking radiation is measured in water by means of induced waves which hit a suitably designed obstacle. Finally, a discussion will be devoted to the Technion experiment involving a system which is associated with a so-called black hole laser.

11.1 The Como experiment

We will now shortly describe an experiment realized at the laboratories of the University of Insubria in Como, Italy, in 2010, [Belgiorno *et al.* (2010a); Rubino *et al.* (2011)]. See also criticisms and reply in [Schutzhold and Unruh (2011); Belgiorno *et al.* (2011a)]. In such experiment, it has been used ultrashort laser pulse filaments to create a traveling RIP in a transparent dielectric medium, constituted by fused silica glass, getting experimental evidence of photon emission that bears many characteristics we have previously described for Hawking radiation in a dispersive dielectric. As we will see, the setting of the experiment has been cured in order to make the signal distinguishable from other known photon emission mechanisms.

Let us recall [Faccio *et al.* (2010)] that ultrashort laser pulse filaments are intense laser pulses in a transparent Kerr medium, which are characterized by a high-intensity spike propagating apparently without diffraction over distances much longer than the Rayleigh length associated to the spike dimensions. These have been proposed for many applications [Couairon and Mysyrowicz (2007); Bergé *et al.* (2007); Gaeta (2003); Kasparian *et al.*

(2003)]. They may either occur spontaneously when a powerful Gaussian shaped beam is loosely focused into the Kerr medium [Couairon and Mysyrowicz (2007); Bergé *et al.* (2007)] or, alternatively, they may be induced by pre-shaping the laser pulse into a Bessel beam [Polesana *et al.* (2007, 2006)].

The experimental layout is shown in Fig. 11.1.

The RIP is created by lasers pulses provided by a regeneratively amplified Nd:glass laser, with repetition rate of 10 Hz. Each pulse has duration 1 ps and a maximum energy of 6 mJ. A 20 degree fused silica axicon (a conical lens) is used for generating a Bessel pulse filament with a cone angle $\theta = 7$ deg. It is placed directly in front of a Kerr sample of fused silica of length 2 cm, where the RIP is generated. The range of the input energy is varied from 100μJ to 1200 μJ.

By means of a lens that images the filament on to the input slit of an imaging spectrometer, the radiation from the filament is collected at 90 degrees with respect to the laser pulse propagation. The spectrum is recorded with a 16 bit, cooled CCD camera.

The choice of taking the measure at 90 degrees has been done in order to suppress and possibly eliminate spurious effects:

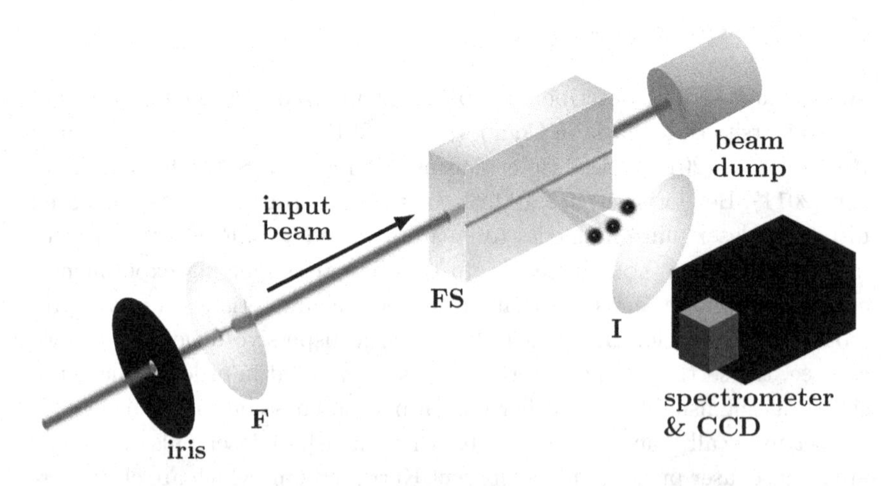

Fig. 11.1 Experimental layout used for detecting analogue Hawking radiation. The input laser pulse is focused into a sample of fused silica (FS) using an axicon or lens (F). An imaging lens (I) collects the photons emitted at 90 deg and sends them to an imaging spectrometer coupled to a cooled CCD camera.

(1) in [Belgiorno *et al.* (2010b)] it has been shown that a superluminal RIP generates spontaneous emission of photons: the Cherenkov-like radiation. This effect is clearly distinguishable from the Hawking like effect, since the emission occurs with no upper bound in the spectral window;
(2) the four wave mixing (FWM) and self phase modulation (SPM) will not occur at 90 degrees. Indeed, the constraints on the phase matching imply that any newly generated frequencies will be emitted at small angles with respect to the propagation axis [Gadonas *et al.* (2001)]. In any case, for the relatively large $\theta = 7$ degrees Bessel cone angle used in the experiments, no FWM, SPM or spectral broadening was observed at any angle, even in the forward direction;
(3) At 90 degrees it is possible to reveal Rayleigh scattering, which surely occurs, but only for vertically polarized light and the scattering process maintains the polarization state. In our experiments we used horizontally polarized light and in any case, in virtue of point (1), there was no input or generated light at the frequencies relevant for the done experiment;
(4) the most relevant problem is the one of fluorescence. It bears many features in common with Hawking radiation but it may still be clearly distinguished from the latter. In figure 11.3(a) is reported the overall spectrum measured at 90 degrees, integrated over 30 laser pulse shots. R, F1 and F2 indicate the laser pulse induced spontaneous Raman, non bridging oxygen hole centre (NBOHC) and oxygen deficient centre (ODC) fluorescences, respectively. For fused silica, these fluorescence peaks are well documented [Zoubir *et al.* (2006); Skuja (1998)], so that they can be subtracted after being fitted with Gaussian functions (in the frequency domain). This leads to a greatly improved contrast and cleaner spectra.

One can also notice that there is a further possibility for avoiding the problem of fluorescences. After looking at the fluorescence spectrum one sees that there are regions, for example between 800nm and 900nm, where there is no fluorescence emission. Thus, one could tune the group velocity of the laser pulse in such a way that the Hawking window falls exactly in this range (where, moreover, the CCD response is maximum).

Figure 11.2 shows the predicted spectral regions for the following three cases: (a) the trailing daughter pulse within a spontaneous filament with v; (b) a Gaussian pulse propagating with $v = v_G$; (c) a Bessel filament propagating with $v = v_G/cos\theta$ and $\theta = 7°$. Two curves are shown in each

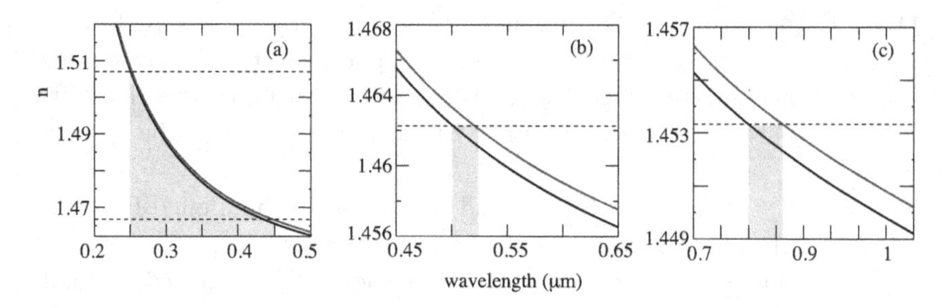

Fig. 11.2 Prediction of the Hawking emission spectral range based on the dielectric perturbation velocity v and the medium refractive index. Lower curves represent the background refractive index n_0 and the upper curves represent $n_0 + \delta n$. (a) Spontaneous filament: the filament perturbation velocity covers a broad range of values, evaluated from numerical simulations. (b) Gaussian pulse: $v = v_G$ is determined by the material dispersion. (c) Bessel filament: $v = v_G/cos\theta$. In all cases $\delta n = 0.001$.

graph: the refractive index n and $n + \delta n$ with $\delta n = 0.001$. We emphasize that the Hawking photons will be emitted only in a bounded spectral window. This is somewhat different from the dispersionless case in which, once v is properly tuned so as to achieve the horizon condition, all frequencies are excited. In the presence of dispersion as in our analogue model, only a limited spectral region is excited and the blackbody spectral shape, typically associated with Hawing radiation, will not be discernible. Regarding the width of the spectral emission window, we note that this is determined by the values of both v and δn. If the perturbation moves with the same velocity as that of a Gaussian pulse centred at 1055 nm, $v = v_G$ (1055 nm) = constant and Hawking emission will be centred around 500 nm with a $\sim$ 20 nm bandwidth. The Bessel filament will have a larger velocity and emission is expected to be around 850 nm with larger bandwidth, $\sim$ 60 nm, due to the lower dispersion at these wavelengths. Conversely, in the case of the spontaneous filament, the variation of v along the propagation direction dominates the emission characteristics and emission is predicted around 350 nm with $\sim$ 200 nm bandwidth. The δn induces only a minor 510 nm broadening at these wavelengths.

Let us enter more the details of the experiment. In fused silica, the Kerr coefficient is $n_2 \sim 3 \cdot 10^{-16}$ cm^2/W. The typical intensity for the pulse is $I \sim 10^{12} - 10^{13}$ W/cm^2, leading to a $\delta n = n_2 I \sim 10^{-2} - 10^{-4}$. Therefore, by taking, for example, a Gaussian pulse centred at 1055nm, with $\delta n = 0.001$ and group velocity, $v_g = d\omega/dk$, determined solely by material dispersion, from the horizon condition

$$\frac{1}{n_0(\omega) + \delta n} < \frac{v}{c} < \frac{1}{n_0(\omega)} \tag{11.1}$$

we would expect Hawking emission between 500 nm and 510 nm. This falls in between the F1 and F2 fluorescence peaks and leads to noisy results even after subtraction of the fluorescence signal. The solution of this problem has been obtained by using Bessel pulses. In Fig. 11.3(b) it is reported the Hawking window as predicted from equation (11.1). The dashed line indicates the value of c/v_B with $\theta = 7$ degrees $v_B = v_g/\cos\theta$ is the velocity of the Bessel pulse. We see that this line intersects the refractive index curves n and $n + \delta n$, with $\delta n = 0.001$ delimiting a window ling in the region between 800 nm and 875 nm, exactly where fluorescences are absent. In Fig. 11.4(a) is shown the resulting spectra, integrated over 3600 laser shots. A Gaussian pulse has been also used to show that no generation corresponds signals notwithstanding the same peak intensity of the Bessel pulses: this is represented by the black line. In particular, it proves the absence of fluorescence or other possible signals in that region, when the RIP velocity does not satisfies the horizon condition. The emitted spectra corresponding to four different input energies of the Bessel filament are evident: the wavelength window is exactly the one predicted by equation (11.1). The dashed lines indicate the Gaussian fitting. It has been also verified that the emitted radiation was unpolarized as it should be for a spontaneous emission.

Notice that the width of the emission band increases with the input energy. In Fig. 11.4(b) are shown the bandwidth and the δn as a function of

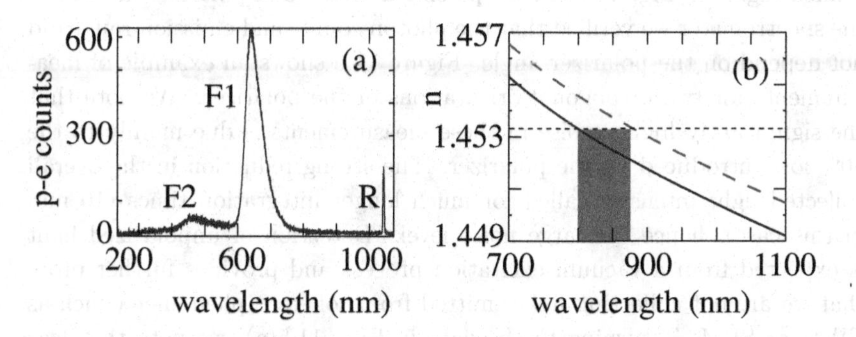

Fig. 11.3 (a) Measured CCD photo-electron (p-e) counts generated by the fused silica fluorescence spectrum. (b) Prediction of the Hawking emission spectral range for the case of fused silica: n_0 (solid line) and $n_0 + \delta n$ (dashed line) with $\delta n = 0.001$. The shaded area indicates the spectral emission region predicted for a Bessel filament.

input energy and Bessel pulse peak intensity, at the position $z = 1$cm where measurements were performed. The linear dependence is qualitatively in agreement with the fact that the emission bandwidth is predicted to depend on δn which in turn is a linear function $\delta n = n_2 I$ of the pulse intensity. By linearly fitting the slope one gets $n_2 = 2.8 \pm 0.5 \times 10^{-16}$ cm^2/W, to be compared with the tabulated value of $n_2 \sim 3 \times 10^{-16}$ cm^2/W [DeSalvo *et al.* (1996); Couairon and Mysyrowicz (2007)]. This shows also a qualitative agreement between the measurements and the model based on Hawking-like radiation emission.

In the same experiment has been also investigated the Hawking production by pulse filamentation in other frequency ranges, further corroborating the thesis of Hawking photon production.

We first show the experimental results obtained from the spontaneous filament. We measured the spectra around 350 nm, emitted from the dielectric perturbation generated by the spontaneous filament and integrated over 3600 laser shots. The measurements were carried out in sample areas that had not been previously irradiated so as to minimize the contribution from the fluorescence signal centred at 470 nm. Moreover, the fluorescence signal has been fitted with a Gaussian function and subtracted out so as to isolate the spectra shown in Fig. 11.5. In particular, Fig. 11.5(a) shows the full spectrum obtained by keeping the input slit of the spectrometer fully open in order to collect photons emitted from the whole filament. The filament, imaged at 90°, is shown in Fig. 11.5(d) and the white lines indicate the region isolated by the spectrometer slit. The measured spectrum agrees well with the predicted emission region that is indicated by the shaded region in Fig. 11.5(a). By placing a broadband polarizer in front of the spectrometer we verified that the photon counts and emission range did not depend on the polarizer angle. Figure 11.6 shows an example of measurements for two orthogonal orientations of the polarizer. We note that the significantly higher noise in these measurements is due mainly to the 50% loss introduced by the polarizer. The strong reduction in the overall collected light intensity called for much longer integration times (10 min in this case), hence the large noise level. Detection of unpolarized light is expected from a vacuum excitation process and provides further proof that we are not collecting light emitted from coherent phenomena such as FWM or SPM. Returning to the data in Fig. 11.5(a), we note the clear increase in the spectral intensity toward longer wavelengths. We attribute this to the fact that the perturbation velocity is varying along the filament length with a non-uniform acceleration, resulting in larger accumulation of

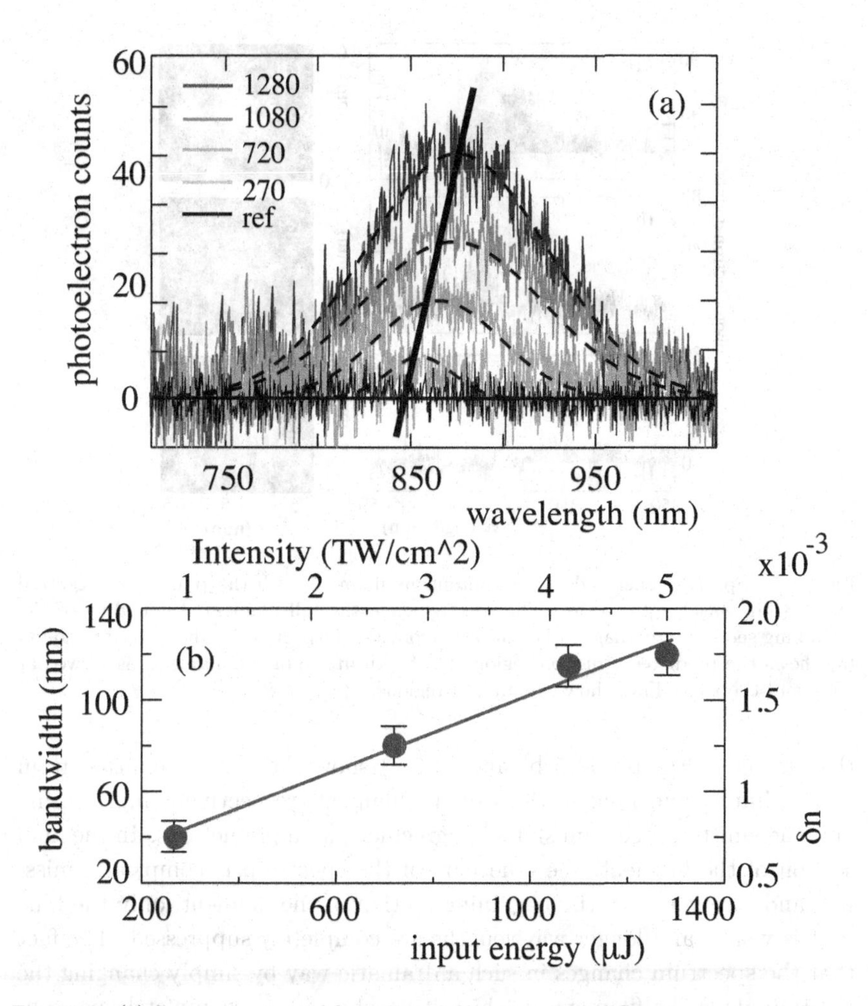

Fig. 11.4 (a) Spectra generated by a Bessel filament. Five different curves are shown corresponding to a reference spectrum obtained with a Gaussian pulse (black line) and indicated Bessel energies (in μJ). Dashed lines are guides for the eye. The solid line connects the spectral peaks, highlighting the $\sim$ 40 nm shift with increasing energy, in close agreement with the predicted 45 nm shift. (b) Bandwidths at FWHM and RIP δn versus input energy and intensity. Solid line: linear fit $\delta n = n_2 I$ with $n_2 = 2.8 \times 10^{-16}$ cm^2/W.

photons for smaller variations of the velocity, i.e. towards the end of the filament. According to equation (11.1), this should result in the accumulation of photons at longer wavelengths. We verified this interpretation by closing the spectrometer input slit so as to select different longitudinal portions of

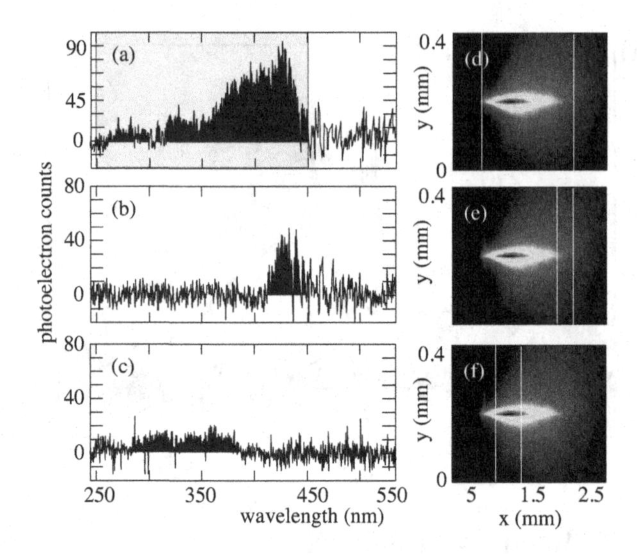

Fig. 11.5 Spectra generated by a spontaneous filament with the pump pulse centred at 1055 nm wavelength. (a)(c) The spectra when the full filament, the ending or the beginning sections are imaged onto the spectrometer, respectively. The shaded region in (a) shows the predicted emission region. (d)(f) An image of the filament, as viewed at 90° , and the white lines show the imaged regions.

the filament. Figures 11.5(b) and 11.5(c) show the spectra emitted from the beginning and final sections of the filament, respectively, highlighting how the emitted spectrum shifts in agreement with predictions. In the final section of the filament, the blue part of the spectrum is completely missing, and vice versa, in the beginning section of the filament, only the blue part is visible and longer wavelengths are completely suppressed. The fact that the spectrum changes in such a dramatic way by simply changing the position along the filament at which it is collected is a completely new observation and, to the best of our knowledge, does not find an explanation in any standard (e.g. four wave mixing of self-phase modulation) mechanism.

We conclude with few final comments: equation (11.1) refers to the existence of phase velocity horizons. However, other analogue systems rely on the existence of group velocity horizons [Rousseaux *et al.* (2010); Schutzhold and Unruh (2011)], so that it remains an open question regarding the relative role of the two different types of horizon. In the present experiment, in the case of the Bessel pulse, a group horizon does not even exist, since there are no frequencies that satisfy $v_g(\omega) = v_B$. Thus, at least in the RIP setting, the phase velocity horizon alone may lead to photon emission in the

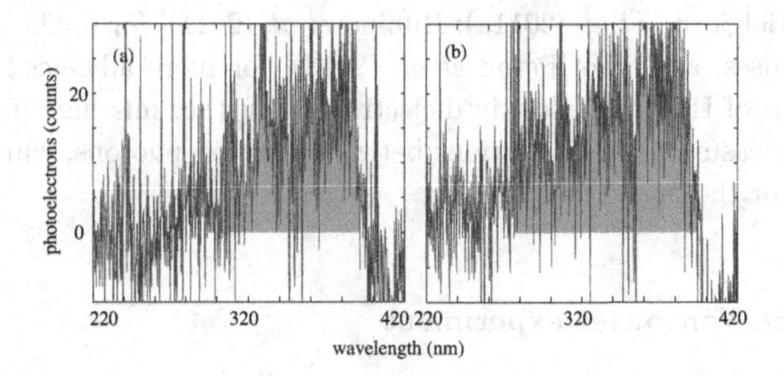

Fig. 11.6 Spectrum generated by a spontaneous filament with a pump-pulse central wavelength of 800 nm. Both (a) and (b) were taken under identical measurement conditions after placing a broadband polarizer in front of the spectrometer. The two graphs show measurements for orthogonal orientations of the polarizer, i.e. of the measured photons.

predicted spectral window. In other words, the superluminal Bessel pulse is subcritical and no blocking horizon associated with the dispersion relation is present. One can still imagine two different scenarios for the photon production associated with the Bessel pulse: from one hand, the dispersion curve at hand is only an approximation which is valid in a restricted range of frequencies; one could hypothesize that the photon production is a long wavelength phenomenon, in the sense that long wave infrared vacuum fluctuations seed the pair emission, and they might be involved with a blocking horizon. On the other hand, a different process of seeding could take place, where the seeding fluctuations start from inside the pulse and, because of their being subluminal with respect to the superluminal Bessel pulse, emerge from the trailing edge of the pulse and therein suffer a mode conversion process. Input and output modes lie on different dispersion curves, as the IN mode lies on the dispersion curve with $\delta n = \delta n_{max}$ and the P, N modes lie on the asymptotic dispersion curve with $\delta n = 0$. Developments of this study are found in [Rubino (2013)]. One could also project analytical calculations for the subcritical case, inspired by analogous studies in a different framework [Coutant and Weinfurtner (2016)].

There are difficulties with the experiment such that no definitive smoking gun has been recognized in the experiment itself for the identification of Hawking radiation. Mainly the absence of a blocking horizon in the case of Bessel pulse, and also the configuration with measures taken at 90° has been realized as involving a scattering process which has not yet reproduced in numerical simulations, as well as the intensity of photocounting is not yet accounted for (see in particular [Finazzi and Carusotto (2014)]). See however [Belgiorno *et al.* (2011a); Rubino *et al.* (2011)] for further analysis and responses, and also [Petev *et al.* (2013)] for more advanced numerical analysis of Hawking effect in dielectrics. A further missing information concerns measurement correlations between emitted photons, which is not available for the present experiment.

11.2 The Vancouver experiment

The theory underlying the feasibility of the experiment is contained in [Schutzhold and Unruh (2002)]. Therein, is shown that surface waves in a open channel flow can be a robust benchmark for Hawking radiation in labs. indeed, by varying in particular the height of the fluid, both the speed of gravity waves and the subluminal/superluminal character of the dispersion relation. Also in the present case, a white hole horizon is created. This is achieved as follows. The experimental setup involves a 6.2 m long, 0.15 m wide and 0.48 m deep flume. A spatial obstacle 1.55 m long and 0.106 m high is posed in the flume, and is suitably designed so as to minimize any flow separation. The obstacle function is to generate spatial inhomogeneities in the velocity of the flow, as required by the theory in order to be involved with an analogue spacetime geometry. The flume flow is generated by a pump, and, schematically, from left to right, at the level of the flume one meets an intake reservoir, the flume, the obstacle, the wave generator, an adjustable weir. At the level of the flume (see Fig. 2 in the arXiv version 1008.1911 of [Weinfurtner *et al.* (2011)]), there is a intake reservoir which provides the background flow with a reduced noise associated with inhomogeneities, turbulence and surface waves generated by the pump. The background flow is perturbed by counterpropagating waves generated downstream with respect to the obstacle by a wave generator, an oscillating mesh 2 m downstream the obstacle. These surface waves represent the ingoing waves of the process. If a blocking horizon (group horizon) is present, they undergo a mode conversion process and are transformed in a pair of large wavenumber modes associated with the analogue Hawking process. By means of laser-induced fluorescence water surface is illuminated, and also photographed by a high resolution camera. We refer to [Weinfurtner *et al.* (2011, 2013)] for more details about the experimental setup.

The background flow is assumed to be steady and incompressible, and the spatial velocity profile is

$$v(x) = \frac{q}{h(x)}, \tag{11.2}$$

where q represents the (fixed) two-dimensional flow rate, and $h(x)$ is the depth of the fluid, varying with x, i.e. the coordinate along the direction parallel to the fluid flow. The dispersion relation one may refer to [Schutzhold and Unruh (2002)] is

$$(f + vk)^2 = \left(\frac{gk}{2\pi}\right) \cdot \tanh(2\pi kh), \tag{11.3}$$

with f the frequency, k the wavenumber, g the gravitational acceleration, v the fluid flow velocity and h as above. For $2\pi kh < 1$, (11.3) can be approximated by

$$(f + vk) = \sqrt{ghk}, \tag{11.4}$$

and one speaks of 'shallow water waves'. If, instead, $2\pi kh > 1$, (11.3) can be approximated by

$$(f + vk) = \sqrt{\frac{gk}{2\pi}}, \tag{11.5}$$

and deep water waves are obtained. The IN wave, whose wavenumber is k^+_{IN}, is a shallow wave which is sent upstream. There are two emerging waves OUT which are characterized by much larger wavenumbers k^+_{OUT} and k^-_{OUT}. These outgoing states correspond to deep water waves swept downstream and resulting from mode conversion of the IN wave due to the presence of a blocking horizon. k^+_{IN} is a positive norm state, with positive group velocity and positive phase velocity; k^+_{OUT} corresponds to a positive norm state, with positive phase velocity and negative group velocity, whereas k^-_{OUT} is a negative norm state with negative phase and group velocities. If thermality holds true, it is expected that

$$\frac{|J_x(k^-_{OUT})|}{|J_x(k^+_{OUT})|} = \exp(-\beta f), \tag{11.6}$$

which would prove thermality of the stimulated Hawking effect taking place in the experiment.

In the experiment, nine different frequencies were utilized, between 0.02 Hz and 0.67 Hz. q is 0.045 m^2/s and $h = 0.194m$. The change of variables

$$d\xi = \frac{dx}{v(x)}, \tag{11.7}$$

allows to rewrite the convective derivative $\partial_t + v(x)\partial_x$ as $\partial_t + \partial_\xi$, and one may introduce κ as the wavenumber associated with ξ. In any event, the conserved norm (cf. also Chapter 8) has the form

$$\int d\kappa \frac{|A(f,\kappa)|^2}{f+\kappa}, \tag{11.8}$$

where $A(f,\kappa)$ is the t,ξ-Fourier transform of the vertical displacement of the wave. Measured amplitudes confirmed that negative norm states were actually produced and also that thermality occurs, with a temperature $T = 6 \cdot 10^{-12} K$. Note that spontaneous Hawking emission is not measurable because of the low temperature involved.

The Vancouver experiment is very interesting and well-grounded from a physical point of view. Criticisms have been raised because of the subcritical character of five of the nine explored experimental frequencies (see the discussion in [Michel and Parentani (2014); Euvé *et al.* (2015); Michel and Parentani (2015b); Robertson *et al.* (2016)]), to which a further study about the imprints of Hawking effect in the subcritical case was dedicated [Coutant and Weinfurtner (2016)].

11.3 The Technion experiment

Hawking effect in BEC has been considered in several papers [Balbinot *et al.* (2005, 2008); Carusotto *et al.* (2008); Macher and Parentani (2009a); Recati *et al.* (2009); Mayoral *et al.* (2011); Zapata *et al.* (2011); Larre *et al.* (2012); Michel and Parentani (2013)], and particular analysis of the possible signatures of this phenomenon in BECs have been studied in [Balbinot *et al.* (2008)]. The main focus herein is on the so-called black hole lasers [Corley and Jacobson (1999)].

Starting from these theoretical results, J. Steinhauer realized some experiments at the Technion laboratories in order to prove the existence of both the analogue Hawking and the lasing effects. For a full account see [Steinhauer (2014, 2016)]. See also [Steinhauer (2015); Steinhauer and de Nova (2017)].

The experiment has been realized with a Bose Einstein condensate of $^{87}R_b$ atoms in the state $F = 2$, $m_F = 2$. The $1D$-dimensional confinement has been imposed with a transversal potential realized using a magnetic trap by means of a laser beam of $123Hz$ of radial trap frequency. The thickness of the beam was of $5\mu m$ and the laser wavelength of $812nm$. The energy level spacing of the transversal confinement was of about $6nK$, non harmonic. The axial potential has been shaped to be less than parabolic.

The horizon is created by a step potential in three different regimes. The height of the steps in the three cases are respectively of $3.2nK$, $6.4nK$, and $9.6nK$. The coherence length λ_c was about of $2\mu m$ in the lasing region, and the width of the steps was approximately equal to ξ. The number of atoms involved has not been declared, but the chemical potential was of $8nK$, from which the number of atoms can be estimated.

The step potential generate a waterfall which accelerates the atoms at the left of the step to superluminal velocity. This fixes a black hole horizon. The step potential is moved with constant velocity along the one dimensional reduction axes, whereas all the remaining potentials are kept at rest. This has the effect to get a slow motion of the atoms on the right of the step, and a superluminal motion on the left. The confining potential however makes the left atoms to be slowed down far on the left so that at a certain point a white hole horizon also appears. The supersonic region between the two horizons resulted to have a depth of about 10ξ.

The experiment has been performed by leaving the system to evolve for $120ms$. The condensate density was imaged at seven given times along each evolution, by means of a phase contrast method. At each time the collection of 80 images was averaged in order to reduce noise.

In the correspondence of the trapped region, in the supersonic region, 6/7 peaks with growing amplitude in time appeared and have been interpreted by the author of the experiment as the result of the interference between Hawking particles and negative modes, giving rise to an Hawking triggered lasing effect. Therefore, the resulting observed data have been interpreted by Steinhauer as an evidence of Hawking effect.

We are not interested in entering further into details and providing a complete critical analysis here. The Nature papers have been obtaining a lot of interest and international resonance. Still, there is not yet a full convergence of experts of the matter on the experimental results. We limit ourselves to mention that doubts have been arose in different papers about the original interpretation of Steinhauer, since accurate numerical simulations show that the data measured in the experimental level can be reproduced by means of purely classical effects not related to the Hawking process [Tettamanti *et al.* (2016); Wang *et al.* (2017a)] and, moreover even the lasing effect can be questioned [Wang *et al.* (2017a,b)].

Appendix A

Algebraic methods in QFT

As we will see, certain results about Hawking radiation (and not only) can be made rigorous by means of the algebraic formulation of quantum field theory. In this chapter we will shortly review the main tools which will be applied to compute quantum effects on black holes backgrounds.

A.1 Araki-Haag-Kastler algebraic formulation of QFT

In this section, for convenience, we will shortly review the main steps in the algebraic formulation of quantum field theory of Haag and Kastler [Haag and Kastler (1964)]. We will follow [Bogolubov *et al.* (1975)], part VI, to which, and the corresponding bibliography, we refer for further details. The main idea is to work with "operations on a physical system" in place of working with the physical observables, typically described by unbounded operators. This is because any measurement in a laboratory is performed locally in space and time, and allow for a knowledge of observables up to a given precision.

We first recall that a $*$-algebra $\mathcal{U}$ is an algebra endowed with an anti-linear involution

$$* : \mathcal{U} \longrightarrow \mathcal{U}, \qquad A \mapsto A^*, \qquad *^2(A) = A. \tag{A.1}$$

A $*$-algebra is a Banach algebra if is endowed with a norm

$$\| \ \| : \mathcal{U} \longrightarrow \mathbb{R}_{\geq}, \tag{A.2}$$

such that

$$\|AB\| \leq \|A\| \ \|B\|, \tag{A.3}$$

$$\|A^*\| = \|A\|, \tag{A.4}$$

$$||\mathbb{I}|| = 1, \tag{A.5}$$

and $\mathcal{U}$ is closed w.r.t. the norm.

A Banach algebra satisfying

$$||A|| = \sqrt{||A^* \, A||} \tag{A.6}$$

is said a C^*-algebra. For example, the set of all bounded operators on a separable Hilbert space, with the Hermitian conjugation as involution and the operator norm as a norm, is a C^*-algebra.

We also recall that a *directed set* is a set X with a partial ordering relation $<$, such that for any $U, V \in X$ it exist at least one $Z \in X$ such that $U < Z$ and $V < Z$. For example, note that any topological set is direct set partially ordered by inclusion.

A quantum field theory on the Minkowski spacetime can be algebraically axiomatized as follows. If $\mathcal{O}$ is any open subset in the spacetime $\mathbb{R}^{1,3}$, to the local observables and local operations in $\mathcal{O}$, one associates an abstract C^*-algebra $\mathcal{U}(\mathcal{O})$ with unit $\mathbb{I}$, satisfying the following five conditions:

(1) Let

$$\mathcal{U} = \overline{\bigcup_{\mathcal{O}} \mathcal{U}(\mathcal{O})}, \tag{A.7}$$

where the closure is understood w.r.t. the norm, the algebra of quasilocal observables. Then, the $\mathcal{U}(\mathcal{O})$ form a net from the Minkowski spacetime to the C^*-algebra $\mathcal{U}$, such that

$$\mathcal{O} \subset \mathcal{O}' \qquad \Rightarrow \qquad \mathcal{U}(\mathcal{O}) \subset \mathcal{U}(\mathcal{O}'), \tag{A.8}$$

where the second inclusion means subalgebra;

(2) *Relativistic covariance*: for any Poincaré transformation $g \equiv (\Lambda, a)$, $\Lambda \in SL(1,3)$, $a \in \mathbb{R}^4$, it exists an automorphism of $\mathcal{U}$ such that

$$\mathcal{U}(\mathcal{O}) \longmapsto \mathcal{U}(g\mathcal{O}), \tag{A.9}$$

$g\mathcal{O}$ being the punctual transformation of $\mathcal{O}$;

(3) *Local commutativity*: if $\mathcal{O}$ and $\mathcal{O}'$ are spatially separated, t.i. if

$$\eta(x, x') > 0$$

for any $(x, x') \in \mathcal{O} \times \mathcal{O}'$, then $\mathcal{U}(\mathcal{O})$ and $\mathcal{U}(\mathcal{O}')$ do commute;

(4) *Primitivity*: $\mathcal{U}$ admits a faithful irreducible representation on a Hilbert space $\mathcal{H}$,

$$\rho : \mathcal{U} \longrightarrow B(\mathcal{H}), \tag{A.10}$$

where $B(\mathcal{H})$ is the C^*-algebra of bounded operators on $\mathcal{H}$;

(5) *Causality*: define the future (past) shadow of $\mathcal{O}$ as the set of all point x such that any past (future) directed timelike or lightlike ray originating from x passes through $\mathcal{O}$. Moreover, define the causal envelop $C(\mathcal{O})$ of $\mathcal{O}$, as the union of the past and future shadows of $\mathcal{O}$. Then we require

$$\mathcal{U}(\mathcal{O}) = \mathcal{U}(C(\mathcal{O})). \tag{A.11}$$

Indeed, one requires two further conditions, but before doing this some comments are in order. First we note that the Causality requirement is necessary to define the Hamiltonian evolution of the system. The time evolution would require a definition of an initial condition of certain Cauchy data on a local set with compact support in an open subset $\mathcal{O}$ of a Cauchy surface in $\mathbb{R}^{1,3}$. Such a set, however, is not an open set in spacetime, but $C(\mathcal{O})$ is, so that one can define in general $\mathcal{U}(\mathcal{O}_t) := \mathcal{U}(C(\mathcal{O}_t))$, where $\mathcal{O}_t$ is a bounded region at time t. Second, we have said that one defines the quasilocal observables in terms of an abstract C^*-algebra. As such we require for it to be representable faithfully on a Hilbert space in terms of bounded operators. Among these operators, only the selfadjoint ones, associated to elements $A^* = A$ of the algebra, will correspond to observables. Thus, differently from the usual picture, observables are associated only to bounded operators. We note that the requirement of restricting the image of the representation into $B(\mathcal{H})$ is superfluous since one can prove the following

Proposition A.1. *All representations $(\rho, \mathcal{H})$ of a Banach $*$-algebra $\mathcal{U}$ over a Hilbert space $\mathcal{H}$ are continuous and satisfy $\|\rho(A)\|_{\mathcal{H}} \leq \|A\|_{\mathcal{U}}$, for any $A \in \mathcal{U}$.*

Moreover, it is well known that different representations can be inequivalent in the sense that are not connected by unitary transformations. However, the equivalence between two representations should take into account that they cannot be distinguished if the measurements give the same results up to arbitrarily small discrepancies. For this reason in [Haag and Kastler (1964)] two representations $(\rho_i, \mathcal{H}_i)$, $i = 1, 2$ are said to be physically equivalent if for any positive integer n and any choice of local observables $A_1, \ldots, A_n \in \mathcal{U}$, and any positive operator of trace class $B_1 \in B(\mathcal{H}_1)$, and for any $\epsilon > 0$, it exists a positive operator of trace class $B_2 \in B(\mathcal{H}_2)$ such that

$$|\mathrm{Tr}(B_1\rho_1(A_i)) - \mathrm{Tr}(B_2\rho_2(A_i))| < \epsilon, \qquad i = 1, \ldots, n. \tag{A.12}$$

This definition coincides with the notion of weak equivalence of representations, and has been characterized by Fell [Fell (1960)] who has shown that two representations of a C^*-algebra $\mathcal{U}$ are weakly equivalent if and only if they have the same kernel in $\mathcal{U}$. In particular, it follows that all the faithful representations are weakly and then physically equivalent (even though they are not unitarily equivalent) [Haag and Kastler (1964)].

Third, not all the elements of $\mathcal{U}$ correspond to observables, since most of them are not selfadjoint. As we said, in general they correspond to operations that change the state of a physical system. A state is defined in terms of positive linear functionals over $\mathcal{U}$. Thus a state is a map

$$\omega : \mathcal{U} \longrightarrow \mathbb{C}, \tag{A.13}$$

such that

$$\omega(A^*A) \geq 0. \tag{A.14}$$

Strictly speaking, a state should be normalized so that $\omega(\mathbb{I}) = 1$. However, it is often useful to work with non normalized states and next normalize all computed quantities. In this sense, positive linear operators which differs by a positive multiplicative constant individuate the same state. A state is said *pure* if it cannot be written as a sum of two linearly independent positive operators. A transformation is a linear map

$$T : \mathcal{U} \longrightarrow \mathcal{U}, \tag{A.15}$$

sending positive operators into positive operators. In particular, a *pure transformation* is a transformation that maps pure states in pure states. These are obtained by the unit ball $B_1(0) = \{A \in \mathcal{U} | \|A\| \leq 1\}$, via the map

$$\omega \longmapsto \omega_A, \quad \omega_A(B) := \omega(A^\dagger BA). \tag{A.16}$$

In this case one can define the probability of having a transition of the system under the action of A as

$$P(\omega, A) = \frac{\omega(A^\dagger A)}{\omega(I)}. \tag{A.17}$$

One can take a contact with a more familiar formalism thanks to the GNS construction of a representation due to Gel'fand, Neimark and Segal:

Theorem A.1. *Let $\mathcal{U}$ be a Banach $*$-algebra with unit and ω be a positive linear functional over $\mathcal{U}$. Then, it exists a representation $(\rho, \mathcal{H})$ uniquely determined (up to unitary equivalences) by ω, with a cyclic vector $\chi \in \mathcal{H}$ such that*

$$\omega(A) = (\chi | \rho(A)\chi), \tag{A.18}$$

for all $A \in \mathcal{U}$.

The fact that χ is cyclic means that

$$\mathcal{H} = \overline{\rho(\mathcal{U})\chi}. \tag{A.19}$$

It also happens that such representation is irreducible iff ω is a pure state, or equivalently iff any vector $x \neq 0$ in H is cyclic.

From the above statements, one can prove the existence of a cyclic vector which is invariant under translations [Doplicher (1965)]. However, to define a relativistic QFT one needs a Poincaré invariant vacuum state, so that, following [Bogolubov *et al.* (1975)] we add the axiom

6. *Spectral condition*: It exists a faithful irreducible representation $(R, \mathcal{H})$ over a Hilbert space, with a unique non vanishing vector $\chi_0 \in \mathcal{H}$ which is invariant under the Poincaré group action. Moreover, the spectrum of the Hermitian generators of the nontrivial unitary representation of the Poincaré translations belongs to the future lightcone.

In particular, it follows by the previous considerations, that the vector χ_0 is a cyclic vector. This representation is not unique since there can be several unitarily inequivalent representations, but all physically equivalent.

Finally, one would include the fields, which again are expected to be defined by a net of algebras

$$\mathcal{O} \longmapsto \mathcal{F}(\mathcal{O}). \tag{A.20}$$

However, in general one cannot expect for the fields to be all observable, so that the algebras of local fields $\mathcal{F}(\mathcal{O})$ will contain properly the subalgebras of local observables. The only requirements one adds for $\mathcal{F}(\mathcal{O})$ is to be weakly closed, which means that they are Von Neumann algebras. Such algebra may contain several different fields, for example, bosonic and fermionic fields. Because of the different behavior under rotations, the relative phases among fermionic and bosonic fields is completely irrelevant. Thus, in any representation of the algebra, the corresponding Hilbert space will be separated in *coherent subspaces* each one corresponding to an independent field (in the sense of the global phase), so that pure states can belong in any coherent subspace but not in a combination of them. This leads to the last axiom:

7. The algebra of local observables must leave invariant any coherent subspace.

For our purposes, we don't need to develop here the consequences of this formulation, for which we refer to the literature [Haag (1996)], and [Halvorson and Muger (2007)], and [Brunetti *et al.* (2015)] for a more modern treatment.

A.2 Haag-Hugenholz-Winnik formulation of quantum statistical systems

The problem of studying the equilibrium states of a quantum system with an infinite number of degrees of freedom in a direct way, in place of takeing the thermodynamic limit of a finite theory, has been first considered by R. Kubo [Kubo (1957)] in 1957 and by P. C. Martin and J. Schwinger [Martin and Schwinger (1959)] in 1959. A general formulation describing equilibrium states in quantum statistical mechanics, valid for general quantum systems, has been developed by R. Haag, N. M. Hugenholz and M. Winnik in 1967 [Haag *et al.* (1967)] .

A.2.1 *Structure of the statistical system*

Observables and operations acting on a physical system are described by a C^*-algebra $\mathcal{U}$. A positive linear functional over this algebra gives the expectation values and, in this way, it determines a state of the system. Thus, among all possible states, one wants to determine the equilibrium states. One can think to define these by taking the limit of infinite volume of a sequence of Gibbs grand canonical ensembles with increasing volumes. Let us adopt the convention of indicating with R_V a spatial region of volume V. In particular, we will consider a sequence $\{R_{V_n}\}_{n=1}^{\infty}$, with $R_{V_i} \subset R_{V_{i+1}}$, with $\bigcup_n R_{V_n} = \mathbb{R}^3$, which also mean $V_i < V_{i+1}$ and $\lim_{n\to\infty} V_i = \infty$.

Let us then start with a fixed representation of $\mathcal{U}$ over an Hilbert space. In such a representation there are well defined an Hamiltonian operator $\hat{H}$ and a number operator $\hat{N}$. At the time $t = 0$ consider the creator field operator $\boldsymbol{a}^\dagger$ and the destructor linear field operator $\boldsymbol{a}$ acting on a Hilbert space $\mathcal{H} = L^2(\mathbb{R}^3, \mu)$, where μ is the Lebesgue measure. The commutator algebra is then described by

$$[\boldsymbol{a}(f), \boldsymbol{a}^\dagger(g)] = (f|g), \qquad [\boldsymbol{a}(f), \boldsymbol{a}(g)] = [\boldsymbol{a}^\dagger(f), \boldsymbol{a}^\dagger(g)] = 0. \quad \text{(A.21)}$$

One usually thinks about these as defining the one particle creators and destructors $\hat{a}^\dagger(x)$ and $\hat{a}(x)$ in the point $x \in \mathbb{R}^3$ by writing

$$\boldsymbol{a}^\dagger(f) = \int_{\mathbb{R}^3} \hat{a}^\dagger(x) f(x) d\mu(x), \qquad \boldsymbol{a}(f) = \int_{\mathbb{R}^3} \hat{a}(x) f^*(x) d\mu(x), \text{ (A.22)}$$

so that

$$[\hat{a}(x), \hat{a}^\dagger(y)] = \delta^3(x - y), \quad \text{(A.23)}$$

and so on. In $\mathcal{H}$ there is a distinguished state Ω_0, the 0-particle state or vacuum, such that $\boldsymbol{a}(f)\Omega_0 = 0 \ \forall \ f \in \mathcal{H}$. Despite the fact that the support of the function in $\mathcal{H}$ may be the whole $\mathbb{R}^3$, this is not the suitable representation for describing the equilibrium state for the system in the infinite volume region (the whole $\mathcal{R}^3$). Indeed, on $\mathcal{H}$, as we said, the Hamiltonian operator and the number operator are well defined separately, whereas in a thermal equilibrium configuration for a system with infinite degrees of freedom, only the combination $\hat{H} - \mu\hat{N}$, μ being the chemical potential, can be well defined. For this reason one must proceed further as follows, in order to get the right representation of $\mathcal{U}$.

Now, fix a region R_V of finite volume. The set

$$\mathcal{R}_V = \{\boldsymbol{a}(f), \boldsymbol{a}^\dagger(f)\}_{f\in\mathcal{H},\text{supp}(f)\subseteq R_V}$$

generates a Von Neumann algebra $\mathcal{U}(R_V)$, which acts in a natural way on $\mathcal{H}$. Thus one can define

$$\mathcal{H}_{R_V} := \mathcal{U}(R_V)\Omega_0. \tag{A.24}$$

Clearly, $\mathcal{H}_{R_V}$ is the support space for an irreducible representation for $\mathcal{U}(R_V)$. It results that $\mathcal{U}(V)$ is closed w.r.t. the operators norm, so that it is a C^*-algebra. Moreover, if we consider the above sequences of regions, it can be shown that

$$\mathcal{U} = \overline{\bigcup_n \mathcal{U}(R_{V_n})}, \tag{A.25}$$

where the closure is in the operators norm. Note also that for non intersecting regions V, W, $V \cap W = \emptyset$, the corresponding Von Neumann algebras do commute.

We now want to proceed in order to define a representation of $\mathcal{U}$ on an Hilbert space H_ω obtained as a limit of the representations $\mathcal{H}_{R_{V_n}}$ when $n \to \infty$. In such a limit we want to have a well defined mean number density of particles. In this way, neither the number operator $\hat{N}$ or the Hamiltonian $\hat{H}$, which are well defined on $\mathcal{H}$, can be well defined on $\mathcal{H}_\omega$, whereas the combination

$$\hat{H}' = \hat{H} - \mu\hat{N} \tag{A.26}$$

must be well defined in order to apply the formalism of the grand canonical ensemble. However, if we restrict to the Von Neumann algebra on any finite volume region R_V, $\mathcal{U}(R_V)$, then both operators will be well defined. For this reason one expects that the representations of $\mathcal{U}(R_V)$ over $\mathcal{H}$ and over $\mathcal{H}_\omega$ should be essentially equivalent (and, indeed, they are quasiequivalent). Thus, one works with the operator $\hat{H}'$, and the corresponding unitary one parameter group operator $U(t) = \exp(i\hat{H}'t)$. In order to be able to define the grand canonical ensemble, one has to require the following three conditions [Haag *et al.* (1967)]:

(1) For any region R_V with finite volume, the *partial Hamiltonian* $\hat{H}'_{R_V}$ is well defined, and

$$U_{R_V}(t) = e^{i\hat{H}'_{R_V} t} \in \mathcal{U}(R_V).$$

We require for the partial Hamiltonians to be quasiadditive, which means that, if one has a decomposition $R_V = R_{V'} \cup R_{V''}$ and $S = R_{V'} \cap R_{V''}$ is a surface, then

$$\hat{H}'_S := \hat{H}'_{R_V} - \hat{H}'_{R_{V'}} - \hat{H}'_{R_{V''}}$$

is just a surface term. For each $A^{R_V} \in \mathcal{U}(R_V)$ one also defines $A_t^{R_V} = U_{R_V}(t)A^{R_V}U_{R_V}(t)^{-1}$. Then, if R_{V_n} is the increasing sequence of regions, we assume that, for any fixed t, $A_t^{R_{V_n}} = U_{R_{V_n}}(t)A^{R_{V_1}}U_{R_{V_n}}(t)^{-1}$ is a Cauchy sequence in the operators norm topology, so that it exists

$$\lim_{n\to\infty} A_t^{R_{V_n}} \in \mathcal{U}.$$

This continues to be true if we take $A \in \mathcal{U}$ in place of $A^{R_{V_1}}$ since we can always find an approximating sequence $A^{R_{V_n}} \in \mathcal{U}(R_{V_n})$, strongly convergent to A. This also defines the approximation of $U(t)$ as

$$A_t = U(t)AU(t)^{-1} = \lim_{n\to\infty} U_{R_{V_n}}(t)AU_{R_{V_n}}(t)^{-1}.$$

(2) In order to define the Gibbs states for all positive V and inverse temperature β and for a given range for the chemical potential μ, one needs to require

$$\mathrm{Tr}_{\mathcal{H}_{R_V}} e^{-\beta \hat{H}'_{R_V}} < \infty$$

for any finite volume region R_V. This also implies that for any $A \in \mathcal{U}(R_V)$ is well defined the expectation value

$$\omega_{R_V}(A) := \mathrm{Tr}_{\mathcal{H}_{R_V}}(\rho_{R_V} A),$$

with density operator

$$\rho_{R_V} := (\mathrm{Tr}_{\mathcal{H}_{R_V}} e^{-\beta \hat{H}'_{R_V}})^{-1} e^{-\beta \hat{H}'_{R_V}}.$$

(3) Given the sequence $\{R_{V_n}\}_{n=1}^{\infty}$ increasing to $\mathbb{R}^3$ and $A \in \mathcal{U}(R_{V_1})$, we assume always to exist the limit

$$\omega(A) := \lim_{n\to\infty} \omega_{R_{V_n}}(A).$$

This implies that $\omega(A^\dagger\, A) > 0$, $|\omega(A)| \leq \|A\|$, and ω is then uniquely well defined, since, by strong convergence, as before it can be defined for any $A \in \mathcal{U}$.

Note that the last condition is required to ensure the existence of the infinite volume limit. However, it would be interesting to be able to weaken such condition enough to include the possibility of phase transitions for particular values of β and μ.

A.2.2 *The Gibbs states*

In order to define the equilibrium states for the infinite system, one first defines the Gibbs states for the finite volume systems and then takes the $V \to \infty$ limit. Since ρ_{R_V} is positive with finite trace, one can uniquely define

$$\chi_{R_V} = (\rho_{R_V})^{\frac{1}{2}}, \tag{A.27}$$

so that it is Hilbert-Schmidt and that if $A \in \mathcal{U}(R_V)$ then

$$A\chi_{R_V} = 0 \qquad \Rightarrow \qquad A = 0. \tag{A.28}$$

Note that we then can write

$$\omega_{R_A}(A) = \mathrm{Tr}_{\mathcal{H}_{R_V}}(\chi_{R_V} A \chi_{R_V}), \tag{A.29}$$

from which we also see that

$$\omega_{R_V}(A^\dagger\, A) = 0 \qquad A = 0. \tag{A.30}$$

Now, we can represent $\mathcal{U}(R_V)$ on a particular Hilbert space $\mathcal{H}$, the bilateral ideal of Hilbert-Schmidt operators. Indeed, this is a linear space endowed with a natural sesquilinear product

$$(\chi|\chi') = \mathrm{Tr}(\chi^\dagger\, \chi'). \tag{A.31}$$

This generates a norm w.r.t. which $\mathcal{H}$ is closed. On this Hilbert space one can define two representations of $\mathcal{U}(R_V)$: a linear one R, and an antilinear one S. The linear one is defined by

$$R(A)\chi := A\chi, \tag{A.32}$$

the natural action of $\mathcal{U}(R_V)$ over the Hilbert-Schmidt operators, whereas the antilinear one is defined by

$$S(A)\chi := \chi A^\dagger. \tag{A.33}$$

It is easy to see that these are C^*-algebra representations. If with a prime we indicate the centralizer of the algebra (t.i. the set of all elements commuting with the whole algebra) then it is not difficult to show that

$$R(\mathcal{U}(R_V))'' = S(\mathcal{U}(R_V))'. \tag{A.34}$$

If we introduce the conjugation operator J defined by $J(\chi) := \chi^\dagger$, so that $J = J^{-1}$ and $J\chi_{R_V} = \chi_{R_V}$, then one gets the following theorem generalizing a result of Araki and Woods [Araki and Woods (1963)] for the ideal Bose gas:

Theorem A.2. *The vector χ_{R_V} generating $\mathcal{H}$ is cyclic for both representations, and one has*

$$\omega_{R_V}(A) = (\chi_{R_V}|R(A)\chi_{R_V}) = \overline{(\chi_{R_V}|S(A)\chi_{R_V})}. \tag{A.35}$$

Moreover, R and S are connected by the antiunitary conjugation by

$$JR(A) = S(A)J. \tag{A.36}$$

For the proof see [Haag *et al.* (1967)]. However, it is worth to note that the proof of the theorem is based mainly on the fact that $A\chi_{R_V} = 0$ implies $A = 0$. Since χ_{R_V} will define the equilibrium state, it is important to remark that the mentioned property is true for a large class of Hilbert-Schmidt operators, which can then be used to define equilibrium states for all possible values of β and μ.

The time evolution $A_t = U(t)AU(t)^{-1}$ can be represented by R, via unitary operators $\hat{U}(t)$ over $\mathcal{H}$ such that

$$R(A_t) = \hat{U}(t)R(A)\hat{U}(t)^{-1}. \tag{A.37}$$

The general solution of this equation is $\hat{U}(t) = R(U(t))V(t)$, where $V(t)$ is a unitary element in the commutant $R(\mathcal{U}(R_V))'$. However, there is only a possible choice of $V(t)$ such that $\hat{U}(t)\chi_{R_V} = \chi_{R_V}$. Since this is the property we want to require for an equilibrium state (which indeed means $\omega(A_t) = \omega(A)$), we get

$$\hat{U}(t) = R(U(t))S(U(t)). \tag{A.38}$$

The corresponding infinitesimal generator is thus

$$\hat{H}' = R(H') - S(H'). \tag{A.39}$$

A.2.3 *KMS condition and the infinite volume limit*

In order to pass to the infinite volume limit one needs first to look at the analytic properties of the complex time evolution. Fixing $z \in \mathbb{C}$ such that $\mathrm{Re}(z) = t$, and setting

$$A_z = e^{iH'z}Ae^{-iH'z}, \tag{A.40}$$

we see that $A_z e^{-\beta H'}$ is bounded and of trace class for any $0 \leq \mathrm{Im}(z) \leq \beta$, whereas $e^{-\beta H'}A_z$ is bounded and of trace class for any $-\beta \leq \mathrm{Im}(z) \leq 0$. It then follows that

$$F_{AB}(z) := \mathrm{Tr}(BA_z e^{-\beta H'}) \tag{A.41}$$

is well defined in the strip $0 \leq \mathrm{Im}(z) \leq \beta$, analytic in $0 < \mathrm{Im}(z) < \beta$ and continuous at the boundaries. Similarly

$$G_{AB}(z) := \mathrm{Tr}(e^{-\beta H'}A_z B) \tag{A.42}$$

is well defined in the strip $-\beta \leq \mathrm{Im}(z) \leq 0$, analytic in $-\beta < \mathrm{Im}(z) < 0$ and continuous at the boundaries. If we restrict to $\mathrm{Im}(z) = 0$ then $z = t$ and we see that

$$F_{AB}(t) = \omega_{R_V}(BA_t), \qquad G_{AB}(t) = \omega_{R_V}(A_t B) \tag{A.43}$$

and using the cyclic property of the trace we see that

$$F_{AB}(t+i\beta) = G_{AB}(t). \tag{A.44}$$

This properties of analyticity and periodicity are known as the KMS conditions. However, in order to take the limit, HHW reformulate such conditions in a more synthetic way. If $\hat{f}$ is a Schwarzian test function of one real variable, t.i. a smooth function with compact support $\mathcal{D}(\mathbb{R})$, then

$$f(z) = \int_{\mathbb{R}} \hat{f}(s)e^{izs}ds$$

is an analytic function such that $f(t+iy)t^n$ is bounded for any fixed y, n. Then one can rewrite the periodicity condition in the form

$$\int_{\mathbb{R}} f(t-i\beta)\omega_{R_V}(BA_t)dt = \int_{\mathbb{R}} f(t)\omega_{R_V}(A_tB)dt. \tag{A.45}$$

Indeed, this result to be equivalent to the KMS conditions, since if this equality holds for any $A, B \in \mathcal{U}(R_V)$ and $\hat{f} \in \mathcal{D}(\mathbb{R})$, then one can prove not only the periodicity condition but also the analyticity properties. We will call this the *equivalent KMS condition.*

By using properties 1 and 3 of section A.2.1, one can prove that

$$\lim_{n\to\infty} \int_{\mathbb{R}} f(t)\omega_{R_{V_n}}(BA_t^{R_{V_n}})dt = \int_{\mathbb{R}} f(t)\omega(BA_t)dt \tag{A.46}$$

converges for any rapidly decreasing smooth function $f \in \mathcal{S}(\mathbb{R})$. This shows that the equivalent KMS condition continue to hold in the infinite volume limit. Thus, it completely characterizes the properties of the limit state ω, and under the GNS construction allows to construct the corresponding representation of $\mathcal{U}$. However, in order to do this, some further requirements must be imposed.

First, note that the C^*-algebra $\mathcal{U}$ contains a $*$-algebra $\tilde{\mathcal{U}}$ generated by all elements $A \in \mathcal{U}$ such the Fourier transform $\hat{A}(s)$ of A_t is a distribution with compact support. If $A \in \tilde{\mathcal{U}}$ one can define the automorphism (which is not a $*$-automorphism)

$$A \longmapsto A_{ix} := \int_{\mathbb{R}} \hat{A}(s)e^{xs}ds, \tag{A.47}$$

for any $x \in \mathbb{R}$. Note that the invariance of ω implies $\omega(A_{ix}) = \omega(A)$ for all $A \in \tilde{\mathcal{U}}$. On the other hand, the equivalent KMS condition can be rewritten in the form

$$\int_{\mathbb{R}} [\omega(BA_{t+i\beta}) - \omega(A_tB)]f(t)dt = 0, \qquad \forall \hat{f}(s) \in \mathcal{D}(\mathbb{R}),\ A \in \tilde{\mathcal{U}}. \tag{A.48}$$

This is equivalent to

$$\omega(BA_{i\beta}) = \omega(AB), \qquad \forall A \in \tilde{\mathcal{U}},\ B \in \mathcal{U}. \tag{A.49}$$

If also $B \in \tilde{\mathcal{U}}$, this becomes

$$\omega(A^{\dagger}_{i\frac{\beta}{2}} B_{i\frac{\beta}{2}}) = \omega(BA^{\dagger}), \qquad \forall A, B \in \tilde{\mathcal{U}}. \tag{A.50}$$

This is a very simple form, equivalent to the KMS conditions. One further requires

$$\omega(A^{\dagger}A) = 0 \quad \Rightarrow \quad A = 0,\ \forall A \in \mathcal{U}. \tag{A.51}$$

From physics, in place of this condition, one would require a weaker condition, t.i. $\omega(A^{\dagger}A) = 0$ implies $A = 0$ for all $A \in \mathcal{U}(R_V)$ for some finite V. But in [Haag *et al.* (1967)] it is shown that the stronger condition must be required if the C^*-algebra must be simple (t.i. does not contains nontrivial bilateral ideals). On the other side, a C^*-algebra of commutation relations is necessarily simple [Segal *et al.* (1967)].With these instruments one can construct the thermal representation associated to the equilibrium state ω.

A.2.4 *The thermal representations*

We take as representation space for $\mathcal{U}$, the algebra $\mathcal{U}$ itself, seen as a linear space. Following [Haag *et al.* (1967)] we use the notation $\psi_A = A$ to indicate A as an element of the linear space in place of an element of the algebra. In particular, $\mathcal{U}$ contains the unit element 1 so that one defines

$$\Omega := \psi_1. \tag{A.52}$$

In $\mathcal{U}$ we introduce the scalar product

$$(\psi_A|\psi_B) := \omega(A^{\dagger}B), \tag{A.53}$$

which is well defined thanks to the positivity of ω and the condition (A.51). The closure of $\mathcal{U}$ w.r.t. this scalar product defines the Hilbert space

$$\mathcal{H}_{\omega} := \overline{\mathcal{U}}_{(|)}. \tag{A.54}$$

On the other hand, one can introduce in $\mathcal{U}$ a second scalar product

$$(\psi_A|\psi_B)' := \omega(BA^{\dagger}). \tag{A.55}$$

This defines a second Hilbert space

$$\mathcal{H}'_{\omega} := \overline{\mathcal{U}}_{(|)'}. \tag{A.56}$$

We will use the symbol ψ'_A for its vectors, in order to distinguish them from the elements of $\mathcal{H}$.

We then quote some important results [Haag-KMS]:

Theorem A.3. *$\tilde{\mathcal{U}}$ is dense in $\mathcal{H}_{\omega}$ and in $\mathcal{H}'_{\omega}$.*

Theorem A.4. *The set* $\{\psi_A + \psi_{A_{i\frac{\beta}{2}}}\}_{A \in \tilde{\mathcal{U}}}$ *is dense in* $\mathcal{H}_\omega$.

A similar statement holds for $\mathcal{H}'_\omega$.

For $A \in \tilde{\mathcal{U}}$ one can define the map

$$S : \tilde{\mathcal{U}} \subset \mathcal{H}_\omega \longrightarrow \tilde{\mathcal{U}} \subset \mathcal{H}'_\omega, \tag{A.57}$$

by $\psi'_A = S\psi_{A_{i\frac{\beta}{2}}}$. Since this map is isometric and $\tilde{\mathcal{U}}$ is dense, both the range and domain of S extend to the whole Hilbert spaces. Then, for all $A \in \mathcal{U}$ one defines the element $\phi_A \in \mathcal{H}$ by

$$\phi_A := S^{-1}\psi'_A. \tag{A.58}$$

In particular, $\phi_A = \psi_{A_{i\frac{\beta}{2}}}$ if $A \in \tilde{A}$, and then, by (A.50), $(\phi_A|\phi_B) = \omega(BA^\dagger)$. This allow to define the linear representation R over $\mathcal{H}_\omega$ by

$$R(A)\psi_B := \psi_{AB}, \qquad \forall\, A, B \in \mathcal{U}, \tag{A.59}$$

and the antilinear representation S over $\mathcal{H}_\omega$ by

$$S(A)\phi_B := \phi_{BA^\dagger}, \qquad \forall\, A, B \in \mathcal{U}, \tag{A.60}$$

and it holds

$$\omega(A) = (\Omega|R(A)\Omega) = \overline{(\Omega|S(A)\Omega)}. \tag{A.61}$$

One also define a conjugation J by

$$J\psi_A := \phi_{A^\dagger}, \qquad \forall A \in \mathcal{U}. \tag{A.62}$$

J is clearly antiunitary and extends to the whole $\mathcal{H}_\omega$. Moreover, $J^2 = 1$ and $S(A) = JR(A)J$. One can show

Theorem A.5. $R(\mathcal{U})'' = S(\mathcal{U})'$.

The time evolution is described by the operator

$$\hat{U}(t)\psi_A = \psi_{A_t}. \tag{A.63}$$

Then, the operator $\hat{U}$ is unitary, commutes with J, and satisfies

$$\hat{U}(t)R(A)\hat{U}(t)^{-1} = R(A_t), \qquad \hat{U}(t)S(A)\hat{U}(t)^{-1} = S(A_t), \tag{A.64}$$

$$\hat{U}(t)\Omega = \Omega. \tag{A.65}$$

Finally, by defining the operator

$$T : \mathcal{H}_\omega \longrightarrow \mathcal{H}_\omega, \qquad \psi_A \longmapsto \phi_A. \tag{A.66}$$

Then, one can show that T and T^{-1} are both selfadjoint, positive and unlimited, and satisfy

$$TR(A)T^{-1} = R(A_{i\frac{\beta}{2}}), \tag{A.67}$$

$$TS(A)T^{-1} = S(A_{-i\frac{\beta}{2}}) \tag{A.68}$$

for all $A \in \tilde{\mathcal{U}}$, and one has

$$T = e^{-\frac{\beta}{2}H'}. \tag{A.69}$$

Summarizing, the equilibrium state ω satisfies the properties first characterized by Kubo [Kubo (1957)] and Araki and Wyss [Araki and Wyss (1964)]:

- ω is an impure state, so that R (and S) is reducible;
- there is an antiunitary conjugation operator J such that $J^2 = 1$ and $JR(A)J = S(A)$. Moreover $S(\mathcal{U})' = R(\mathcal{U})''$, the bicommutant being isomorphic to the weak closure of $R(\mathcal{U})$;
- if $\hat{h}(\vec{x})$ is the energy density operator, the generator of time translation in $\mathcal{H}_\omega$ is formally

$$\hat{H} = \int_{\mathbb{R}} (R(\hat{h}(\vec{x})) - S(\hat{h}(\vec{x})))d^3\vec{x}. \tag{A.70}$$

Formally means that indeed $\hat{H}$ is not well defined, but one must consider $\hat{H}' = \hat{H} - \mu\hat{N}$.

A.3 The Tomita-Takesaki theorem

The KMS construction of Haag, Hugenholz and Winnik find a natural place in the mathematical theory Tomita and Takesaki of Von Neumann algebras [Tomita (1967); Takesaki (1970); Haag (1996)]. An important ingredient to understand a fundamental result of Tomita and Takesaki is the concept of polar decomposition.

A.3.1 *Polar decomposition*

Let

$$\hat{O} : \mathcal{H} \longrightarrow \mathcal{H}', \tag{A.71}$$

a closed operator with dense domain in $\mathcal{H}$. Then, one can show that there is a unique decomposition

$$\hat{O} = \hat{\Phi}\hat{R}, \tag{A.72}$$

where $\hat{R}$ is a positive selfadjoint operator in $\mathcal{H}$ with the same domain of $\hat{O}$, and

$$\hat{\Phi} : D(\hat{O}) \longrightarrow \overline{\text{Image}(\hat{O})},$$

is isometric on the orthogonal complement of the kernel of $\hat{O}$ (which coincides with the kernel of $\hat{\Phi}$). Formula (A.72) is called the *polar decomposition* of $\hat{O}$. Indeed, it is not difficult to prove that under these conditions the operator $\hat{A} := \hat{O}^\dagger \hat{O}$ is selfadjoint and positive, so that it admits a unique positive square root which is just $\hat{R}$.

An analogue result is obtained for an antilinear closed operator

$$\hat{K} : \mathcal{H} \longrightarrow \mathcal{H}', \tag{A.73}$$

with dense domain $D(\hat{K}) \subset \mathcal{H}$. In this case the polar decomposition takes the form

$$\hat{K} = \hat{J}\hat{R}, \tag{A.74}$$

where $\hat{R} = \sqrt{\hat{K}^\dagger \hat{K}}$ is positive and selfadjoint, with domain $D(\hat{R}) = D(\hat{K})$, and

$$\hat{J} : D(\hat{O}) \longrightarrow \overline{\text{Image}(\hat{O})},$$

is an anti-isometry on the orthogonal complement of the kernel $\ker(\hat{K}) = \ker(\hat{J})$.

Before stating the famous result of Tomita and Takesaki, let us recall some facts about Von Neumann algebras.

A.3.2 *Some simple facts about Von Neumann algebras*

A Von Neumann algebra $\mathcal{V}$ is a $*$-algebra closed w.r.t. the weak topology. If $\mathcal{V}$ is a subalgebra of $B(\mathcal{H})$, the set of bounded operators over an Hilbert space, then the *commutant* $\mathcal{V}'$ of $\mathcal{V}$ is the subset of all bounded operators that commute with the whole $\mathcal{V}$. The intersection between $\mathcal{V}$ and its commutant is called the *center* $\mathcal{V}_a$ of $\mathcal{V}$:

$$\mathcal{V}_a := \mathcal{V} \cap \mathcal{V}'. \tag{A.75}$$

A Von Neumann algebra $\mathcal{V}$ is called a *factor* if its center consists only in the multiples of the unit element:

$$\mathcal{V}_a = \{\lambda \mathbb{I}\}_{\lambda \in \mathbb{C}}. \tag{A.76}$$

All factors can be classified in terms of the properties of a function called the *relative dimension function* D, defined via the following theorem ([Bogolubov *et al.* (1975)], Th. 22.2):

Theorem A.6. *It exists a function $D(\Pi)$, defined on any projector Π belonging to a given factor $\mathcal{V}$, determined apart from a normalization constant by the following three properties:*

(1) $D(\Pi) > 0$, *if* $\Pi \neq 0$, *and* $D(0) = 0$;
(2) *if* Π_1 *and* Π_2 *are equivalent w.r.t. the factor* $\mathcal{V}$ *(t.i* $\Pi_2 = S\Pi_1 S^{-1}$, $S \in \mathcal{V}$*) then* $D(\Pi_1) = D(\Pi_2)$;
(3) *if* $\Pi_1 = \Pi_2$ *then* $D(\Pi_1 + \Pi_2) = D(\Pi_1) + D(\Pi_2)$.

Then, a classification of factors follows:

Theorem A.7. *After a suitable choice of normalization, the range of the relative dimension can be reduced to one of the following possible sets:*

(I_n) *the set of integers* $F_n = \{1, \ldots, n\}$;
(I_∞) *the set of all non negative integers including* ∞;
(II_1) *the set* $[0, 1]$;
(II_∞) *the set* $[0, \infty]$;
(III) *the numbers* 0 *and* ∞.

For this reason the corresponding factors are called of type $I_n, I_\infty, II_1, II_\infty, III$.

Recall that for a subalgebra $\mathcal{V}$ of $B(\mathcal{H})$, a vector $x \in \mathcal{H}$ is said cyclic if $\mathcal{V}x$ is dense in $\mathcal{H}$, and is said to be *separating* if Ax, $A \in \mathcal{V}$ implies $A = 0$. We call *normalized state* over a C^*-algebra with unity, a linear positive functional ω such that $\omega(\mathbb{I}) = 1$. In particular, it is said to be a *faithful state* if $\omega(A^\dagger A) = 0$ implies $A = 0$. We end this short review with the following results:

Theorem A.8. *Let* $\mathcal{V}$ *be a Von Neumann algebra,* ω *a normal state and* ρ_ω *the associated GNS representation. Then,* $\rho(\mathcal{V})$ *is a Von Neumann algebra. Moreover, if* ω *is faithful, then* ρ_ω *is a* $*$*-isomorphism.*

Theorem A.9. *A Von Neumann algebra contains a faithful state iff it is isomorphic to a Von Neumann subalgebra of* $B(\mathcal{H})$*, endowed with a separating cyclic vector.*

Equivalently,

Theorem A.10. *A Von Neumann algebra contains a faithful state iff it is of countable gender.*

Recall that a Von Neumann algebra is of countable gender iff any family of orthogonal projector is countable.

A.3.3 *The Tomita-Takesaki theorem and the KMS condition*

Let us consider a Von Neumann algebra $\mathcal{V}$ possessing a faithful normal state. Then we can identify it with the image of its GNS representation $\mathcal{V} \simeq \rho_\omega(\mathcal{V})$. Thus, it also has a separating cyclic vector χ. Let us consider the map

$$\sigma : D(\sigma) \subset \mathcal{H} \longrightarrow \mathcal{H}, \quad A\chi \longmapsto A^\dagger \chi. \tag{A.77}$$

This map is well defined because if $A\chi = A'\chi$ then $A = A'$, since χ is separating. Moreover, it is densely defined because of cyclicity. It results that σ is a closable operator, and its closure

$$\Sigma := \overline{\sigma} \tag{A.78}$$

is an antilinear operator, which admits a polar decomposition

$$\Sigma = J\Delta^{\frac{1}{2}}, \qquad \Delta = \Sigma^\dagger \Sigma. \tag{A.79}$$

In particular, Δ being strictly positive, we can find a selfadjoint operator H such that $\Delta = \exp(-H)$, so that

$$\Delta^{it} = \exp(-iHt) \tag{A.80}$$

is a strongly continuous one parameter subgroup.

Now, we can state the main result of Tomita and Takesaki:

Theorem A.11. *Let $\mathcal{V}$ a Von Neumann algebra with a separating cyclic vector, and let be $\Sigma = J\Delta^{\frac{1}{2}}$ defined as above. Then*

(1) $\Delta^{-it}\mathcal{V}\Delta^{it} = \mathcal{V} \qquad \forall t \in \mathbb{R}$;
(2) $J\mathcal{V}J^{-1} = \mathcal{V}'$,

where the prime indicates the commutant.

Notice that the map

$$m_t(A) := A \longmapsto \Delta^{-it} A \Delta^{it} \tag{A.81}$$

is weakly continuous in $B(\mathcal{H})$, and then in $\mathcal{V}$. The maps m_t generate the *group of modular automorphisms* of the Von Neumann algebra. The main interesting fact is that such results are strictly connected with the KMS condition and are indeed strongly characterized by it. Indeed, we can state the following theorems:

Theorem A.12. *Let $\mathcal{V}$ be a Von Neumann algebra with a separating cyclic vector χ. Let ω_χ the state defined by $\omega_\chi(A) = (\chi|A\chi)$ for all $A \in \mathcal{V}$. Then ω_χ satisfies the KMS condition w.r.t. the time evolution described by the modular automorphisms, with inverse temperature $\beta = 1$.*

On the opposite, the KMS condition characterizes the modular group:

Theorem A.13. *Let $\mathcal{V}$ be a C^*-algebra of operators on a Hilbert space $\mathcal{H}$ with a cyclic vector ξ. Let ω_χ the state defined by $\omega_\chi(A) = (\chi|A\chi)$ for all $A \in \mathcal{V}$, such that it satisfies the KMS condition w.r.t. a group $U(t)$ of automorphisms of $\mathcal{V}$. Then ω_χ is invariant under evolution, t. i.*

$$\omega_\chi(U(t)(A)) = \omega_\chi(A), \qquad \forall A \in \mathcal{V},\ t \in \mathbb{R}. \tag{A.82}$$

Moreover, χ is separating for the weak closure $\mathcal{V}''$ of $\mathcal{V}$. In particular, $U(t)$ is just the restriction to $\mathcal{V}$ of the modular automorphism m_t of the Von Neumann algebra $\mathcal{V}''$ defined by the vector χ.

Bibliography

Akhmedov, E. T., Akhmedova, V., Pilling, T. and Singleton, D. (2007). Thermal radiation of various gravitational backgrounds, *Int. J. Mod. Phys.* **A22**, pp. 1705–1715, doi:10.1142/S0217751X07036130.

Akhmedov, E. T., Akhmedova, V. and Singleton, D. (2006). Hawking temperature in the tunneling picture, *Phys. Lett.* **B642**, pp. 124–128, doi:10.1016/j.physletb.2006.09.028.

Akhmedov, E. T., Pilling, T. and Singleton, D. (2008). Subtleties in the quasi-classical calculation of Hawking radiation, *Int. J. Mod. Phys.* **D17**, pp. 2453–2458, doi:10.1142/S0218271808013947.

Akhmedova, V., Pilling, T., de Gill, A. and Singleton, D. (2008). Temporal contribution to gravitational WKB-like calculations, *Phys. Lett.* **B666**, pp. 269–271, doi:10.1016/j.physletb.2008.07.017.

Akhmedova, V., Pilling, T., de Gill, A. and Singleton, D. (2009). Comments on anomaly versus WKB/tunneling methods for calculating Unruh radiation, *Phys. Lett.* **B673**, pp. 227–231, doi:10.1016/j.physletb.2009.02.022.

Araki, H., Kastler, D., Takesaki, M. and Haag, R. (1977). Extension of KMS states and chemical potential, *Communications in Mathematical Physics* **53**, 2, pp. 97–134.

Araki, H. and Woods, E. J. (1963). Representations of the canonical commutation relations describing a nonrelativistic infinite free Bose gas, *J. Math. Phys.* **4**, pp. 637–662, doi:10.1063/1.1704002.

Araki, H. and Wyss, W. (1964). Representations of canonical anticommutation relations, *Helv. Phys. Acta* **37**, pp. 136–159.

Arnold, V. (1989). *Mathematical Methods of Classical Mechanics*, Graduate Texts in Mathematics 60 (Springer, Berlin, Germany), ISBN 9783642839405.

Audretsch, J. and Muller, R. (1992). Amplification of the black hole Hawking radiation by stimulated emission, *Phys. Rev.* **D45**, pp. 513–519, doi:10.1103/PhysRevD.45.513.

Aurégan, Y., Fromholz, P., Michel, F., Pagneux, V. and Parentani, R. (2015). Slow sound in a duct, effective transonic flows, and analog black holes, *Phys. Rev.* **D92**, 8, p. 081503, doi:10.1103/PhysRevD.92.081503.

Bachelot, A. (1999). The Hawking effect, *Ann. Inst. H. Poincare Phys. Theor.* **70**, pp. 41–99.

Balasz, N. L. (1955). The Propagation of Light Rays in Moving Media, *JOSA* **45**, pp. 63–64.

Balbinot, R., Carusotto, I., Fabbri, A. and Recati, A. (2010). Testing Hawking particle creation by black holes through correlation measurements, *Int. J. Mod. Phys.* **D19**, pp. 2371–2377, doi:10.1142/S0218271810018463.

Balbinot, R., Fabbri, A., Fagnocchi, S. and Parentani, R. (2005). Hawking radiation from acoustic black holes, short distance and back-reaction effects, *Riv. Nuovo Cim.* **28**, 3, pp. 1–55, doi:10.1393/ncr/i2006-10001-9.

Balbinot, R., Fabbri, A., Fagnocchi, S., Recati, A. and Carusotto, I. (2008). Nonlocal density correlations as signal of Hawking radiation in BEC acoustic black holes, *Phys. Rev.* **A78**, p. 021603, doi:10.1103/PhysRevA.78.021603.

Balbinot, R., Fabbri, A., Frolov, V. P., Nicolini, P., Sutton, P. and Zelnikov, A. (2001). Vacuum polarization in the Schwarzschild space-time and dimensional reduction, *Phys. Rev.* **D63**, p. 084029, doi:10.1103/PhysRevD.63.084029.

Banerjee, R. (2009). Covariant Anomalies, Horizons and Hawking Radiation, *Int. J. Mod. Phys.* **D17**, pp. 2539–2542, doi:10.1142/S0218271808014175.

Banerjee, R. and Kulkarni, S. (2009). Hawking Radiation, Covariant Boundary Conditions and Vacuum States, *Phys. Rev.* **D79**, p. 084035, doi:10.1103/PhysRevD.79.084035.

Banerjee, R. and Majhi, B. R. (2009). Connecting anomaly and tunneling methods for Hawking effect through chirality, *Phys. Rev.* **D79**, p. 064024, doi:10.1103/PhysRevD.79.064024.

Barcelo, C., Liberati, S., Sonego, S. and Visser, M. (2004). Causal structure of acoustic spacetimes, *New J. Phys.* **6**, p. 186, doi:10.1088/1367-2630/6/1/186.

Barcelo, C., Liberati, S. and Visser, M. (2005). Analogue gravity, *Living Rev. Rel.* **8**, p. 12, doi:10.12942/lrr-2005-12, [Living Rev. Rel.14,3(2011)].

Barton, G. (1963). Introduction to advanced field theory, *Interscience Tracts Phys. Astron.* **22**, pp. 1–163.

Bekenstein, J. D. and Meisels, A. (1977). Einstein A and B Coefficients for a Black Hole, *Phys. Rev.* **D15**, pp. 2775–2781, doi:10.1103/PhysRevD.15.2775.

Belgiorno, F. (1998). Black hole thermodynamics and spectral analysis, *J. Math. Phys.* **39**, pp. 4608–4633, doi:10.1063/1.532527.

Belgiorno, F. and Cacciatori, S. L. (2016). Stimulated emission in black holes and in analogue gravity, *Gen. Rel. Grav.* **48**, 11, p. 145, doi:10.1007/s10714-016-2140-4.

Belgiorno, F., Cacciatori, S. L., Clerici, M., Gorini, V., Ortenzi, G., Rizzi, L., Rubino, E., Sala, V. G. and Faccio, D. (2010a). Hawking radiation from ultrashort laser pulse filaments, *Phys. Rev. Lett.* **105**, p. 203901, doi:10.1103/PhysRevLett.105.203901.

Belgiorno, F., Cacciatori, S. L., Clerici, M., Gorini, V., Ortenzi, G., Rizzi, L., Rubino, E., Sala, V. G. and Faccio, D. (2011a). Reply to Comment on: Hawking radiation from ultrashort laser pulse filaments, *Phys. Rev. Lett.* **107**, p. 149402, doi:10.1103/PhysRevLett.107.149402.

Belgiorno, F., Cacciatori, S. L. and Dalla Piazza, F. (2014). Perturbative photon production in a dispersive medium, *Eur. Phys. J.* **D68**, p. 134, doi:10.1140/epjd/e2014-40803-6.

Belgiorno, F., Cacciatori, S. L. and Dalla Piazza, F. (2015). Hawking effect in dielectric media and the Hopfield model, *Phys. Rev.* **D91**, 12, p. 124063, doi:10.1103/PhysRevD.91.124063.

Belgiorno, F., Cacciatori, S. L. and Dalla Piazza, F. (2016a). The Hopfield model revisited: Covariance and Quantization, *Phys. Scripta* **91**, 1, p. 015001, doi:10.1088/0031-8949/91/1/015001.

Belgiorno, F., Cacciatori, S. L. and Dalla Piazza, F. (2017a). Tunneling method for Hawking radiation in the Nariai case, *General Relativity and Gravitation* **49**, 8, p. 109.

Belgiorno, F., Cacciatori, S. L., Dalla Piazza, F. and Doronzo, M. (2016b). Exact quantisation of the relativistic Hopfield model, *Annals Phys.* **374**, pp. 338–365, doi:10.1016/j.aop.2016.09.001.

Belgiorno, F., Cacciatori, S. L., Dalla Piazza, F. and Doronzo, M. (2016c). Path integral quantization of the relativistic Hopfield model, *Phys. Rev.* **D93**, 6, p. 065020, doi:10.1103/PhysRevD.93.065020.

Belgiorno, F., Cacciatori, S. L., Dalla Piazza, F. and Doronzo, M. (2016d). $\Phi - \Psi$ model for electrodynamics in dielectric media: exact quantisation in the Heisenberg representation, *Eur. Phys. J.* **C76**, 6, p. 308, doi:10.1140/epjc/s10052-016-4146-1.

Belgiorno, F., Cacciatori, S. L., Dalla Piazza, F. and Doronzo, M. (2017b). Hopfield-Kerr model and analogue black hole radiation in dielectrics, *Phys. Rev. D* **96**, p. 096024, doi:10.1103/PhysRevD.96.096024, URL `https://link.aps.org/doi/10.1103/PhysRevD.96.096024`, arXiv preprint hep-th 1707.01663.

Belgiorno, F., Cacciatori, S. L., Ortenzi, G., Rizzi, L., Gorini, V. and Faccio, D. (2011b). Dielectric black holes induced by a refractive index perturbation and the Hawking effect, *Phys. Rev.* **D83**, p. 024015, doi:10.1103/PhysRevD.83.024015.

Belgiorno, F., Cacciatori, S. L., Ortenzi, G., Sala, V. G. and Faccio, D. (2010b). Quantum Radiation from Superluminal Refractive-Index Perturbations, *Phys. Rev. Lett.* **104**, p. 140403, doi:10.1103/PhysRevLett.104.140403.

Belgiorno, F., Cacciatori, S. L. and Viganò, A. (2017c). Spectral boundary conditions and solitonic solutions in a classical sellmeier dielectric, *The European Physical Journal C* **77**, 6, p. 404, doi:10.1140/epjc/s10052-017-4975-6, URL `https://doi.org/10.1140/epjc/s10052-017-4975-6`.

Bergé, L., Skupin, S., Nuter, R., Kasparian, J. and Wolf, J.-P. (2007). Ultrashort filaments of light in weakly ionized, optically transparent media, *Reports on Progress in Physics* **70**, 10, p. 1633, URL `http://stacks.iop.org/0034-4885/70/i=10/a=R03`.

Bergmann, P. G. (1961). "Gauge-invariant" variables in general relativity, *Phys. Rev. (2)* **124**, pp. 274–278.

Bermudez, D. and Leonhardt, U. (2016). Hawking spectrum for a fiber-optical analog of the event horizon, *Phys. Rev.* **A93**, 5, p. 053820, doi:10.1103/PhysRevA.93.053820.

Bertlmann, R. A. (1996). *Anomalies in quantum field theory* (Oxford, UK: Clarendon (1996) 566 p. (International series of monographs on physics: 91)).

Besse, A. L. (1987). *Einstein Manifolds* (Springer-Verlag, Berlin, Heidelberg, New York), ISBN 0387152792.

Birrell, N. D. and Davies, P. C. W. (1984). *Quantum Fields in Curved Space*, Cambridge Monographs on Mathematical Physics (Cambridge Univ. Press, Cambridge, UK), ISBN 0521278589, 9780521278584, 9780521278584, doi:10.1017/CBO9780511622632, URL `http://www.cambridge.org/mw/academic/subjects/physics/theoretical-physics-and-mathematical-physics/quantum-fields-curved-space?format=PB`.

Bisognano, J. J. and Wichmann, E. H. (1975). On the duality condition for a hermitian scalar field, *Journal of Mathematical Physics* **16**, 4, pp. 985–1007, doi:10.1063/1.522605, URL `https://doi.org/10.1063/1.522605`.

Bisognano, J. J. and Wichmann, E. H. (1976). On the duality condition for quantum fields, *Journal of Mathematical Physics* **17**, 3, pp. 303–321, doi:10.1063/1.522898, URL `http://aip.scitation.org/doi/abs/10.1063/1.522898`.

Bjorken, J. and Drell, S. (1964). *Relativistic quantum mechanics*, International series in pure and applied physics (McGraw-Hill).

Bjorken, J. and Drell, S. (1965). *Relativistic quantum fields*, International series in pure and applied physics (McGraw-Hill).

Bogolubov, N. N., Logunov, A. A. and Todorov, I. T. (1975). *Introduction to axiomatic quantum field theory*, Mathematical Physics Monograph Series, No. 18. (Benjamin, Inc.).

Bondi, H., van der Burg, M. G. J. and Metzner, A. W. K. (1962). Gravitational waves in general relativity. 7. Waves from axisymmetric isolated systems, *Proc. Roy. Soc. Lond.* **A269**, pp. 21–52, doi:10.1098/rspa.1962.0161.

Borchers, H. J. (1996). Translation group and particle representations in quantum field theory, *Lect. Notes Phys. Monogr.* **40**, pp. 1–131, doi:10.1007/978-3-540-49954-1.

Boulware, D. G. (1975). Quantum Field Theory in Schwarzschild and Rindler Spaces, *Phys. Rev.* **D11**, p. 1404, doi:10.1103/PhysRevD.11.1404.

Brout, R., Massar, S., Parentani, R. and Spindel, P. (1995a). A Primer for black hole quantum physics, *Phys. Rept.* **260**, pp. 329–454, doi:10.1016/0370-1573(95)00008-5.

Brout, R., Massar, S., Parentani, R. and Spindel, P. (1995b). Hawking radiation without transPlanckian frequencies, *Phys. Rev.* **D52**, pp. 4559–4568, doi:10.1103/PhysRevD.52.4559.

Brunetti, R., Dappiaggi, C., Fredenhagen, K. and Yngvason, J. (eds.) (2015). *Advances in algebraic quantum field theory*, Mathematical Physics Studies (Springer), ISBN 9783319213521, 9783319213538, doi:10.1007/978-3-319-21353-8.

Bruschi, D. E., Friis, N., Fuentes, I. and Weinfurtner, S. (2013). On the robustness of entanglement in analogue gravity systems, *New J. Phys.* **15**, p. 113016, doi:10.1088/1367-2630/15/11/113016.

Cacciatori, S. L., Belgiorno, F., Gorini, V., Ortenzi, G., Rizzi, L., Sala, V. G. and Faccio, D. (2010). Space-time geometries and light trapping in travelling refractive index perturbations, *New J. Phys.* **12**, p. 095021, doi:10.1088/1367-2630/12/9/095021.

Calmet, X. (2014). *Quantum Aspects of Black Holes*, Fundamental Theories of Physics (Springer International Publishing), ISBN 9783319108520, URL `https://books.google.it/books?id=VESgBQAAQBAJ`.

Cardoso, V., Coutant, A., Richartz, M. and Weinfurtner, S. (2016). Detecting Rotational Superradiance in Fluid Laboratories, *Phys. Rev. Lett.* **117**, 27, p. 271101, doi:10.1103/PhysRevLett.117.271101.

Carusotto, I., Fagnocchi, S., Recati, A., Balbinot, R. and Fabbri, A. (2008). Numerical observation of Hawking radiation from acoustic black holes in atomic BECs, *New J. Phys.* **10**, p. 103001, doi:10.1088/1367-2630/10/10/103001.

Chowdhury, B. D. (2008). Problems with Tunneling of Thin Shells from Black Holes, *Pramana* **70**, pp. 593–612, doi:10.1007/s12043-008-0001-8, [Pramana70,3(2008)].

Christensen, S. M. and Fulling, S. A. (1977). Trace Anomalies and the Hawking Effect, *Phys. Rev.* **D15**, pp. 2088–2104, doi:10.1103/PhysRevD.15.2088.

Corda, C. (2011). Effective temperature for black holes, *JHEP* **08**, p. 101, doi:10.1007/JHEP08(2011)101.

Corley, S. (1998). Computing the spectrum of black hole radiation in the presence of high frequency dispersion: An Analytical approach, *Phys. Rev.* **D57**, pp. 6280–6291, doi:10.1103/PhysRevD.57.6280.

Corley, S. and Jacobson, T. (1996). Hawking spectrum and high frequency dispersion, *Phys. Rev.* **D54**, pp. 1568–1586, doi:10.1103/PhysRevD.54.1568.

Corley, S. and Jacobson, T. (1999). Black hole lasers, *Phys. Rev.* **D59**, p. 124011, doi:10.1103/PhysRevD.59.124011.

Couairon, A. and Mysyrowicz, A. (2007). Femtosecond filamentation in transparent media, *Physics Reports* **441**, 2, pp. 47 – 189, doi:http://dx.doi.org/10.1016/j.physrep.2006.12.005, URL `http://www.sciencedirect.com/science/article/pii/S037015730700021X`.

Coutant, A., Fabbri, A., Parentani, R., Balbinot, R. and Anderson, P. (2012a). Hawking radiation of massive modes and undulations, *Phys. Rev.* **D86**, p. 064022, doi:10.1103/PhysRevD.86.064022.

Coutant, A. and Parentani, R. (2010). Black hole lasers, a mode analysis, *Phys. Rev.* **D81**, p. 084042, doi:10.1103/PhysRevD.81.084042.

Coutant, A. and Parentani, R. (2014a). Hawking radiation with dispersion: The broadened horizon paradigm, *Phys. Rev.* **D90**, 12, p. 121501, doi:10.1103/PhysRevD.90.121501.

Coutant, A. and Parentani, R. (2014b). Undulations from amplified low frequency surface waves, *Phys. Fluids* **26**, p. 044106, doi:10.1063/1.4872025.

Coutant, A., Parentani, R. and Finazzi, S. (2012b). Black hole radiation with short distance dispersion, an analytical S-matrix approach, *Phys. Rev.* **D85**, p. 024021, doi:10.1103/PhysRevD.85.024021.

Coutant, A. and Weinfurtner, S. (2016). The imprint of the analogue Hawking effect in subcritical flows, *Phys. Rev.* **D94**, 6, p. 064026, doi:10.1103/PhysRevD.94.064026.

Coutant, A. and Weinfurtner, S. (2018a). Low-frequency analogue Hawking radiation: The Bogoliubov-de Gennes model, *Phys. Rev. D* **97**, p. 025006, doi:10.1103/PhysRevD.97.025006, URL `https://link.aps.org/doi/10.1103/PhysRevD.97.025006`.

Coutant, A. and Weinfurtner, S. (2018b). Low-frequency analogue Hawking radiation: The Korteweg-de Vries model, *Phys. Rev. D* **97**, p. 025005, doi:10.1103/PhysRevD.97.025005, URL `https://link.aps.org/doi/10.1103/PhysRevD.97.025005`.

Damour, T. (1975). Klein paradox and vacuum polarization, in *Proc. first Marcel Grossmann Meeting on General Relativity* (Trieste, Italy), pp. 459–482.

Damour, T. and Ruffini, R. (1976). Black Hole Evaporation in the Klein-Sauter-Heisenberg-Euler Formalism, *Phys. Rev.* **D14**, pp. 332–334, doi:10.1103/PhysRevD.14.332.

Dappiaggi, C., Moretti, V. and Pinamonti, N. (2011). Rigorous construction and Hadamard property of the Unruh state in Schwarzschild spacetime, *Adv. Theor. Math. Phys.* **15**, 2, pp. 355–447, doi:10.4310/ATMP.2011.v15.n2.a4.

Dappiaggi, C., Moretti, V. and Pinamonti, N. (2017). Hadamard States From Light-like Hypersurfaces, ArXiv preprint math-ph 1706.09666.

Das, A. K. (1997). *Finite Temperature Field Theory* (World Scientific, New York), ISBN 9789810228569, 9789814498234.

Das, S., Robinson, S. P. and Vagenas, E. C. (2008). Gravitational anomalies: A Recipe for Hawking radiation, *Int. J. Mod. Phys.* **D17**, pp. 533–539, doi:10.1142/S0218271808012218.

Davies, P. C. W. and Fulling, S. A. (1977). Radiation from Moving Mirrors and from Black Holes, *Proc. Roy. Soc. Lond.* **A356**, pp. 237–257, doi:10.1098/rspa.1977.0130.

Davies, P. C. W., Fulling, S. A. and Unruh, W. G. (1976). Energy Momentum Tensor Near an Evaporating Black Hole, *Phys. Rev.* **D13**, pp. 2720–2723, doi:10.1103/PhysRevD.13.2720.

Davydov, A. (1984). *Teoria del solido* (Mir, Moscow).

de Gill, A., Singleton, D., Akhmedova, V. and Pilling, T. (2010). A WKB-Like Approach to Unruh Radiation, *Am. J. Phys.* **78**, pp. 685–691, doi:10.1119/1.3308568.

de Gosson, M. (2016). *Born-Jordan Quantization*, Fundamental Theories of Physics (Springer, Berlin, Germany), ISBN 9783642839405.

De Lorenci, V. A. and Klippert, R. (2002). Analog gravity from electrodynamics in nonlinear media, *Phys. Rev.* **D65**, p. 064027, doi:10.1103/PhysRevD.65.064027.

DeSalvo, R., Said, A. A., Hagan, D. J., Van Stryland, E. W. and Sheik-Bahae, M. (1996). Infrared to ultraviolet measurements of two-photon absorption and n_2 in wide bandgap solids, *IEEE Journal of Quantum Electronics* **32**, 8, pp. 1324–1333, doi:10.1109/3.511545.

DeWitt, B. (2003). *The Global Approach to Quantum Field Theory*, no. v. 1 in International series of monographs on physics (Clarendon Press), ISBN 9780198527909, URL https://books.google.it/books?id=mSKzXTEusz0C.

DeWitt, B. S. (1975). Quantum Field Theory in Curved Space-Time, *Phys. Rept.* **19**, pp. 295–357, doi:10.1016/0370-1573(75)90051-4.

Dimock, J. (1980). Algebras of local observables on a manifold, *Comm. Math. Phys.* **77**, 3, pp. 219–228.

Dimock, J. (2011). *Quantum Mechanics and Quantum Field Theory: A Mathematical Primer* (Cambridge University Press), ISBN 9781139497480, URL https://books.google.it/books?id=Y4m8V7-83swC.

Dirac, P. A. M. (1947). *The Principles of Quantum Mechanics* (Oxford, at the Clarendon Press), 3d ed.

Dirac, P. A. M. (1950). Generalized hamiltonian dynamics, *Canadian J. Math.* **2**, pp. 129–148.

Doplicher, S. (1965). An Algebraic Spectrum Condition, *Commun. Math. Phys.* **1**, pp. 1–5.

Eastham, M. (1989). *The Asymptotic Solution of Linear Differential Systems: Applications of the Levinson Theorem* (Oxford University Press).

Ehlers, J., Prasanna, A. R. and Breuer, R. A. (1987). Propagation of Gravitational Waves Through Pressureless Matter, *Class. Quant. Grav.* **4**, pp. 253–264, doi:10.1088/0264-9381/4/2/009.

Euvé, L. P., Michel, F., Parentani, R., Philbin, T. G. and Rousseaux, G. (2016). Observation of noise correlated by the Hawking effect in a water tank, *Phys. Rev. Lett.* **117**, 12, p.121301, doi:10.1103/PhysRevLett.117.121301.

Euvé, L.-P., Michel, F., Parentani, R. and Rousseaux, G. (2015). Wave blocking and partial transmission in subcritical flows over an obstacle, *Phys. Rev.* **D91**, 2, p. 024020, doi:10.1103/PhysRevD.91.024020.

Fabbri, A. and Navarro-Salas, J. (2005). *Modeling black hole evaporation* (London, UK: Imp. Coll. Pr. (2005) 334 p).

Faccio, D. (2012). Laser pulse analogues for gravity and analogue Hawking radiation, *Contemporary Physics* **53**, 2, pp. 97–112.

Faccio, D., Arane, T., Lamperti, M. and Leonhardt, U. (2012). Optical black hole lasers, *Class. Quant. Grav.* **29**, p. 224009, doi:10.1088/0264-9381/29/22/224009.

Faccio, D., Belgiorno, F., Cacciatori, S., Gorini, V., Liberati, S. and Moschella, U. (2013). Proceedings, 9th SIGRAV Graduate School in Contemporary Relativity and Gravitational Physics on Analogue Gravity Phenomenology, *Lect. Notes Phys.* **870**, pp. pp.1–439.

Faccio, D., Cacciatori, S., Gorini, V., Sala, V. G., Averchi, A., Lotti, A., Kolesik, M. and Moloney, J. V. (2010). Analogue gravity and ultrashort laser pulse filamentation, *EPL (Europhysics Letters)* **89**, 3, p. 34004, URL http://stacks.iop.org/0295-5075/89/i=3/a=34004.

Faccio, D. and Wright, E. M. (2017a). Nonlinear Zel'dovich effect: Parametric amplification from medium rotation, *Phys. Rev. Lett.* **118**, p. 093901, doi:10.1103/PhysRevLett.118.093901, URL https://link.aps.org/doi/

10.1103/PhysRevLett.118.093901.

Faccio, D. and Wright, E. M. (2017b). Erratum: Nonlinear Zel'dovich effect: Parametric amplification from medium rotation [phys. rev. lett. 118, 093901 (2017)], *Phys. Rev. Lett.* **118**, p. 169902, doi: 10.1103/PhysRevLett.118.169902, URL https://link.aps.org/doi/10.1103/PhysRevLett.118.169902.

Fano, U. (1956). Atomic theory of electromagnetic interactions in dense materials, *Phys. Rev.* **103**, pp. 1202–1218.

Fedoryuk, M. (1993). *Asymptotic Analysis* (Springer Verlag-Berlin).

Fell, J. (1960). The dual spaces of C^*-algebras, *Trans. Amer. Math. Soc.* **94**, pp. 365–403.

Fewster, C. J. and Verch, R. (2013). The necessity of the Hadamard condition, *Class. Quant. Grav.* **30**, p. 235027, doi:10.1088/0264-9381/30/23/235027.

Feynman, R. P. (1949). The theory of positrons, *Phys. Rev.* **76**, pp. 749–759, doi:10.1103/PhysRev.76.749.

Finazzi, S. and Carusotto, I. (2013). Quantum vacuum emission in a nonlinear optical medium illuminated by a strong laser pulse, *Phys. Rev.* **A87**, 2, p. 023803, doi:10.1103/PhysRevA.87.023803.

Finazzi, S. and Carusotto, I. (2014). Spontaneous quantum emission from analog white holes in a nonlinear optical medium, *Phys. Rev.* **A89**, 5, p. 053807, doi:10.1103/PhysRevA.89.053807.

Finazzi, S. and Parentani, R. (2011). Spectral properties of acoustic black hole radiation: broadening the horizon, *Phys. Rev.* **D83**, p. 084010, doi:10.1103/PhysRevD.83.084010.

Finazzi, S. and Parentani, R. (2012). Hawking radiation in dispersive theories, the two regimes, *Phys. Rev.* **D85**, p. 124027, doi:10.1103/PhysRevD.85.124027.

Fre, P., Gorini, V., Magli, G. and Moschella, U. (1999). *Classical and Quantum Black Holes*, Series in High Energy Physics, Cosmology and Gravitation (CRC Press), ISBN 9781420050684, URL https://books.google.it/books?id=FYDpUmBCILIC.

Frolov, V. and Zelnikov, A. (2011). *Introduction to Black Hole Physics* (OUP Oxford), ISBN 9780199692293, URL https://books.google.it/books?id=nZsNduVxmB8C.

Frolov, V. P., Fursaev, D. V. and Zelnikov, A. I. (1996). Black hole entropy: Off-shell versus on-shell, *Phys. Rev.* **D54**, pp. 2711–2731, doi: 10.1103/PhysRevD.54.2711.

Frolov, V. P. and Novikov, I. D. (eds.) (1998). *Black hole physics: Basic concepts and new developments* (Dordrecht, Netherlands: Kluwer Academic (1998) 770 p).

Fujikawa, K. and Suzuki, H. (2004). *Path integrals and quantum anomalies* (Oxford, UK: Clarendon (2004) 284 p), doi:10.1093/acprof:oso/9780198529132.001.0001.

Fulling, S. A. (1973). Nonuniqueness of canonical field quantization in Riemannian space-time, *Phys. Rev.* **D7**, pp. 2850–2862, doi:10.1103/PhysRevD.7.2850.

Fulling, S. A. (1977). Alternative Vacuum States in Static Space-Times with Horizons, *J. Phys.* **A10**, pp. 917–951, doi:10.1088/0305-4470/10/6/014.

Fulling, S. A. (1989). *Aspects of Quantum Field Theory in Curved Spacetime*, London Mathematical Society St (Cambridge University Press), ISBN 9780521377683, URL https://books.google.it/books?id=Zo5_3cmtFEUC.

Fulling, S. A. and Ruijsenaars, S. N. M. (1987). Temperature, periodicity and horizons, *Physics Reports* **152**, 3, pp. 135 – 176, doi:http://dx.doi.org/10.1016/0370-1573(87)90136-0, URL http://www.sciencedirect.com/science/article/pii/0370157387901360.

Fursaev, D. V. and Solodukhin, S. N. (1995). On the description of the Riemannian geometry in the presence of conical defects, *Phys. Rev.* **D52**, pp. 2133–2143, doi:10.1103/PhysRevD.52.2133.

Gadonas, R., Jarutis, V., Pakauskas, R., Smilgeviius, V., Stabinis, A. and Vaiaitis, V. (2001). Self-action of Bessel beam in nonlinear medium, *Optics Communications* **196**, 1, pp. 309 – 316, doi:http://dx.doi.org/10.1016/S0030-4018(01)01386-4, URL http://www.sciencedirect.com/science/article/pii/S0030401801013864.

Gaeta, A. L. (2003). Collapsing light really shines, *Science* **301**, 5629, pp. 54–55, doi:10.1126/science.1083629, URL http://science.sciencemag.org/content/301/5629/54.

Gamelin, T. (2001). *Complex Analysis* (New York, United States: Springer (2001) 478p).

Garay, L. J., Anglin, J. R., Cirac, J. I. and Zoller, P. (2000). Sonic analog of gravitational black holes in Bose-Einstein condensates, *Phys. Rev. Lett.* **85**, pp. 4643–4647, doi:10.1103/PhysRevLett.85.4643, URL https://link.aps.org/doi/10.1103/PhysRevLett.85.4643.

Gavrilov, S. P., Gitman, D. M. and Tomazelli, J. L. (2008). Density matrix of a quantum field in a particle-creating background, *Nucl. Phys.* **B795**, pp. 645–677, doi:10.1016/j.nuclphysb.2007.11.029.

Gel'fand, I. and Shilov, G. (1964). *Generarized Functions. Volume I. Properties and Operations* (New York, United States: Academic Press (1964) 423p).

Gerace, D. and Carusotto, I. (2012). Analog Hawking radiation from an acoustic black hole in a flowing polariton superfluid, *Phys. Rev.* **B86**, p. 144505, doi:10.1103/PhysRevB.86.144505.

Gerlach, U. H. (1976). The Mechanism of Black Body Radiation from an Incipient Black Hole, *Phys. Rev.* **D14**, pp. 1479–1508, doi:10.1103/PhysRevD.14.1479.

Gibbons, G. W. (1975). Vacuum Polarization and the Spontaneous Loss of Charge by Black Holes, *Commun. Math. Phys.* **44**, pp. 245–264, doi:10.1007/BF01609829.

Gibbons, G. W. and Hawking, S. W. (1977a). Action Integrals and Partition Functions in Quantum Gravity, *Phys. Rev.* **D15**, pp. 2752–2756, doi:10.1103/PhysRevD.15.2752.

Gibbons, G. W. and Hawking, S. W. (1977b). Cosmological Event Horizons, Thermodynamics, and Particle Creation, *Phys. Rev.* **D15**, pp. 2738–2751, doi:10.1103/PhysRevD.15.2738.

Gibbons, G. W. and Perry, M. J. (1976). Black Holes in Thermal Equilibrium, *Phys. Rev. Lett.* **36**, p. 985, doi:10.1103/PhysRevLett.36.985.

Gibbons, G. W. and Perry, M. J. (1978). Black Holes and Thermal Green's Functions, *Proc. Roy. Soc. Lond.* **A358**, pp. 467–494, doi:10.1098/rspa.1978.0022.

Ginzburg, V. L. and Frolov, V. P. (1987). Vacuum in a homogeneous gravitational field and excitation of a uniformly accelerated detector, *Soviet Physics Uspekhi* **30**, 12, p. 1073, URL http://stacks.iop.org/0038-5670/30/i=12/a=A04.

Gitman, D. M. and Tyutin, I. V. (1990). *Quantization of Fields with Constraints*, Springer Series in Nuclear and Particle Physics (Springer, Berlin, Germany), ISBN 9783642839405.

Gordon, W. (1923). Zur Lichtfortpflanzung nach der Relativitätstheorie, *Annalen der Physik* **377**, 22, pp. 421–456, doi:10.1002/andp.19233772202, URL http://dx.doi.org/10.1002/andp.19233772202.

Greiner, W. and Reinhardt, J. (2013). *Quantum Electrodynamics* (Springer Berlin Heidelberg), ISBN 9783662052464, URL https://books.google.it/books?id=6y3rCAAAQBAJ.

Grib, A., Mamayev, S. and Mostepanenko, V. (1994). *Vacuum Quantum Effects in Strong Fields* (Friedmann Laboratory Pub.), URL https://books.google.it/books?id=azBdcgAACAAJ.

Haag, R. (1996). *Local Quantum Physics. Fields, Particles, Algebras* (Berlin, Germany: Springer, (1996) 392 p).

Haag, R., Hugenholtz, N. M. and Winnink, M. (1967). On the equilibrium states in quantum statistical mechanics, *Commun. Math. Phys.* **5**, pp. 215–236, doi:10.1007/BF01646342.

Haag, R. and Kastler, D. (1964). An algebraic approach to quantum field theory, *J. Math. Phys.* **5**, pp. 848–861, doi:10.1063/1.1704187.

Haag, R., Narnhofer, H. and Stein, U. (1984). On Quantum Field Theory in Gravitational Background, *Commun. Math. Phys.* **94**, p. 219, doi:10.1007/BF01209302.

Hafner, D. (2009). Creation of fermions by rotating charged black-holes, *Mémoires de la SOCIÉTÉ MATHÉMATIQUE DE FRANCE* **117**, pp. 1–158.

Halvorson, H. and Muger, M. (2007). Algebraic quantum field theory, in *Philosophy of Physics: Handbook of the philosophy of science. Pt. A. 2 , Pt. A.* (Elsevier), ISBN 9780444515605, URL https://books.google.it/books?id=UGpJEFA5zXcC, arXiv preprint math-ph/0602036.

Hartle, J. B. and Hawking, S. W. (1976). Path Integral Derivation of Black Hole Radiance, *Phys. Rev.* **D13**, pp. 2188–2203, doi:10.1103/PhysRevD.13.2188.

Hawking, S. W. (1974). Black hole explosions, *Nature* **248**, pp. 30–31, doi:10.1038/248030a0.

Hawking, S. W. (1975). Particle Creation by Black Holes, *Commun. Math. Phys.* **43**, pp. 199–220, doi:10.1007/BF02345020, [Erratum: ibid. 46, 206 (1976)].

Hawking, S. W. (1980). THE PATH INTEGRAL APPROACH TO QUANTUM GRAVITY, in *General Relativity: An Einstein Centenary Survey* (Cambridge University Press), pp. 746–789.

Hawking, S. W. and Ellis, G. F. R. (2011). *The Large Scale Structure of Space-Time*, Cambridge Monographs on Mathematical Physics (Cambridge University Press), ISBN 9780521200165, 9780521099066, 9780511826306, 9780521099066, doi:10.1017/CBO9780511524646.

Hehl, F., Kiefer, C. and Metzler, R. (1998). *Black Holes: Theory and Observation: Proceedings of the 179th W.E. Heraeus Seminar Held at Bad Honnef, Germany, 18–22 August 1997*, Lecture Notes in Physics (Springer Berlin Heidelberg), ISBN 9783540651581, URL `https://books.google.it/books?id=khM219HEmdOC`.

Himemoto, Y. and Tanaka, T. (2000). A Generalization of the model of Hawking radiation with modified high frequency dispersion relation, *Phys. Rev.* **D61**, p. 064004, doi:10.1103/PhysRevD.61.064004.

Hopfield, J. J. (1958). Theory of the Contribution of Excitons to the Complex Dielectric Constant of Crystals, *Phys. Rev.* **112**, pp. 1555–1567, doi:10.1103/PhysRev.112.1555.

Huttner, B. and Barnett, S. M. (1992). Quantization of the electromagnetic field in dielectrics, *Phys. Rev.* **A46**, 7, p. 4306, doi:10.1103/PhysRevA.46.4306.

Huttner, B., Baumberg, J. J. and Barnett, S. M. (1991). Canonical quantization of light in a linear dielectric, *EPL (Europhysics Letters)* **16**, 2, pp. 177–182, URL `http://stacks.iop.org/0295-5075/16/i=2/a=010`.

Isham, C. J. (1977). Quantum Field Theory in Curved Space Times: An Overview, in *8th Texas Symposium on Relativistic Astrophysics Boston, Mass., December 13-17, 1976*, p. 114.

Iso, S., Umetsu, H. and Wilczek, F. (2006a). Anomalies, Hawking radiations and regularity in rotating black holes, *Phys. Rev.* **D74**, p. 044017, doi:10.1103/PhysRevD.74.044017.

Iso, S., Umetsu, H. and Wilczek, F. (2006b). Hawking radiation from charged black holes via gauge and gravitational anomalies, *Phys. Rev. Lett.* **96**, p. 151302, doi:10.1103/PhysRevLett.96.151302.

Israel, W. (1976). Thermo field dynamics of black holes, *Phys. Lett.* **A57**, pp. 107–110, doi:10.1016/0375-9601(76)90178-X.

Jacobson, T. (1991). Black hole evaporation and ultrashort distances, *Phys. Rev.* **D44**, pp. 1731–1739, doi:10.1103/PhysRevD.44.1731.

Jacobson, T. (1993). Black hole radiation in the presence of a short distance cutoff, *Phys. Rev.* **D48**, pp. 728–741, doi:10.1103/PhysRevD.48.728.

Jacobson, T. (1994). A Note on Hartle-Hawking vacua, *Phys. Rev.* **D50**, pp. R6031–R6032, doi:10.1103/PhysRevD.50.R6031.

Jacobson, T. (1996). On the origin of the outgoing black hole modes, *Phys. Rev.* **D53**, pp. 7082–7088, doi:10.1103/PhysRevD.53.7082.

Jacobson, T. (1999). Trans Planckian redshifts and the substance of the space-time river, *Prog. Theor. Phys. Suppl.* **136**, pp. 1–17, doi:10.1143/PTPS.136.1.

Jacobson, T. and Kang, G. (1993). Conformal invariance of black hole temperature, *Class. Quant. Grav.* **10**, pp. L201–L206, doi:10.1088/0264-9381/10/11/002.

Jacquet, M. and König, F. (2015). Quantum vacuum emission from a refractive-index front, *Phys. Rev. A* **92**, p. 023851, doi:10.1103/PhysRevA.92.023851, URL `https://link.aps.org/doi/10.1103/PhysRevA.92.023851`.

Kasparian, J., Rodriguez, M., Méjean, G., Yu, J., Salmon, E., Wille, H., Bourayou, R., Frey, S., André, Y.-B., Mysyrowicz, A., Sauerbrey, R., Wolf, J.-P. and Wöste, L. (2003). White-light filaments for atmospheric analysis, *Science* **301**, 5629, pp. 61–64, doi:10.1126/science.1085020, URL `http://science.sciencemag.org/content/301/5629/61`.

Kay, B. S. (1985). The Double Wedge Algebra for Quantum Fields on Schwarzschild and Minkowski Space-times, *Commun. Math. Phys.* **100**, p. 57, doi:10.1007/BF01212687.

Kay, B. S. and Wald, R. M. (1991). Theorems on the Uniqueness and Thermal Properties of Stationary, Nonsingular, Quasifree States on Space-Times with a Bifurcate Killing Horizon, *Phys. Rept.* **207**, pp. 49–136, doi: 10.1016/0370-1573(91)90015-E.

Kerner, R. and Mann, R. B. (2006). Tunnelling, temperature and Taub-NUT black holes, *Phys. Rev.* **D73**, p. 104010, doi:10.1103/PhysRevD.73.104010.

Khanna, F. C., Malbouisson, A. P. C., Malbouisson, J. M. C. and Santana, A. R. (2009). *Thermal quantum field theory - Algebraic aspects and applications* (Singapore, Singapore: World Scientific Publishing Company (2009) 484 p).

Kim, S. P. and Hwang, W.-Y. P. (2011). Vacuum Polarization and Persistence on the Black Hole Horizon, ArXiv preprint, hep-th 1103.5264.

Kittel, C. (1987). *Quantum Theory of Solids* (Wiley, New York).

Kolesik, M. and Moloney, J. V. (2004). Nonlinear optical pulse propagation simulation: From Maxwell's to unidirectional equations, *Phys. Rev. E* **70**, p. 036604, doi:10.1103/PhysRevE.70.036604, URL `https://link.aps.org/doi/10.1103/PhysRevE.70.036604`.

Kolesik, M., Moloney, J. V. and Mlejnek, M. (2002). Unidirectional optical pulse propagation equation, *Phys. Rev. Lett.* **89**, p. 283902, doi:10.1103/PhysRevLett.89.283902, URL `https://link.aps.org/doi/10.1103/PhysRevLett.89.283902`.

Kraus, P. and Wilczek, F. (1995). Self-interaction correction to black hole radiance, *Nucl. Phys.* **B433**, pp. 403–420, doi:10.1016/0550-3213(94)00411-7.

Kravtsov, Y. A., Ostrovsky, L. A. and Stepanov, N. S. (1974). Geometrical optics of inhomogeneous and nonstationary dispersive media, *Proceedings of the IEEE* **62**, 11, pp. 1492–1510, doi:10.1109/PROC.1974.9656.

Kubo, R. (1957). Statistical mechanical theory of irreversible processes. 1. General theory and simple applications in magnetic and conduction problems, *J. Phys. Soc. Jap.* **12**, pp. 570–586, doi:10.1143/JPSJ.12.570.

Laflamme, R. (1989). Geometry and Thermofields, *Nucl. Phys.* **B324**, pp. 233–252, doi:10.1016/0550-3213(89)90191-0.

Larre, P. E., Recati, A., Carusotto, I. and Pavloff, N. (2012). Quantum fluctuations around black hole horizons in Bose-Einstein condensates, *Phys. Rev.* **A85**, p. 013621, doi:10.1103/PhysRevA.85.013621.

Leonhardt, U. and Piwnicki, P. (2000). Relativistic effects of light in moving media with extremely low group velocity, *Phys. Rev. Lett.* **84**, pp. 822–825, doi:10.1103/PhysRevLett.84.822.

Leonhardt, U. and Robertson, S. (2012). Analytical theory of Hawking radiation in dispersive media, *New J. Phys.* **14**, p. 053003, doi:10.1088/1367-2630/14/5/053003.

Linder, M. F., Schutzhold, R. and Unruh, W. G. (2016). Derivation of Hawking radiation in dispersive dielectric media, *Phys. Rev.* **D93**, 10, p. 104010, doi:10.1103/PhysRevD.93.104010.

Luks, A. and Perinová, V. (2009). *Quantum aspects of light propagation*, Lecture Notes in Mathematics, Vol. 128 (Springer, Berlin-New York).

Macher, J. and Parentani, R. (2009a). Black hole radiation in Bose-Einstein condensates, *Phys. Rev.* **A80**, p. 043601, doi:10.1103/PhysRevA.80.043601.

Macher, J. and Parentani, R. (2009b). Black/White hole radiation from dispersive theories, *Phys. Rev.* **D79**, p. 124008, doi:10.1103/PhysRevD.79.124008.

Martellini, M., Sodano, P. and Vitiello, G. (1978). Vacuum Structure for a Quantum Field Theory in Curved Space-time, *Nuovo Cim.* **A48**, p. 341, doi:10.1007/BF02781601.

Martin, P. C. and Schwinger, J. S. (1959). Theory of many particle systems. 1. *Phys. Rev.* **115**, pp. 1342–1373, doi:10.1103/PhysRev.115.1342.

Massar, S. and Parentani, R. (2000). How the change in horizon area drives black hole evaporation, *Nucl. Phys.* **B575**, pp. 333–356, doi:10.1016/S0550-3213(00)00067-5.

Mayoral, C., Recati, A., Fabbri, A., Parentani, R., Balbinot, R. and Carusotto, I. (2011). Acoustic white holes in flowing atomic Bose-Einstein condensates, *New J. Phys.* **13**, p. 025007, doi:10.1088/1367-2630/13/2/025007.

Melnyk, F. (2004). The Hawking effect for spin 1/2 fields, *Commun. Math. Phys.* **244**, pp. 483–525, doi:10.1007/s00220-003-0999-x.

Michel, F. and Parentani, R. (2013). Saturation of black hole lasers in Bose-Einstein condensates, *Phys. Rev.* **D88**, 12, p. 125012, doi:10.1103/PhysRevD.88.125012.

Michel, F. and Parentani, R. (2014). Probing the thermal character of analogue Hawking radiation for shallow water waves? *Phys. Rev.* **D90**, 4, p. 044033, doi:10.1103/PhysRevD.90.044033.

Michel, F. and Parentani, R. (2015a). Mode mixing in sub- and trans-critical flows over an obstacle: When should Hawking's predictions be recovered? ArXiv preprint gr-qc 1508.02044.

Michel, F. and Parentani, R. (2015b). Nonlinear effects in time-dependent transonic flows: An analysis of analog black hole stability, *Phys. Rev.* **A91**, 5, p. 053603, doi:10.1103/PhysRevA.91.053603.

Miller, P. (2006). *Applied asymptotic analysis* (Graduate Studies in Mathematics, Volume 75. American Mathematical Society).

Milonni, P. (1995). Field quantization and radiative processes in dispersive dielectric media, *Journal of Modern Optics* **42**, 10, pp. 1991–2004.

Misner, C., Thorne, K. and Wheeler, J. (1973). *Gravitation* (W. H. Freeman), ISBN 9780716703440, URL `https://books.google.it/books?id=ExAbAQAAIAAJ`.

Mitra, P. (2007). Hawking temperature from tunnelling formalism, *Phys. Lett.* **B648**, pp. 240–242, doi:10.1016/j.physletb.2007.03.002.

Moretti, V. and Pinamonti, N. (2012). State independence for tunneling processes through black hole horizons and Hawking radiation, *Commun. Math. Phys.* **309**, pp. 295–311, doi:10.1007/s00220-011-1369-8.

Moschella, U. and Schaeffer, R. (2009a). A note on canonical quantization of fields on a manifold, *Journal of Cosmology and Astroparticle Physics* **2009**, 02, p. 033, URL http://stacks.iop.org/1475-7516/2009/i=02/a=033.

Moschella, U. and Schaeffer, R. (2009b). Quantum fields on curved spacetimes and a new look at the Unruh effect, *AIP Conf. Proc.* **1132**, pp. 303–332, doi:10.1063/1.3151844.

Mukhanov, V. and Winitzki, S. (2007). *Introduction to Quantum Effects in Gravity* (Cambridge University Press), ISBN 9780521868341, URL https://books.google.it/books?id=vmwHoxf2958C.

Nakanishi, N. and Ojima, I. (1990). Covariant operator formalism of gauge theories and quantum gravity, *World Sci. Lect. Notes Phys.* **27**, pp. 1–434.

Nikishov, A. I. (1969). Pair production by a constant external field, *Zh. Eksp. Teor. Fiz.* , pp. 1210–1216.

Novello, M. and Bergliaffa, S. E. P. (2003). Effective geometry, *AIP Conference Proceedings* **668**, 1, pp. 288–300, doi:10.1063/1.1587103, URL http://aip.scitation.org/doi/abs/10.1063/1.1587103.

Novello, M., Perez Bergliaffa, S. E., Salim, J., De Lorenci, V. and Klippert, R. (2003). Analog black holes in flowing dielectrics, *Class. Quant. Grav.* **20**, pp. 859–872, doi:10.1088/0264-9381/20/5/306.

O'Neill, B. (2014). *The Geometry of Kerr Black Holes*, Dover Books on Physics (Dover Publications), ISBN 9780486783116, URL https://books.google.it/books?id=jHXCAgAAQBAJ.

Ostrovskii, L. A. (1972). Some General Relations for Waves at the Moving Boundary Between Two Media, *Soviet Journal of Experimental and Theoretical Physics* **34**, p. 293.

Padmanabhan, T. (2005). Gravity and the thermodynamics of horizons, *Phys. Rept.* **406**, pp. 49–125, doi:10.1016/j.physrep.2004.10.003.

Page, D. N. (1976a). Particle Emission Rates from a Black Hole. 2. Massless Particles from a Rotating Hole, *Phys. Rev.* **D14**, pp. 3260–3273, doi:10.1103/PhysRevD.14.3260.

Page, D. N. (1976b). Particle Emission Rates from a Black Hole: Massless Particles from an Uncharged, Nonrotating Hole, *Phys. Rev.* **D13**, pp. 198–206, doi:10.1103/PhysRevD.13.198.

Page, D. N. (1977). Particle Emission Rates from a Black Hole. 3. Charged Leptons from a Nonrotating Hole, *Phys. Rev.* **D16**, pp. 2402–2411, doi:10.1103/PhysRevD.16.2402.

Page, D. N. (2005). Hawking radiation and black hole thermodynamics, *New J. Phys.* **7**, p. 203, doi:10.1088/1367-2630/7/1/203.

Parentani, R. and Brout, R. (1992). Physical interpretation of black hole evaporation as a vacuum instability, *Int. J. Mod. Phys.* **D1**, pp. 169–191, doi:10.1142/S0218271892000082.

Parikh, M. K. (2004). A Secret tunnel through the horizon, *Int. J. Mod. Phys.* **D13**, pp. 2351–2354, doi:10.1142/S0218271804006498, [Gen. Rel. Grav.36,2419(2004)].

Parikh, M. K. and Wilczek, F. (2000). Hawking radiation as tunneling, *Phys. Rev. Lett.* **85**, pp. 5042–5045, doi:10.1103/PhysRevLett.85.5042.

Parker, L. (1975). Probability Distribution of Particles Created by a Black Hole, *Phys. Rev.* **D12**, pp. 1519–1525, doi:10.1103/PhysRevD.12.1519.

Parker, L. and Toms, D. (2009). *Quantum Field Theory in Curved Spacetime: Quantized Fields and Gravity*, Cambridge Monographs on Mathematical Physics (Cambridge University Press), ISBN 9780521877879, URL `https://books.google.it/books?id=5nNuGMBBTjMC`.

Penrose, R. (1963). Asymptotic properties of fields and space-times, *Phys. Rev. Lett.* **10**, pp. 66–68, doi:10.1103/PhysRevLett.10.66.

Petev, M., Westerberg, N., Moss, D., Rubino, E., Rimoldi, C., Cacciatori, S. L., Belgiorno, F. and Faccio, D. (2013). Blackbody emission from light interacting with an effective moving dispersive medium, *Phys. Rev. Lett.* **111**, p. 043902, see also the associated supplementary information.

Philbin, T. (2016). An exact solution for the Hawking effect in a dispersive fluid, *Phys. Rev.* **D94**, 6, p. 064053, doi:10.1103/PhysRevD.94.064053.

Philbin, T. G., Kuklewicz, C., Robertson, S., Hill, S., Konig, F. and Leonhardt, U. (2008). Fiber-optical analogue of the event horizon, *Science* **319**, pp. 1367–1370, doi:10.1126/science.1153625.

Pitaevskii, L. and Stringari, S. (2003). *Bose-Einstein Condensation*, International Series of Monographs on Physics (Clarendon Press), ISBN 9780198507192, URL `https://books.google.it/books?id=rIobbOxC4j4C`.

Poisson, E. (2004). *A Relativist's Toolkit: The Mathematics of Black-Hole Mechanics* (Cambridge University Press), ISBN 9781139451994, URL `https://books.google.it/books?id=bk2XEgz\_ML4C`.

Polesana, P., Couairon, A., Faccio, D., Parola, A., Porras, M. A., Dubietis, A., Piskarskas, A. and Di Trapani, P. (2007). Observation of conical waves in focusing, dispersive, and dissipative Kerr media, *Phys. Rev. Lett.* **99**, p. 223902, doi:10.1103/PhysRevLett.99.223902, URL `https://link.aps.org/doi/10.1103/PhysRevLett.99.223902`.

Polesana, P., Dubietis, A., Porras, M. A., Kučinskas, E., Faccio, D., Couairon, A. and Di Trapani, P. (2006). Near-field dynamics of ultrashort pulsed Bessel beams in media with Kerr nonlinearity, *Phys. Rev. E* **73**, p. 056612, doi:10.1103/PhysRevE.73.056612, URL `https://link.aps.org/doi/10.1103/PhysRevE.73.056612`.

Quan, P. M. (1957). Inductions électromagnétiques en relativité générale et principe de Fermat, *Archive for Rational Mechanics and Analysis* **1**, 1, pp. 54–80, doi:10.1007/BF00297996, URL `http://dx.doi.org/10.1007/BF00297996`.

Radzikowski, M. J. (1996a). A Local to global singularity theorem for quantum field theory on curved space-time, *Commun. Math. Phys.* **180**, pp. 1–22, doi:10.1007/BF02101180.

Radzikowski, M. J. (1996b). Micro-local approach to the Hadamard condition in quantum field theory on curved space-time, *Commun. Math. Phys.* **179**, pp. 529–553, doi:10.1007/BF02100096.

Raine, D. and Thomas, E. (2014). *Black Holes: A Student Text* (Imperial College Press), ISBN 9781783264841, URL `https://books.google.it/books?id=reQ7DQAAQBAJ`.

Recati, A., Pavloff, N. and Carusotto, I. (2009). Bogoliubov Theory of acoustic Hawking radiation in Bose-Einstein Condensates, *Phys. Rev.* **A80**, p. 043603, doi:10.1103/PhysRevA.80.043603.

Richartz, M., Prain, A., Liberati, S. and Weinfurtner, S. (2015). Rotating black holes in a draining bathtub: Superradiant scattering of gravity waves, *Phys. Rev. D* **91**, p. 124018, doi:10.1103/PhysRevD.91.124018, URL `https://link.aps.org/doi/10.1103/PhysRevD.91.124018`.

Robertson, S. and Leonhardt, U. (2010). Frequency shifting at fiber-optical event horizons: The effect of Raman deceleration, *Phys. Rev. A* **81**, p. 063835.

Robertson, S., Michel, F. and Parentani, R. (2016). Scattering of gravity waves in subcritical flows over an obstacle, *Phys. Rev.* **D93**, 12, p. 124060, doi: 10.1103/PhysRevD.93.124060.

Robertson, S. and Parentani, R. (2015). Hawking radiation in the presence of high-momentum dissipation, *Phys. Rev.* **D92**, 4, p. 044043, doi:10.1103/PhysRevD.92.044043.

Robertson, S. J. (2012). The theory of Hawking radiation in laboratory analogues, *J. Phys.* **B45**, p. 163001, doi:10.1088/0953-4075/45/16/163001.

Robinson, S. P. (2005). *Two quantum effects in the theory of gravitation*, Ph.D. thesis, MIT, URL `http://wwwlib.umi.com/dissertations/fullcit?p0808106`.

Robinson, S. P. and Wilczek, F. (2005). A Relationship between Hawking radiation and gravitational anomalies, *Phys. Rev. Lett.* **95**, p. 011303, doi: 10.1103/PhysRevLett.95.011303.

Rousseaux, G., Maissa, P., Mathis, C., Coullet, P., Philbin, T. G. and Leonhardt, U. (2010). Horizon effects with surface waves on moving water, *New J. Phys.* **12**, p. 095018, doi:10.1088/1367-2630/12/9/095018.

Rousseaux, G., Mathis, C., Maissa, P., Philbin, T. G. and Leonhardt, U. (2008). Observation of negative phase velocity waves in a water tank: A classical analogue to the Hawking effect? *New J. Phys.* **10**, p. 053015, doi:10.1088/1367-2630/10/5/053015.

Rubino, E. (2013). *Hawking radiation and mode conversion at optically induced horizons*, Phd thesis in physics, Università degli Studi dell'Insubria, available at url http://hdl.handle.net/10277/517.

Rubino, E., Belgiorno, F., Cacciatori, S. L., Clerici, M., Gorini, V., Ortenzi, G., Rizzi, L., Sala, V. G., Kolesik, M. and Faccio, D. (2011). Experimental evidence of analogue Hawking radiation from ultrashort laser pulse filaments, *New J. Phys.* **13**, p. 085005, doi:10.1088/1367-2630/13/8/085005.

Rubino, E., Lotti, A., Belgiorno, F., Cacciatori, S. L., Couairon, A., Leonhardt, U. and Faccio, D. (2012.). Soliton-induced relativistic-scattering and amplification, *Scientific Reports* **2**, p. 932, doi:10.1038/srep00932, see also the associated supplementary information.

Rubino, E., McLenaghan, J., Kehr, S. C., Belgiorno, F., Townsend, D., Rohr, S., Kuklewicz, C. E., Leonhardt, U., König, F. and Faccio, D. (2012). Negative-frequency resonant radiation, *Phys. Rev. Lett.* **108**, p. 253901, doi:10.1103/PhysRevLett.108.253901, URL `https://link.aps.org/doi/10.1103/PhysRevLett.108.253901`.

Sachs, R. (1962). Asymptotic symmetries in gravitational theory, *Phys. Rev.* **128**, pp. 2851–2864, doi:10.1103/PhysRev.128.2851.

Saida, H. and Sakagami, M.-a. (2000). Black hole radiation with high frequency dispersion, *Phys. Rev.* **D61**, p. 084023, doi:10.1103/PhysRevD.61.084023.

Sannan, S. (1988). Heuristic Derivation of the Probability Distributions of Particles Emitted by a Black Hole, *Gen. Rel. Grav.* **20**, pp. 239–246, doi:10.1007/BF00759183.

Schulman, L.S. (1981). *Techniques and Applications of Path Integration* (New York, USA: Wiley (1981) 358p).

Schutzhold, R. and Unruh, W. G. (2002). Gravity wave analogs of black holes, *Phys. Rev.* **D66**, p. 044019, doi:10.1103/PhysRevD.66.044019.

Schutzhold, R. and Unruh, W. G. (2005). Hawking radiation in an electromagnetic wave-guide? *Phys. Rev. Lett.* **95**, p. 031301, doi:10.1103/PhysRevLett.95.031301.

Schutzhold, R. and Unruh, W. G. (2008). Origin of the particles in black hole evaporation, *Phys. Rev.* **D78**, p. 041504(R), doi:10.1103/PhysRevD.78.041504.

Schutzhold, R. and Unruh, W. G. (2011). Comment on: Hawking Radiation from Ultrashort Laser Pulse Filaments, *Phys. Rev. Lett.* **107**, p. 149401, doi:10.1103/PhysRevLett.107.149401.

Schutzhold, R. and Unruh, W. G. (2013). Hawking radiation with dispersion versus breakdown of the WKB approximation, *Phys. Rev.* **D88**, 12, p. 124009, doi:10.1103/PhysRevD.88.124009.

Schwinger, J. (1993a). Casimir light: a glimpse. *Proceedings of the National Academy of Sciences* **90**, 3, pp. 958–959, doi:10.1073/pnas.90.3.958.

Schwinger, J. (1993b). Casimir light: photon pairs. *Proceedings of the National Academy of Sciences* **90**, 10, pp. 4505–4507, doi:10.1073/pnas.90.10.4505.

Schwinger, J. (1993c). Casimir light: pieces of the action, *Proceedings of the National Academy of Sciences* **90**, 15, pp. 7285–7287.

Schwinger, J. (1993d). Casimir light: the source. *Proceedings of the National Academy of Sciences* **90**, 6, pp. 2105–2106, doi:10.1073/pnas.90.6.2105.

Schwinger, J. (1994). Casimir light: field pressure, *Proceedings of the National Academy of Sciences* **91**, 14, pp. 6473–6475.

Sciama, D. W., Candelas, P. and Deutsch, D. (1981). Quantum Field Theory, Horizons and Thermodynamics, *Adv. Phys.* **30**, pp. 327–366, doi:10.1080/00018738100101457.

Segal, I. E., Bongaarts, P. J. M. and Niemeijer, T. (1967). Representations of the canonical commutations relations, in *Application of Mathematics to Problems in Theoretical Physics: Proceedings, Summer School of Theoretical Physics: Cargese, France, Sep 1965*, Vol. 6, pp. 107–170.

Sewell, G. L. (1982). Quantum fields on manifolds: PCT and gravitationally induced thermal states, *Annals of Physics* **141**, 2, pp. 201 – 224, doi:https://doi.org/10.1016/0003-4916(82)90285-8, URL `http://www.sciencedirect.com/science/article/pii/0003491682902858`.

Shankaranarayanan, S., Padmanabhan, T. and Srinivasan, K. (2002). Hawking radiation in different coordinate settings: Complex paths approach, *Class. Quant. Grav.* **19**, pp. 2671–2688, doi:10.1088/0264-9381/19/10/310.

Shankaranarayanan, S., Srinivasan, K. and Padmanabhan, T. (2001). Method of complex paths and general covariance of Hawking radiation, *Mod. Phys. Lett.* **A16**, pp. 571–578, doi:10.1142/S0217732301003632.

Skuja, L. (1998). Optically active oxygen-deficiency-related centers in amorphous silicon dioxide, *Journal of Non-Crystalline Solids* **239**, 1, pp. 16 – 48, doi:http://dx.doi.org/10.1016/S0022-3093(98)00720-0, URL `http://www.sciencedirect.com/science/article/pii/S0022309398007200`.

Solodukhin, S. N. (1995). The Conical singularity and quantum corrections to entropy of black hole, *Phys. Rev.* **D51**, pp. 609–617, doi:10.1103/PhysRevD.51.609.

Sorokin, Y. M. (1972). On a certain energy relationship for waves in systems having traveling parameters, *Radiophysics and Quantum Electronics* **15**, 1, pp. 36–39, doi:10.1007/BF02209239, URL `https://doi.org/10.1007/BF02209239`.

Srinivasan, K. and Padmanabhan, T. (1999). Particle production and complex path analysis, *Phys. Rev.* **D60**, p. 024007, doi:10.1103/PhysRevD.60.024007.

Starobinsky, A. A. (1973). Amplification of waves during reflection from a rotating "black hole". *Sov. Phys. JETP* **37**, 1, pp. 28–32, [Zh. Eksp. Teor. Fiz.64,48(1973)].

Steinhauer, J. (2014). Observation of self-amplifying Hawking radiation in an analog black hole laser, *Nature Phys.* **10**, p. 864, doi:10.1038/NPHYS3104.

Steinhauer, J. (2015). Measuring the entanglement of analogue Hawking radiation by the density-density correlation function, *Phys. Rev.* **D92**, 2, p. 024043, doi:10.1103/PhysRevD.92.024043.

Steinhauer, J. (2016). Observation of quantum Hawking radiation and its entanglement in an analogue black hole, *Nature Phys.* **12**, p. 959, doi:10.1038/nphys3863.

Steinhauer, J. and de Nova, J. R. M. (2017). Self-amplifying Hawking radiation and its background: a numerical study, *Phys. Rev.* **A95**, 3, p. 033604, doi:10.1103/PhysRevA.95.033604.

Stephens, C. R. (1989). The Hawking Effect in Abelian Gauge Theories, *Annals Phys.* **193**, pp. 255–286, doi:10.1016/0003-4916(89)90001-8.

Stueckelberg, E. C. G. (1941). Remarks on the creation of pairs of particles in the theory of relativity, *Helv. Phys. Acta* **14**, pp. 588–594.

Summers, S. J. and Verch, R. (1996). Modular inclusion, the Hawking temperature and quantum field theory in curved space-time, *Lett. Math. Phys.* **37**, pp. 145–158, doi:10.1007/BF00416017.

Suttorp, L. G. (2007). Field quantization in inhomogeneous anisotropic dielectrics with spatio-temporal dispersion, *Journal of Physics A: Mathematical and Theoretical* **40**, 13, p. 3697, URL `http://stacks.iop.org/1751-8121/40/i=13/a=025`.

Suttorp, L. G. and van Wonderen, A. J. (2004). Fano diagonalization of a polariton model for an inhomogeneous absorptive dielectric, *EPL (Europhysics Letters)* **67**, 5, pp. 766–772.

Swanson, D. (1998). *Theory of Mode Conversion and Tunneling in Inhomogeneous Plasmas*, A Wiley-interscience publication (Wiley), ISBN 9780471247760, URL `https://books.google.it/books?id=ZakemvM28ssC`.

Synge, J. L. (1956). Geometrical optics in moving dispersive media, *Commun. Dublin Inst. Ser. A* **12**.

Taflove, A. and Hagness, S. (2005). *Computational Electrodynamics: The Finite-difference Time-domain Method*, Artech House antennas and propagation library (Artech House), ISBN 9781580538329.

Takagi, S. (1986). Vacuum noise and stress induced by uniform acceleration Hawking-Unruh effect in Rindler manifold of arbitrary dimension, *Progress of Theoretical Physics Supplement* **88**, pp. 1–142, doi:10.1143/PTP.88.1, URL `+http://dx.doi.org/10.1143/PTP.88.1`.

Takahashi, Y. and Umezawa, H. (1975). Thermo field dynamics, *Collect. Phenom.* **2**, pp. 55–80.

Takesaki, M. (1970). *Tomita's theory of modular Hilbert algebras and its applications*, Lecture Notes in Mathematics, Vol. 128 (Springer-Verlag, Berlin-New York).

Tao, J., Wang, P. and Yang, H. (2017). Black hole radiation with modified dispersion relation in tunneling paradigm: Static frame, *Nuclear Physics B* **922**, Supplement C, pp. 346 – 383, doi:https://doi.org/10.1016/j.nuclphysb.2017.06.022, URL `http://www.sciencedirect.com/science/article/pii/S0550321317302171`.

Temme, N. M. (2014). *Asymptotic Methods for Integrals* (World Scientific Publishing Company).

Tettamanti, M., Cacciatori, S. L., Parola, A. and Carusotto, I. (2016). Numerical study of a recent black hole lasing experiment, *Europhys. Lett.* **114**, 6, p. 60011, doi:10.1209/0295-5075/114/60011.

Thorne, K., Price, R. and MacDonald, D. (1986). *Black Holes: The Membrane Paradigm*, Silliman Memorial Lectures (Yale University Press), ISBN 9780300037708, URL `https://books.google.it/books?id=T94hD5rR8oYC`.

Tomita, M. (1967). On canonical forms of von Neumann algebras, in *Fifth Functional Analysis Sympos. (Tôhoku Univ., Sendai, 1967) (Japanese)* (Math. Inst., Tôhoku Univ., Sendai), pp. 101–102.

Torres, T., Patrick, S., Coutant, A., Richartz, M. and Tedford, S., Edmund W.and Weinfurtner (2017). Rotational superradiant scattering in a vortex flow, *Nature Physics* **13**, p. 833, doi:10.1038/nphys4151, URL `http://dx.doi.org/10.1038/nphys4151`.

Troyanov, M. (1986). Les surfaces euclidiennes à singularités coniques, *L'Enseignement Mathèmatique.* **32**, pp. 79–94.

Umezawa, H. (1993). *Advanced Field Theory: Micro, Macro, and Thermal Physics* (New York, USA: AIP).

Umezawa, H., Matsumoto, H. and Tachiki, M. (1982). *Thermo Field Dynamics and Condensed States* (Amsterdam, Netherlands: North-Holland).

Unruh, W. G. (1974). Second quantization in the Kerr metric, *Phys. Rev.* **D10**, pp. 3194–3205, doi:10.1103/PhysRevD.10.3194.

Unruh, W. G. (1976). Notes on black hole evaporation, *Phys. Rev.* **D14**, p. 870, doi:10.1103/PhysRevD.14.870.

Unruh, W. G. (1981). Experimental black hole evaporation, *Phys. Rev. Lett.* **46**, pp. 1351–1353, doi:10.1103/PhysRevLett.46.1351.

Unruh, W. G. (1995). Sonic analog of black holes and the effects of high frequencies on black hole evaporation, *Phys. Rev.* **D51**, pp. 2827–2838, doi: 10.1103/PhysRevD.51.2827.

Unruh, W. G. and Schutzhold, R. (2003). On slow light as a black hole analog, *Phys. Rev.* **D68**, p. 024008, doi:10.1103/PhysRevD.68.024008.

Unruh, W. G. and Schutzhold, R. (2005). On the universality of the Hawking effect, *Phys. Rev.* **D71**, p. 024028, doi:10.1103/PhysRevD.71.024028.

Vanzo, L. (2011). On tunneling across horizons, *Europhys. Lett.* **95**, p. 20001, doi:10.1209/0295-5075/95/20001.

Vanzo, L., Acquaviva, G. and Di Criscienzo, R. (2011). Tunnelling Methods and Hawking's radiation: achievements journal and prospects, *Class. Quant. Grav.* **28**, p. 183001, doi:10.1088/0264-9381/28/18/183001.

Vermeil, H. (1917). Notiz über das mittlere krümmungsmaß einer n-fach ausgedehnten riemann'schen mannigfaltigkeit, *Nachrichten von der Gesellschaft der Wissenschaften zu Gttingen, Mathematisch-Physikalische Klasse* **1917**, pp. 334–344, URL `http://eudml.org/doc/58997`.

Visser, M. (1998a). Acoustic black holes: Horizons, ergospheres, and Hawking radiation, *Class. Quant. Grav.* **15**, pp. 1767–1791, doi:10.1088/0264-9381/15/6/024.

Visser, M. (1998b). Hawking radiation without black hole entropy, *Phys. Rev. Lett.* **80**, pp. 3436–3439, doi:10.1103/PhysRevLett.80.3436.

Visser, M. (2003). Essential and inessential features of Hawking radiation, *Int. J. Mod. Phys.* **D12**, pp. 649–661, doi:10.1142/S0218271803003190.

Wald, R. (1984). *General Relativity* (University of Chicago Press), ISBN 9780226870335.

Wald, R. M. (1975). On Particle Creation by Black Holes, *Commun. Math. Phys.* **45**, pp. 9–34, doi:10.1007/BF01609863.

Wald, R. M. (1976). Stimulated Emission Effects in Particle Creation Near Black Holes, *Phys. Rev.* **D13**, pp. 3176–3182, doi:10.1103/PhysRevD.13.3176.

Wald, R. M. (1977). The Back Reaction Effect in Particle Creation in Curved Space-Time, *Commun. Math. Phys.* **54**, pp. 1–19, doi:10.1007/BF01609833.

Wald, R. M. (1978). Trace Anomaly of a Conformally Invariant Quantum Field in Curved Space-Time, *Phys. Rev.* **D17**, pp. 1477–1484, doi:10.1103/PhysRevD.17.1477.

Wald, R. M. (1995). *Quantum Field Theory in Curved Space-Time and Black Hole Thermodynamics*, Chicago Lectures in Physics (University of Chicago Press, Chicago, IL), ISBN 9780226870274.

Wang, Y.-H., Jacobson, T., Edwards, M. and Clark, C. W. (2017a). Induced density correlations in a sonic black hole condensate, *SciPost Phys.* **3**, p. 022, doi:10.21468/SciPostPhys.3.3.022, URL `https://scipost.org/10.21468/SciPostPhys.3.3.022`.

Wang, Y.-H., Jacobson, T., Edwards, M. and Clark, C. W. (2017b). Mechanism of stimulated hawking radiation in a laboratory bose-einstein condensate, *Phys. Rev. A* **96**, p. 023616, doi:10.1103/PhysRevA.96.023616, URL `https://link.aps.org/doi/10.1103/PhysRevA.96.023616`.

Weinfurtner, S., Tedford, E. W., Penrice, M. C. J., Unruh, W. G. and Lawrence, G. A. (2011). Measurement of stimulated Hawking emission in an analogue system, *Phys. Rev. Lett.* **106**, p. 021302, doi:10.1103/PhysRevLett.106.021302.

Weinfurtner, S., Tedford, E. W., Penrice, M. C. J., Unruh, W. G. and Lawrence, G. A. (2013). *Classical Aspects of Hawking Radiation Verified in Analogue Gravity Experiment* (Springer International Publishing, Cham), ISBN 978-3-319-00266-8, pp. 167–180, doi:10.1007/978-3-319-00266-8_8, URL `https://doi.org/10.1007/978-3-319-00266-8_8`.

Wong, R. (1989). *Asymptotic Approximation of Integrals* (Academic Press, New York).

Zapata, I., Albert, M., Parentani, R. and Sols, F. (2011). Resonant Hawking radiation in Bose-Einstein condensates, *New J. Phys.* **13**, p. 063048, doi:10.1088/1367-2630/13/6/063048.

Zel'dovich, Y. B. (1972). Amplification of cylindrical electromagnetic waves reflected from a rotating body. *Sov. Phys. JETP* **35**, 6, pp. 1085–1087, [Zh. Eksp. Teor. Fiz.62,2076(1972)].

Zoubir, A., Rivero, C., Grodsky, R., Richardson, K., Richardson, M., Cardinal, T. and Couzi, M. (2006). Laser-induced defects in fused silica by femtosecond IR irradiation, *Phys. Rev. B* **73**, p. 224117, doi:10.1103/PhysRevB.73.224117, URL `https://link.aps.org/doi/10.1103/PhysRevB.73.224117`.

Index

$\varphi\psi$–model, 217

anomalous Ward identity, 95
anomaly
  consistent, 95
  covariant, 95

B-field, 200, 207
BEC models, 147
bimetric, 155
black hole laser, 150
blocking horizon, 142
Bogolubov transformation, 76
Born approximation, 229
Boulware state, 68, 70, 92, 129
branch cut, 147, 242, 245
branch points, 240

Cauchy formula, 186, 187, 190, 223, 224, 237, 239, 256, 261
Como experiment, 267
comoving frame, 157
conformal factor, 103
conformally invariant field, 90
conical singularity, 103, 104, 107
Constrained systems
  Dirac approach, 200
constraint
  first class, 200
  primary, 201
  second class, 200
  second-stage, 201

decreasing mode, 241
dielectric metric
  acoustic form, 173
  black hole temperature, 175, 176, 184
  horizons, 173
  static, 172
  stationary, 157
Dirac brackets, 202
dispersion
  optical, 185
  subluminal, 143
  superluminal, 143
dispersion relation
  asymptotic, 223
dispersion relation
  geometrical optics approximation, 146
distributional curvature, 105

eikonal approximation, 236
ergoregion, 158
ergosurface, 158, 159, 161, 162
Euclidean metric, 102
Euclidean section, 101, 104
Euclidean section
  extendability, 103, 106
  polar coordinates, 103
extremal black holes, 109

Feynman-Stueckelberg interpretation, 58, 84

finite difference time domain, 229
Frobenius theorem, 172
Froude number, 143

Gibbs state, 289
Gordon metric, 156, 157, 163
Gross-Pitaevskii equation, 148
group horizon, 142, 146, 192, 261

Hadamard condition, 133
Hartle-Hawking state, 57, 65, 66, 68, 83, 92, 93, 100, 102, 127
Hartle-Hawking-Israel state, 57
Hopfield model
  Hawking temperature, 243

Kerr medium, 155
KMS condition, 64, 116, 286
KMS condition
  infinite volume limit, 290
  Tomita-Takesaki theorem, 297
KMS state, 133

linear region, 250, 254
logarithmic branch point, 75

Manley-Rowe relations
  generalized, 227
mode conversion, 142

near horizon approximation, 147, 236

Penrose diagram
  L-region, 69
  R-region, 69
phase horizon, 189, 192, 261
polarization field, 196, 197, 206
principle of local definiteness, 112, 118
principle of local stability, 112, 118
propagator
  analyticity properties, 63

quantum anomalies, 89
quantum anomalies
  chiral anomaly, 94
  trace anomaly, 89
quantum field theory
  algebraic formulation, 281
quantum fluctuations
  tearing apart of, 141

Ree-Schlieder property, 112
reflection coefficient, 84
Rindler vacuum, 68

Sachs-Bondi energy, 111
saddle points, 239, 249
saddle points
  coalescence, 240
scalar product
  conserved, 204, 219
steepest descent approximation, 238
straddling mode, 74, 75
sub-critical case, 143
super-critical case, 143
surface gravity
  in dispersive models, 144

Technion experiment, 278
thermality
  case of analogous systems, 142
thermofield dynamics, 66
thermofield dynamics
  fictitious Hilbert space, 68
Tomita-Takesaki theory, 112, 131
trans-critical case, 143
transmission coefficient, 84
transplanckian frequencies, 141
trick à la Nikishov, 83
tunneling methods, 73
tunneling methods
  analyticity argument, 81
  Damour and Ruffini approach, 74
  Hamilton-Jacobi approach, 76, 83
  null geodesic method, 78
  Parikh-Wilczek approach, 87
  WKB approximation, 77
turning point, 142, 146, 259, 261

unidirectional pulse propagation equation, 230